3D Imaging in Medicine

Editors

Jayaram K. Udupa, Ph.D.
Adjunct Associate Professor and Director
Medical Image Processing Group
Hospital of the University of Pennsylvania
Philadelphia, Pennsylvania

Gabor T. Herman, Ph.D.
Professor
Medical Image Processing Group
Hospital of the University of Pennsylvania
Philadelphia, Pennsylvania

CRC Press
Boca Raton   Ann Arbor   London   Tokyo

Library of Congress Cataloging-in-Publication Data

3D imaging in medicine / editors, Jayaram K. Udupa, Gabor T. Herman.
       p.   cm.
       Based on a meeting held in Coronado, Calif., Nov. 16–19, 1989.
       Includes bibliographical references.
       Includes index.
       ISBN 0-8493-4294-5
       1. Three-dimensional imaging in medicine--Congresses.   I. Udupa,
Jayaram K.   II. Herman, Gabor T.   III. Title: Three-dimensional imaging in medicine.
       [DNLM: 1. Diagnostic Imaging--congresses.   2. Image Processing.
Computer-Assisted--congresses.   WN 160 Z999 1989]
   R857.T47A13   1991
   616.07′54--dc20
   DNLM/DLC
   for Library of Congress   90-2591
                                           CIP

© 1991 by CRC Press, Inc.

International Standard Book Number 0-8493-4294-5

Library of Congress Card Number 90-2591
Printed in the United States of America   3 4 5 6 7 8 9 0
Printed on acid-free paper

PREFACE

Research in three-dimensional imaging in medicine started in the seventies and the activity in the field became very brisk in the eighties. During this period, the technical developments — design of algorithms, software, and machines — have galloped ahead, accompanied by considerable increase in clinical applications, but clinical validation and the study of clinical usefulness have generally lagged behind. Yet it is the latter aspects that will decide whether or not three-dimensional imaging for medicine will survive and prosper. It is mainly for this reason of fostering acceptance in clinical practice that we organized a meeting entitled "Three-Dimensional Imaging in Medicine" in December 1987 in Philadelphia. Since prior meetings on medically oriented three-dimensional imaging, such as those sponsored by the Society of Photo-Optical Instrumentation Engineers and by the National Computer Graphics Association, catered mainly to the technical aspects, our second prime reason for the 1987 Philadelphia meeting was to provide a forum for dialogue between the technical developers and the clinical end users. Encouraged by the very positive response to that meeting, we decided to organize further meetings of the same type. We also thought that a compendium of articles from leading clinically oriented researchers on their own areas of expertise, supplemented by a technical tutorial article tailored to nonspecialists, would form a useful reference book on 3D imaging for medical researchers and for practicing clinicians, as well as for technical developers. This book is a collection of such articles from the invited speakers of the second meeting on "Three-Dimensional Imaging in Medicine" held in Coronado, California November 16 to 19, 1989.

The first chapter by Udupa describes for the nonspecialist the basic principles and the methodologies commonly used in 3D imaging, in a tutorial form. The chapter by Vannier et al. gives a general introduction to 3D imaging. Hemmy and Brigman survey the product options currently available for 3D imaging and compare them in a general fashion. Gillespie and Herman in their respective articles address issues related to the assessment of clinical efficacy and the validation of the information provided by 3D images. Cutting describes the progress made in surgical planning, mainly for the treatment of craniofacial malformations, using tools for image manipulation. The subsequent two chapters by Wojcik and Harris and by Fishman et al. discuss 3D imaging applications to the musculoskeletal system, the former using surface-rendering techniques and the latter using volume rendering. Robertson and Walker explain how 3D data analysis can lead to more effective custom design of prostheses. The chapter by Levin et al. outlines their approach to 3D imaging of brain anatomy and to the composite imaging and analysis of MR, CT, and PET data. Hoffman illustrates the use of 3D imaging in the study of anatomic, physiologic, and functional information related to cardiopulmonary structures. The use of 3D imaging for radiation therapy planning is described by Rosenman. The last chapter by Hemmy suggests possible future directions for 3D imaging by pointing out some clinical problems for which the existing 3D imaging systems cannot provide satisfactory solutions.

As with any new discipline, 3D imaging is not without its controversies. The views expressed in this book by the individual chapter authors are their own; they do not necessarily agree with those of the editors. In fact, there are some statements in the book which we ourselves would not have made, because we consider that their validity is neither obvious nor has been established scientifically (e.g., the claim in Chapter 8 that volume-rendering techniques are superior to surface-rendering techniques for depicting fine structures, such as sutures and fractures, runs counter to the views expressed in Chapter 1). While we provided all contributors with our editorial comments, we felt that we should leave them to the final decision of what to include in their chapters. Thus we present the state of the art: diverse and vibrant.

Finally, we would like to express our sincere gratitude to all chapter authors as well as to the staff of CRC Press for their cooperation in this project.

Jayaram K. Udupa
Gabor T. Herman

THE EDITORS

Jayaram K. Udupa, Ph.D., is Adjunct Associate Professor in the Department of Radiology at the University of Pennsylvania.

Dr. Udupa received a B.Eng. degree in electronics and communication engineering from Mysore University, India and a Ph.D. in computer science from the Indian Institute of Science, Bangalore. He was a Scientific Officer at the Institute between 1976 and 1978, Research Assistant Professor in the Department of Computer Science, State University of New York at Buffalo from 1978 to 1981, Adjunct Assistant Professor at the Department of Radiology, University of Pennsylvania from 1981 to 1983, Adjunct Associate Professor since 1981 and Director of the Medical Image Processing Group since 1982.

Dr. Udupa is a senior member of the Institute of Electrical and Electronic Engineers and a member of the Mathematical Association of America, the Radiological Society of North America, the Society for Computer Applications in Radiology, the National Computer Graphics Association, the Association for Computing Machinery, and the Society of Photo-Optical Instrumentation Engineers. His current interests are in the development of theory, algorithms, transportable software, and standards for multidimensional biomedical data visualization and analysis.

Gabor T. Herman, Ph.D., is Professor in the Department of Radiology at the University of Pennsylvania.

Dr. Herman received the M.Sc. and Ph.D. degrees in mathematics from the University of London, England, and the M.S. degree in engineering science from the University of California, Berkeley. From 1969 to 1981, he was with the Department of Computer Science, State University of New York at Buffalo, where in 1976 he became the Director of the Medical Image Processing Group. He was Chief of Medical Imaging in the Department of Radiology at the University of Pennsylvania from 1981 to 1988.

Dr. Herman's society memberships include the American Association of Physicists in Medicine, American Society of Neuro-imaging (honorary member), National Computer Graphics Association, and the Radiological Society of North America. He has editorial positions for several journals.

Dr. Herman's research involves the application of computer technology to the processing of radiological and other medical images. Specific topics of his current interest include quantitative three-dimensional medical imaging, quantitative evaluation of image processing methods, stereoscopic viewing of and interaction with three-dimensional data sets, and applications of the above to surgical planning and the analysis of brain disorders.

CONTRIBUTORS

Robert N. Beck, B.S.
Department of Radiology
University of Chicago Hospitals
Chicago, Illinois

Patricia M. Brigman
Wisconsin Medmark
Elm Grove, Wisconsin

Chin-Tu Chen, Ph.D.
Director
Frank Image Analysis Center
Department of Radiology
University of Chicago Hospitals
Chicago, Illinois

George T. Y. Chen, Ph.D.
Director of Radiation Physics
Department of Radiation Oncology
University of Chicago Hospitals
Chicago, Illinois

Malcolm Cooper, M.D.
Director of Nuclear Medicine
Department of Radiology
University of Chicago Hospitals
Chicago, Illinois

Court B. Cutting, M.D.
Assistant Professor of Plastic Surgery
New York University Medical Center
New York, New York

Elliot K. Fishman, M.D.
Associate Professor of Radiology
Director, Computed Body Tomography
Russell H. Morgan Department of
  Radiology and Radiological Science
The Johns Hopkins Medical Institutions
Baltimore, Maryland

Simranjit Galhotra, B.S.
Department of Radiology
University of Chicago Hospitals
Chicago, Illinois

Donald E. Gayou, Ph.D.
Research Associate
Mallinckrodt Institute of Radiology
Washington University School of
  Medicine
St. Louis, Missouri

Anil Gholkar, M.B., B.S., F.R.C.R.
Department of Diagnostic Radiology
University of Manchester
United Kingdom

**James E. Gillespie, M.B., B.Ch.,
  F.R.C.R.**
Department of Diagnostic Radiology
University of Manchester
United Kingdom

John H. Harris, Jr., M.D., D.Sc.
Professor and John S. Dunn Chairman
Department of Radiology
The University of Texas Medical School
  at Houston and
Hermann Hospital
Houston, Texas

David C. Hemmy
Clinical Professor
Department of Neurosurgery
Medical College of Wisconsin
Milwaukee, Misconsin

Gabor T. Herman, Ph.D.
Professor
Medical Image Processing Group
Department of Radiology
University of Pennsylvania
Philadelphia, Pennsylvania

Andreas Herrmann, Ph.D.
Department of Radiology
University of Chicago Hospitals
Chicago, Illinois

Charles F. Hildebolt, D.D.S., Ph.D.
Research Assistant Professor
Mallinckrodt Institute of Radiology
Washington University School of
  Medicine
St. Louis, Missouri

Eric A. Hoffman, Ph.D.
Chief, Section of Cardiothoracic Imaging
  Research
Department of Radiology
Hospital of the University of Pennsylvania
Philadelphia, Pennsylvania

Xiaoping Hu, Ph.D.
Department of Radiology
University of Chicago Hospitals
Chicago, Illinois

Ian Isherwood, M.D., F.R.C.P.,
 F.R.C.R., F.F.R.R.C.S.I.(Hon.)
Professor of Radiology
Department of Diagnostic Radiology
University of Manchester
United Kingdom

Janet E. Kuhlman, M.D.
Assistant Professor of Radiology
Russell H. Morgan Department of
 Radiology and Radiological Science
The Johns Hopkins Medical Institutions
Baltimore, Maryland

David N. Levin, M.D., Ph.D.
Department of Radiology
University of Chicago Hospitals
Chicago, Illinois

Donna Magid, M.D.
Assistant Professor of Radiology
Russell H. Morgan Department of
 Radiology and Radiological Science
The Johns Hopkins Medical Institutions
Baltimore, Maryland

Jeffrey L. Marsh, M.D.
Professor and Director
Cleft Palate and Craniofacial Deformities
 Institute
St. Louis Childrens Hospital
Washington University Medical Center
St. Louis, Missouri

Derek R. Ney, B.S.
Assistant Professor of Radiology
Russell H. Morgan Department of
 Radiology and Radiological Science
The Johns Hopkins Medical Institutions
Baltimore, Maryland

Charles A. Pelizzari, Ph.D.
Director of Computer Section
Department of Radiation Oncology
University of Chicago Hospitals
Chicago, Illinois

Douglas D. Robertson, M.D., Ph.D.
Assistant Professor
Orthopedic Biomechanics Laboratory
Brigham and Women's Hospital
Boston, Massachusetts

Julian Rosenman, Ph.D., M.D.
Associate Professor
Departments of Radiation Oncology and
 Computer Science
The University of North Carolina at
 Chapel Hill
Chapel Hill, North Carolina

Kim K. Tan, Ph.D.
Department of Radiology
University of Chicago Hospitals
Chicago, Illinois

Jayaram K. Udupa, Ph.D.
Adjunct Associate Professor and Director
Medical Image Processing Group
Hospital of the University of
 Pennsylvania
Philadelphia, Pennsylvania

Michael W. Vannier, M.D.
Professor
Mallinckrodt Institute of Radiology
Washington University School of
 Medicine
St. Louis, Missouri

Peter S. Walker, Ph.D.
Professor
Department of Biomedical Engineering
Royal National Orthopaedic Hospital
Brockley Hill, Stanmore
Middlesex, England

W. Gregory Wojcik, M.D., M.S.
Assistant Professor of Radiology
Department of Radiology
The University of Texas Medical School
 at Houston and
Hermann Hospital
Houston, Texas

TABLE OF CONTENTS

Chapter 1

COMPUTER ASPECTS OF 3D IMAGING IN MEDICINE: A TUTORIAL

Jayaram K. Udupa

TABLE OF CONTENTS

I. INTRODUCTION

A. BACKGROUND AND SCOPE

To visualize noninvasively human internal organs in their true form and shape has intrigued mankind for centuries. If the discovery of X-rays gave birth to radiology, the invention of computerized tomography and, recently, magnetic resonance imaging, has revolutionized radiology. Three-dimensional (3D) imaging is another recent development that has brought us closer to fulfilling the age-old quest of noninvasive visualization

3D imaging is evolving as a discipline of its own dealing with the various forms of visualization, manipulation, and analysis of multidimensional medical structures. What used to be the subject of scientific curiosity of a handful of research groups just a few years ago has now grown into a full discipline. Clinical applications of a wide variety and strong commercial interests have been the two major factors contributing to the pace of development. While other chapters in this book give a more detailed account of the various established and potential applications of 3D imaging, evidence of the extent of applications and of commercial interests in this field is best gathered at the annual meetings of the Radiological Society of North America. There has been a steady increase in the representation of both of these factors at this meeting during the past four years. There are other annual meetings such as the Biomedical Applications Conference of the National Computer Graphics Association and the Medical Imaging Conference sponsored by the Society of Photo-Optical Instrumentation Engineers that deal with the technical aspects of 3D imaging.

Though there are several review and survey papers written on 3D imaging techniques,[1-5] there are no published tutorial articles that are easily understood by clinician readers whose main interest is in using 3D imaging tools in their applications but who nevertheless intend to get a grasp of the underlying ideas. In order to know what to expect from a given 3D approach, it is important to understand its shortcomings as well as strong points. Therefore, it becomes necessary to understand the approaches from a phenomenological point of view, although the mathematical and algorithmic details are usually not relevant to this level of understanding. It is for these reasons that we attempt here to present the major results of 3D imaging in a form readily understood by nonspecialists. We will avoid mathematical and technical treatment, yet state related results and explain the ideas, often in a geometric guise, with ample illustrations.

We will discuss only computer-display-technology-based approaches. In other words, we assume that the display device available to us is two-dimensional (2D) in nature and all analysis of multidimensional image data is to be carried out via the 2D screen of the device.

Because of this restriction all 3D imaging approaches considered here do not really create 3D images, but they only give an illusion of three dimensionality via appropriately computed 2D (flat) images. We would like to point out, however, that there are technologies such as holography and vibrating mirror that do provide a "true 3D screen".[1-3] This article will not discuss such approaches. Neither will we describe the numerous clinical applications and results obtained via 3D imaging, except to draw examples to illustrate approaches. While we shall try to describe all major techniques, because of space limitations we will have to leave out the description of some methods. Our selection is based mainly on the popularity of use of the methods.

B. OVERVIEW

A global view of the computer processing done in 3D imaging is best obtained by examining how information in the given multidimensional data set is successively transformed to eventually yield the information that is sought. This sequence of transformations can be schematically represented as shown in Figure 1. The user, of course, is an integral part of these transformations. *Scene space* in this figure represents the space in which the given multidimensional image data are represented. For example, if the data represent a set of CT slice images of a static organ, then the scene space is the 3D space in which the image points together with a (CT) number associated with each image point are defined. When we extract an organ or its surface from image intensities, such information is represented in the *object space,* which is the 3D space in which points of the object or its surface are defined. Hence, such operations represent transformations from scene space to object space. When we rotate, scale, or translate scenes, objects, or surfaces, they are represented in another 3D space called the *image space*. The *view space* available to us is the 2D screen of a computer display device. Because of the disparity in dimensionality between image and view space, the information in the image space has to be subjected to a dimensionality-reduction transformation which usually consists of somehow projecting the object or surface onto the screen. To make up for the loss of dimensionality, usually a variety of illusions of depth are created in the displayed image via shading, showing only visible parts, transparency, stereo display, and displaying in rapid succession images representing successive views of the object (this is commonly known as animation). Often, certain measurements are made on the object based on the image displayed in the view space; for example, the measure may be the volume of a certain organ, the length of a ridge on a surface, or the mean CT number in a specified region. In more sophisticated mensuration, we may be interested in more than one parameter. Such operations, hence, represent a transformation from the view space to the *parameter space.*

There are other possible transformations of information between spaces as indicated in Figure 1. We shall identify them while describing the various 3D imaging operations.

C. CLASSIFICATION OF APPROACHES

To systematize our discussion, we classify the 3D imaging approaches as shown in Figure 2. The first level of grouping is based on how the multidimensional information is displayed and analyzed. In *slice imaging* our aim is to extract certain 2D images (e.g., a coronal cross section) from the given multidimensional image data so that they can be directly displayed in the 2D view space for visualization and analysis. *Projective imaging* deals with techniques for extracting multidimensional information from the given image data and for depicting such information in the 2D view space via a process of projection. *Surface rendering* and *volume rendering* are the two major classes of approaches available under projective imaging. In surface rendering, visualization and analysis of multidimensional data are based on object boundaries. Volume rendering is not based on pre-extracted object boundaries; rather the objective here is to depict pseudo-surfaces and interfaces of various types of tissues, somewhat like in a radiograph but in a highly controlled and more sophisticated way. Finally, *volume*

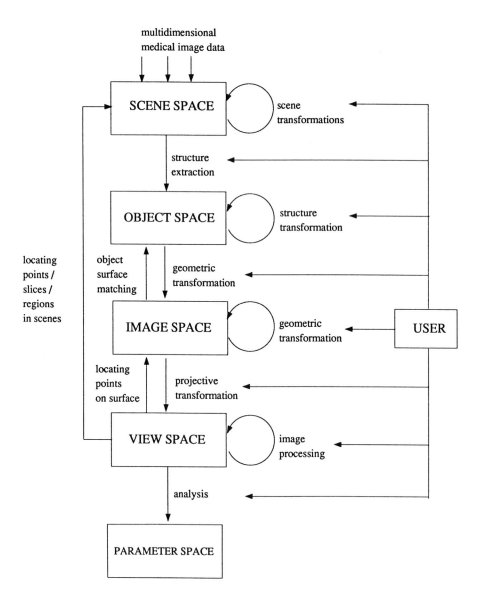

FIGURE 1. A schematic representation of common 3D imaging transformations.

imaging (different from volume rendering) methods, unlike slice and projective imaging, are based on technologies such as holography and varifocal mirror that provide a 3D view space.

It is interesting to note that while slice imaging provides basically a 2D mode of visualization and analysis, projective imaging allows a $2\frac{1}{2}$ D (more than 2D, yet not true 3D) mode. Although volume imaging provides a true 3D mode, this technology is not yet well developed, and, hence, its actual clinical use has been insignificant compared to that of slice and projective imaging. We will begin with a definition of our terminology in the next section, which will also describe certain operations that are common to a majority of the approaches. The third section will be devoted to a discussion of slice imaging methods. Surface and volume rendering will be covered in the fourth and fifth sections, respectively. While up to the fifth section our main concern is visualization, we will discuss some aspects of manipulation and analysis in the sixth section. Aspects of software and hardware related to 3D imaging will be addressed in the seventh section. The eighth section is an outline of the history of developments in 3D imaging. We will spare the last section for some general remarks and criticism on the current state of affairs.

II. TERMINOLOGY AND PREPROCESSING

The purpose of this section is first to define some of the key concepts, phrases, and terminology and then to describe some of the processing operations that are common to a majority of the approaches.

A. BASICS

Medical imaging devices such as computerized tomography (CT) scanners and magnetic resonance imaging (MRI) scanners essentially sample certain properties of tissues within a region inside the human body and represent the sampled values in the form of a set of cross-sectional images. The region within which the property values are estimated is usually cylindrical in the case of CT and cuboid (a rectangular box each of whose six sides are rectangles) in the case of MRI. We assume without loss of generality that this region is always cuboid; if it is of any other shape, we can determine a cuboid region that encloses the given region and assign a property value corresponding to air to the difference region. The sampling process can be regarded as a process of partitioning the cuboid region into a number of small cuboids and then assigning a number to each small cuboid. As is commonly done, we call the small cuboid a *voxel* (abbreviation for volume element along the lines of the abbreviation *pixel* for picture element). We call the value associated with the voxel its *density,* the array of voxels along with the value associated with each voxel a 3D *scene* or simply a *scene,* and the cuboid region in which the scene is defined as the *scene region* (see Figure 3). In short, imaging scanners produce scenes of the scanned body region. The voxel densities of a scene occupy a certain range of values (usually integers) with a minimum denoted by G_L and a maximum G_H. We call the scene for which $G_L = 0$ and $G_H = 1$ a *binary scene.* Of course, binary scenes

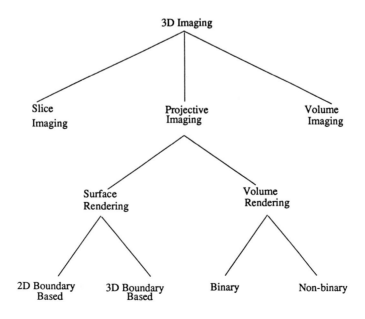

FIGURE 2. A classification of the 3D imaging approaches.

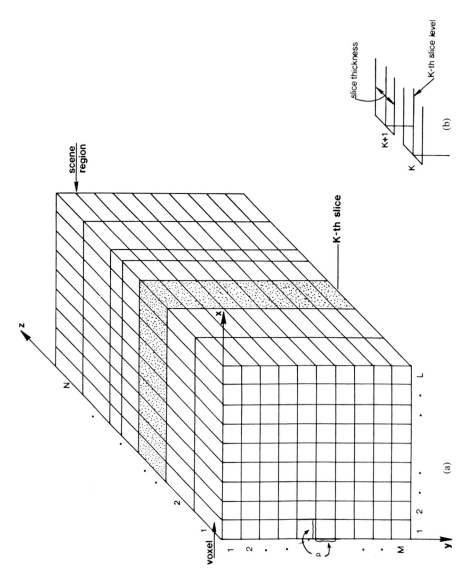

FIGURE 3. (a) Illustration used for the definition of basic terms. (b) The slice location (level) and thickness for the K-th and (K+1)th slice here are so chosen that they neither overlap nor abut. As we increase the thickness for both slices, they first abut and then overlap.

are not generated by scanners but can be created through computer processing. We assign a right-handed xyz coordinate system to the scene. (Arrange the thumb, the forefinger, and the middle finger to point along x, y, and z, respectively. The fingers are now automatically mutually orthogonal. This can be done only with the fingers of the right hand for the figure shown. This is the most natural coordinate system considering how the slices are displayed on the screen.) The xyz space in which the scene is defined represents the scene space we referred to previously.

The coordinate system allows us to index the voxels in the scene by a triple of integers (i, j, k) where the numbers represent respectively the x, y, and z indices of the voxel. In an L × M × N scene (meaning a scene with an L × M × N voxel array, see Figure 3), the subscene defined by the set of all voxels with coordinates (i, j, K), for all i between 1 and L, for all j between 1 and M, and a fixed K between 1 and N, together with the voxel densities, is called the *K-th slice* of the scene. Note that the K-th slice may not always represent a transaxial cross section since we do not assume that the z-axis of the scene space is in the direction of the long axis of the patient. This is often the case in MRI since the cross-sectional images may be acquired in a coronal, sagittal, or even oblique fashion.

Some explanation regarding the z-dimension of the voxel is in order. Each slice image estimated by the scanner has a certain "slice thickness" and "slice location" associated with it. These may be chosen in such a way that the successive slice images may or may not overlap or they may abut (see Figure 3b). Our definition of the z-dimension of the voxel is independent of slice thickness but depends only on slice location. We assume that the K-th slice extends from the K-th slice level (location) to the (K + 1)-th slice level. Thus, for example, the 1st slice extends from the xy plane (level of the 1st slice) to the level of the 2nd slice, and hence, the z-dimension of the voxels in the 1st slice is the difference in the location of the 1st and the 2nd slice. (For the N-th (last) slice, we assume this dimension to be the same as that of the N-1th slice.) Clearly, the voxels in two different slices may have different z-dimension.

The actual meaning of the voxel density depends, of course, on the imaging modality. This should be borne in mind while determining the organ for which 3D imaging and analysis is to be carried out. In CT, for example, the density of a voxel represents the extent to which X-rays passing through the voxel are attenuated by the tissue within the voxel. In MRI, on the other hand, the density represents an aggregate of one of several magnetic properties of the tissue molecules within the voxel. A knowledge of how these properties vary from one tissue to another is critical in determining the structure of interest.

B. SCENE PROCESSING

The operations described in this section process an input scene to produce another scene. They can hence be characterized as transformations from scene space into the scene space (see Figure 1).

1. Volume of Interest (VOI)

Often the structure that we want to display and analyze occupies only a small portion of the scene-region. The *VOI operation* allows us to create another scene whose scene region is again a cuboid which encloses all of the structure of interest but as little as possible of irrelevant structures. The structure of interest may be an entire organ or an organ system or just a part of an organ. The main purpose of the VOI operation is to minimize the computer storage space required for further processing of the scene. As an example, a 512 × 512 × 64 scene contains more than 16.5 million voxels. If the scene region of the structure of interest is 200 × 200 × 64, then the number of voxels, about 2.5 million, is nearly seven times less than that for the whole scene. Since most computers have just about enough (often less) main memory to store the smaller scene, and in many scene processing and other operations the time spent in transferring the scene data from a secondary storage medium (such as a disk) to the

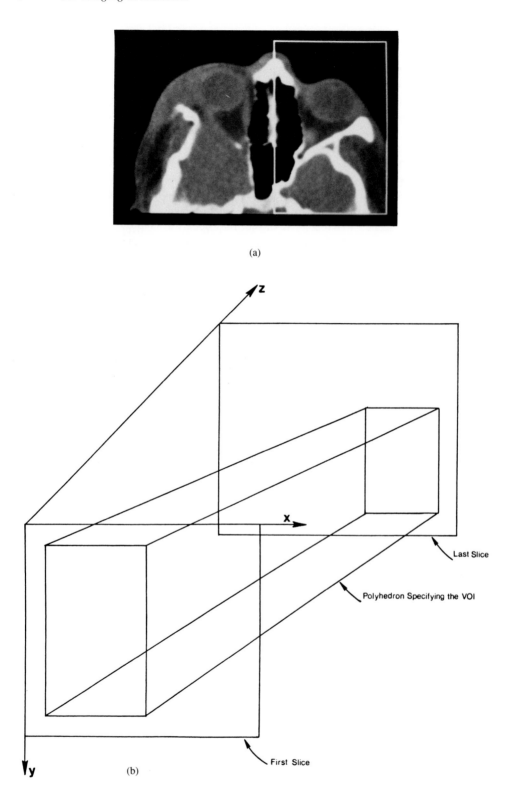

(a)

(b)

FIGURE 4. Illustration of the VOI operation: (a) the volume is a cuboid and (b) the volume is a polyhedron.

main memory becomes an overriding part of the total processing time, the importance of the VOI operation becomes obvious. Another reason for the VOI operation may be to exclude a part of an organ and to image the remainder so as to reveal its interior. For example, if we wish to create a 3D display of a part of the spinal column revealing the spinal canal, then we specify a VOI using a rectangle of fixed size and location on all slices of the scene (see Figure 4a). By specifying a fixed rectangle and a subset of slices in the given scene we have essentially indicated the scene region for the output scene. Creation of the output scene consists of just collecting of voxels (along with their density) that fall inside the rectangle in each slice of the set.

A somewhat more sophisticated VOI operation[6] is to specify a polyhedron instead of a cuboid (of course, the scene region for the output scene should still be a cuboid). A simple way of doing this is to specify two rectangles, one on the first of a set of slices and the other on the last slice. The polyhedron is defined by joining the corresponding vertices of the two rectangles, which of course may have different locations and sizes (see Figure 4b). This operation gives the effect of cutting by oblique planes (planes not orthogonal to the xy plane) which is often useful in the spinal example given above. To compute the output scene, we simply have to collect all voxels that fall inside the polyhedron and then appropriately pad the exterior region with just enough voxels to create a cuboid scene region.

2. Filtering

The main purpose of the filtering operation is either to smooth or to enhance the scene. Although many other forms of scene processing can also be done through filtering, we will concentrate here only on these two forms. The output scene created by the filtering operation has the same scene region as the input scene. In other words, both scenes have the same set of voxels but their densities may be different. The main idea behind scene filtering is to assign a density value to a voxel v in the output scene based on the densities of voxels in a small neighborhood of v in the input scene. Depending on how the density values are made use of and how the neighborhood is defined we get different filtering operations.

For example, if we assign to v a weighted average of the densities, then the local variations are averaged or smoothed. Hence, after the whole scene is filtered this way if we display the same slice from the input and the output scene side by side, the latter appears smoother than the former. Thus, in this filtering operation (commonly known as *low-pass* filtering), the density $g_o(v)$ assigned to a voxel v in the output scene is given by

$$g_o(v) = \frac{\sum_{j=0}^{n} w_j g_i(v_j)}{\sum_{j=0}^{n} w_j} \tag{1}$$

where $g_i(v_j)$ is the density of voxel v_j in the input scene, w_j is the weight (any finite positive number) associated with v_j, $v_0 = v$, and $v_1, v_2,...,v_n$ are the neighbors of v. Some of the commonly used neighborhood definitions are illustrated in Figure 5. The definition in Figure 5(a) is essentially 2D and hence the filtering in a slice is not affected by its adjoining slices. The other two definitions allow 3D operations. The weights are usually chosen such that the voxels closer to v_0 are given more weight than those that are farther away. In Figure 5(a), for example, we may choose $w_0 = 4$, $w_1 = w_3 = w_5 = w_7 = 3$ and $w_2 = w_4 = w_6 = w_8 = 2$. Note that the range of densities in the output scene remains roughly the same as that in the input scene.

In enhancement (also known as *high-pass* filtering), the idea is to emphasize edges in the scene. Clearly, since edges appear where there are large differences in density, to enhance edges we simply have to compute differences of densities of neighboring voxels. Here again, we can get different types of effects depending on how the differences and the neighborhood are defined. We give two examples.

(i) Using the neighborhood of Figure 5(a):

$$g_o(v) = \left| \frac{w_2 g_i(v_2) + w_1 g_i(v_1) + w_8 g_i(v_8)}{(w_2 + w_1 + w_8)} - \frac{w_4 g_i(v_4) + w_5 g_i(v_5) + w_6 g_i(v_6)}{(w_4 + w_5 + w_6)} \right| +$$
$$\left| \frac{w_6 g_i(v_6) + w_7 g_i + w_8 g_i(v_8)}{(w_6 + w_7 + w_8)} - \frac{w_2 g_i(v_2) + w_3 g_i(v_3) + w_4 g_i(v_4)}{(w_2 + w_3 + w_4)} \right| \qquad (2)$$

(ii) Using the neighborhood of Figure 5(b):

$$g_o(v) = \frac{w_1 |g_i(v_1) - g_i(v_3)| + w_2 |g_i(v_4) - g_i(v_2)| + w_3 |g_i(v_6) - g_i(v_5)|}{w_1 + w_2 + w_3} \qquad (3)$$

In these equations |x| represents the absolute value of x, and the weights are any finite, positive numbers. It is readily seen that the voxel density $g_o(v)$ in the output scene is simply the sum of the differences of input densities in the vicinity of v along the principal directions.

A third type of filtering that is often quite useful in 3D imaging is called *median filtering*. In this operation, $g_o(v)$ is taken to be the median value of the densities $g_i(v_1),...,g_i(v_n)$. For example, suppose we use the neighborhood of Figure 5(b) and that the neighbor densities around v [i.e., the values of $g_i(v_1),...,g_i(v_6)$] are 100, 120, 110, 200, 90, 80. To find the median

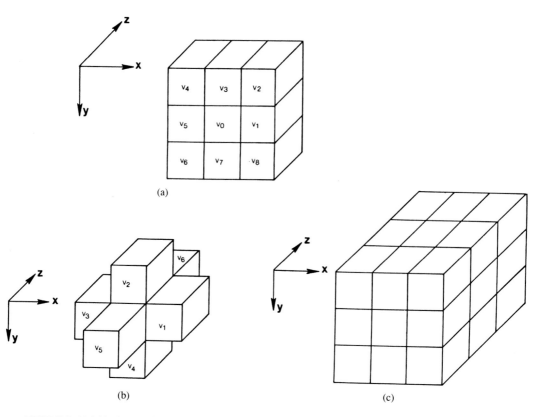

(a)

(b)

(c)

FIGURE 5. Neighborhood definition for the filtering operation. (a) Neighboring voxels are defined in the same slice. (b) Neighboring voxels, total 6, share a face with the voxel v_0 which is in the center. (c) The total number of neighboring voxels is 26, and v_0 is in the center.

value, we arrange these densities in the ascending order 80, 90, 100, 110, 120, 200, and then take the middle value (either 100 or 110) and assign it to $g_o(v)$.

The use of low-pass filtering in 3D imaging is mainly to suppress noise. One undesirable aspect of this operation is that along with noise, the real sharp density variations due to the presence of tissue boundaries are also often smoothed out. Of course, it is possible to detect such discontinuities through high-pass filtering and do low-pass filtering selectively. The median filter on the other hand is quite effective in suppressing noise and at the same time in preserving the sharpness of real boundaries. The common use of high-pass filtering is in determining where tissue boundaries are located by computing the variation in density. This is also used in determining the shading function for surface rendering as we will describe in later sections.

C. INTERPOLATION

Since the separation between successive slice locations in a scene is usually much greater (2 to 15 times) than the dimension of the voxel in the x or y direction [denoted by p in Figure 3(a)], the voxels are in general cuboids and not cubes. Further, the separation itself may not be uniform between successive slice locations. It is often desirable to scan a patient in this fashion for a variety of reasons. For example, in craniofacial surgical planning, if the site of deformity is the region of the orbits, it is desirable to get detailed information in this region because the bony structures in this region are fine and thin. Hence, an optimal CT scanning strategy for a given patient dose constraint is to have slices closely located in this region but more spread out in other regions above and below the orbits. In short, the scenes generated by scanners usually do not represent an isotropic sampling of the scanned body region.

For a variety of reasons, mainly to do with accuracy and sometimes for processing efficiency, it becomes necessary to create a uniformly sampled scene from the given nonuniformly sampled scene — an operation called *interpolation*. We will describe several techniques that are commonly used for interpolating scenes.

To begin with, we assume that the x and y dimensions p of the voxel in the input scene and that in the output scene, p', are identical. Our aim is to create an output scene of cubic voxels of side $p' (= p)$. The interpolation operation can be described by concentrating on one column of voxels in the input scene in the z-direction with coordinates (I, J, k) for some fixed I and J and for $1 \leq k \leq N$, since the same operation is repeated for all columns. We have shown the first two voxels of such a column in Figure 6(a). The simplest way of determining the density to be assigned to the voxels $v_1', v_2', \ldots$ of the output scene in this column is by using the *nearest neighbor* rule: assign to v_m' the density of the voxel v_n nearest to v_m'. One possible definition of nearness is using the distance between the center of v_m' and that of v_n. By this rule, the voxels v_1' and v_2' in Figure 6(a) are both assigned the density of v_1 while $v_3', v_4',$ and v_5' are all assigned the density of v_2. It is clear that nearest-neighbor interpolation is equivalent to duplicating thinner slices. A somewhat better approach, called *linear interpolation,* assigns different values that vary in a linear fashion to the voxels (v_1', v_2') falling within the same voxel (v_1) of the input scene. One possible model of this form of variation assumes that the density varies linearly along z across the voxel v_n from the density of v_n at the front of v_n to that of $v_n + 1$ at the end of v_n (which is also the beginning of $v_n + 1$). In the context of Figure 6(a), since the density within v_1 varies uniformly from its density $g_i(v_1)$ at its front to $g_i(v_2)$ at its back, the density $g_o(v_2')$ at the center of v_2' which is at a distance $p + p/2$ from the front of v_1 is given by

$$g_o(v_2') = g_i(v_1) + \left[\frac{g_i(v_2) - g_i(v_1)}{t_1} \right] (p + p/2) \tag{4}$$

To compute the density of any other voxel v_m', we simply have to determine the voxel v_n of

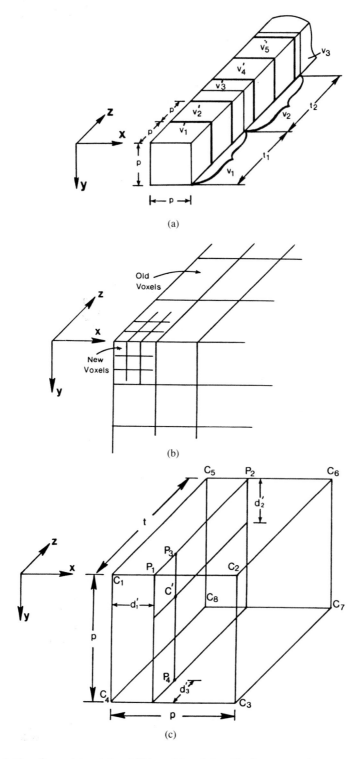

FIGURE 6. Illustration of scene interpolation. (a) Linear interpolation. The first two voxels — v_1, v_2 — of a typical column of voxels along the z-direction in the input scene are shown. The voxels v_1', $v_2',...,v_5'$ of the interpolated (output) scene are indicated by heavy lines. (b and c) Trilinear interpolation. C_1, $C_2,...,C_8$ are the centers of eight voxels that are closest to C', the center of v', whose density we want to estimate. Note that the polyhedron $C_1, C_2,...,C_8$ does not itself correspond to any voxels, new or old.

the input scene in which the center C'_m of v'_m falls and then compute the distance d'_m of C'_m from the front end of v_n. The density of v'_m is given by a formula similar to that in Equation 3:

$$g_o(v'_m) = g_i(v_n) + \left[\frac{g_i(v_{n+1}) - g_i(v_n)}{t_n} \right] d'_m \qquad (5)$$

where t_n is the z-dimension of v_n. Note that we cannot estimate interpolated density at centers which fall inside v_N, where v_N is the voxel in the column under consideration in the N-th slice of the input scene. Hence, the interpolated voxels extend up to the front end of the N-th slice.

A still more sophisticated method that is also commonly used in 3D imaging is the so-called *trilinear interpolation* method. As the name implies, instead of interpolating in only one (z) direction as in the previous method, this allows (linear) interpolation in all three directions. It is not necessary to assume p = p' as done in the previous method. This implies that we can create a scene that is more finely or more coarsely sampled than the input scene. Imagine that we build the new voxels of side p' starting from the origin of the scene space [Figure 6(b)]. Our problem is to determine the density $g_o(v')$ at the center of each new voxel v'.

The principle is somewhat similar to that of linear interpolation except that the step indicated by Equations 4 and 5 are applied a few times to compute $g_o(v')$. The steps involved are illustrated in Figure 6(c). We make a slight refinement to the assumed model of variation of density across the old voxels. In this model, we assume that the density $g_i(v)$ of the old voxel varies linearly from the center of its front end (face) to the centers of the front ends of its six neighbors in the principal directions, where the neighbors of v are defined as in Figure 5(b). Given this model, to determine the density at any point C' (which represents the center of the front of the new voxel v') in the scene space, we find eight old voxels $v_1, v_2, ..., v_8$ such that the centers $C_1, C_2, ..., C_8$ of their front ends are closest to C' as illustrated in Figure 6(c). The trilinear problem then is to determine the density at C' knowing the density $g_i(v_1), ..., g_i(v_8)$ at the centers $C_1, ..., C_8$. Note that since we know the coordinates of C' in the scene space as well as the distance between these centers, we can calculate the distances marked d'_1, d'_2 and d'_3 of C' with respect to the centers $C_1, ..., C_8$. The actual computation of the density at C' now consists of seven calculations similar to that in Equation 5: first we compute the density at P_1 knowing the density at C_1 and C_2, then at P_2 knowing the density at C_5 and C_6, and then at P_3 knowing the density at P_1 and P_2. Similarly, we compute the density at P_4 using three calculations. Finally $g_o(v')$ is computed knowing the density at P_3 and P_4. Clearly, if C' falls on a face of the cuboid $C_1...C_8$, we need only three calculations, and only one if C' is on an edge.

It should be clear by now that, computationally, the nearest-neighbor rule is the simplest interpolation method while the trilinear rule is the costliest of the three methods. Though the linear, and even more so the trilinear, methods lead to subjectively more pleasing visualization than the nearest-neighbor method, how they or other higher order interpolation schemes actually influence the clinical application at hand is largely unknown.

The interpolation problem, in our opinion, has received less attention than it deserves. We recently developed a new technique[7] called *shape-based interpolation* which instead of inter-polating densities first determines the structure and then interpolates it. The main idea in this technique is to first convert the given scene into a binary scene (we will discuss such scene transformations in the following section) such that the voxels with density 1 in this scene represent the structure of interest and those with density 0 represent nonstructure regions. Shape-based interpolation uses this binary scene as input and creates another binary scene for the specified voxel size p'. The method consists of first converting the input binary scene S_b into a scene S_g, both with the same scene region, by assigning to each voxel v in S_g a number that represents the shortest distance between v and the (2D) boundary in S_b within the slice that contains v. The number is positive if v has a density 1 in S_b within the slice that contains v.

The number is positive if v has a density 1 in S_b, else the number is negative. Now we interpolate S_g using linear, trilinear, or other interpolation rules to create another scene S_g'. Remembering that the numbers associated with voxels in S_g' represent distances from boundary, we create the final interpolated binary scene S_b' by assigning to a voxel in S_b' the density 1 if the corresponding voxel in S_g' has positive value. All other voxels are assigned the density 0. Our preliminary evaluation indicates that shape-based interpolation leads to more accurate quantitative analysis, such as estimation of volumes and areas, and to better subjective depiction in surface rendering methods.[7]

D. SEGMENTATION

The purpose of segmentation is to identify the structure of interest in the given scene. There are basically two types of approaches to scene segmentation: Those which directly produce structure boundary information, called *boundary-based* approaches, and those that produce information about the space occupied by the structure, called *region-based* approaches. In the former case, the output of segmentation is the structure boundary, and, in the latter case, the output is a binary scene in which the voxels with density 1 represent the space occupied by the structure. In the remainder of this section, we will describe some of the segmentation techniques commonly used in 3D imaging under both these approaches.

1. Boundary-Based Approaches

If the imaging modality used for generating the input scene is effective in distinguishing various tissue regions within the part of the body scanned, we expect to discern the boundary of the structure of interest with the surrounding tissue regions in a display of the slices of the scene. At such a boundary it is reasonable to expect a discontinuity in the distribution of the density (and, perhaps, of a number of other properties computed from the density). The main idea behind boundary-based approaches is to somehow detect where such discontinuities occur in order to locate the boundary.

A simple way of detecting discontinuities in the distribution of density is to find out how rapidly the density changes at each voxel. At voxels in the vicinity of the boundary this rate is likely to be much higher than at those away from the boundary. If a formula were available to us to express the scene density variation as a function of x, y, and z, we can use a mathematical operator called the gradient to compute the rate of change. The output of this operation at every point x, y, z is a vector whose magnitude indicates the magnitude of this rate and direction indicates the direction in which the change is maximum. Since such a formula is not available, we use local differences computed at each voxel (in a fashion similar to the high-pass filtering operation described earlier) to estimate the magnitude of the rate of change (we ignore the direction of the gradient vector for the time being, but later on explain how to use this information for surface shading).

We first consider a 2D example and then move on to the 3D case to describe gradient-based boundary detection. Figure 7(a) shows a part of a slice of a 3D scene which contains a part of a boundary. Since a slice, and even more so a 3D scene, may contain many boundaries, to simplify the problem we assume that a starting location (a voxel face) on the boundary of interest is known. Our aim is to extract the boundary, starting from this location, as a ribbon of voxel faces [Figure 7(c)]. To find such a starting face, we may display the slice and have the user point to a location near the boundary of interest where it is most unequivocal. Suppose ab represents such a face in Figure 7(a). (Note that line segments such as ab in Figures 7(a), (b), and (c) actually represent a face and a point such as a represents an edge of a face because of the depth of the slice. For simplicity we have not shown this depth in Figures 7(b) and (c).) Our idea is to expand the boundary from the face ab from both of its edges a and b into a full boundary within the slice. The steps in this process are illustrated in Figures 7(b) and (c). We keep a pool of open edges which, to begin with, consists of a and b. At edge a we interrogate which of the three possible edges, a_1, a_2, and a_3, can be chosen so that together with a we get

15

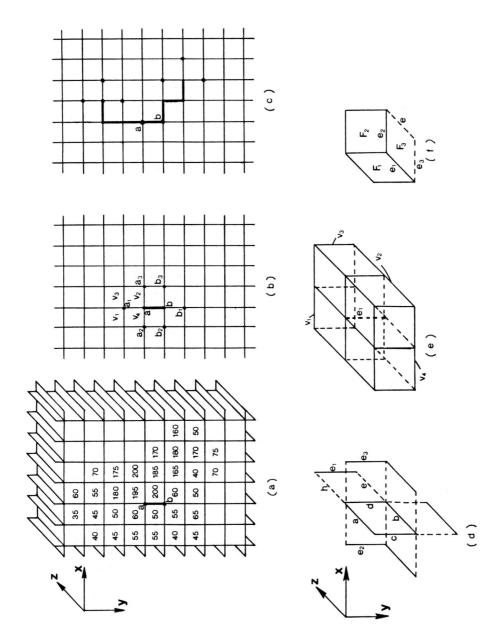

FIGURE 7. Illustration of a gradient-based boundary detection technique. (a) A part of a slice of a scene; ab represents the initial boundary face from which detection starts. (b and c) show how detection proceeds. (d, e, f) Illustration of the 3D case.

one of the three possible faces aa_1, aa_2 and aa_3. It is in this choice that we use the gradient. We estimate the gradient at a_1 by computing the sum $|g(v_1) - g(v_2)| + |g(v_3) - g(v_4)|$ and similarly at a_2 and a_3. That edge among a_1, a_2, a_3 is chosen which gives the maximum gradient. When we choose an edge, say edge a_1, we remove a but add a_1 to the pool to indicate that it is an "open end". We expand from edge b in a similar fashion. It is easy to verify that after a few steps the faces discovered by this algorithm for the example in Figure 7(a) will be as shown in Figure 7(c).

Of course, our aim is to eventually empty the pool, when we will have constructed a closed boundary. During this process it is possible that the boundary we have traced so far crosses itself. Clearly this happens when a previously chosen edge is chosen again. In such a case we discard this choice and accept the second best choice. If all three edges lead to this situation we "backup" one edge (which means a previously formed face should be discarded) and see if alternative choices (those not yielding maximum gradient) can get us out of the situation of self-crossing boundary. Backing up may have to be repeated several times. Eventually, of course, we will succeed and create a closed boundary.

Note that the above algorithm is essentially 2D in nature and our aim was to extract a boundary within a given slice. With a slight modification it can be generalized to detect a complete 3D surface. The idea is illustrated in Figures 7(d) and (e). Suppose our initial face is ab as in Figure 7(d). We now wish to expand from all four edges a, b, c, d instead of just the two edges a and b as done previously, and the pool contains all four edges. For each of these edges we examine which of the three possible next edge selections yields an optimal face. For example, for edge d, the three edges to be examined are e_1, e_2, and e_3. As before, an edge is selected if the gradient at the edge is maximum. The magnitude of the gradient at e_1, for example, is the sum [see Figure 7(e)] $|g(v_1) - g(v_2)| + |g(v_3) - g(v_4)|$. If we select e_1, d is removed from the pool but e_1 and the remaining two edges e and h of the face de_1 are added to the pool since we have to expand from all of these edges. Clearly, the pool grows rapidly, at least initially (note that, in the 2D case, the pool always contains two edges). After generating a few faces, it is possible that some of the edges (such as e_1, e_2 in Figure 7(f) of a face such as F_3 just found) may be already in the pool (because faces F_1 and F_2 were previously generated). In this case, we remove from the pool such edges and add only edges (e_3) not already in the pool. As in the 2D case, our aim is to empty the pool. The backing-up operation in 3D is similar to the 2D case.

Our above description is based on what is perhaps the first published segmentation algorithm[8] in 3D imaging other than thresholding (see References 9 and 10 for other gradient-based techniques). This algorithm is readily extended[11] to detect the boundary of a dynamic organ (which can be considered as a four-dimensional (4D) object in a 4D scene space xyzt where the t axis represents time). Further, the selection of the next edge can be based not just on one criterion such as the gradient magnitude but on a set of constraints. For example, we may insist that the gradient magnitude should be within a certain interval, the density also should be within a certain interval, and that the mean density (and possibly other properties) on both sides of the potential boundary face should agree with previously known values. In addition we may make these constraints themselves vary as different sections of the boundary are detected (this appears reasonable if different boundary segments represent interfaces between different tissue regions). We may then think of the selection to be done by a processor which has three edges as its input and an edge or an error signal as its output. The error signal, of course, indicates the need for backing up on the boundary. Such complex approaches are just beginning to be explored.[12]

2. Region-Based Approaches

The premise in region-based approaches is that the voxels that are inside the structure of interest can be distinguished from those that are outside or those that surround it by a set of common features (properties). The problem of segmentation here can be considered as determining whether or not each voxel in the scene belongs to the structure. The result of such an exercise is a binary scene where a density of 1 indicates that the voxel belongs to the structure and a density of 0 means the voxel is outside.

The simplest of these approaches is *thresholding*:[13] the voxel is considered to belong to the structure if a certain feature value evaluated at the voxel exceeds a fixed threshold value. The feature may be the voxel density itself or it may be an entity derived from the voxel density distribution such as mean density computed from a small neighborhood of each voxel. A more general form of thresholding is to use lower level and an upper level for the threshold and to classify the voxel as belonging to the structure if the feature value associated with the voxel falls between the two levels. Often it helps to use multiple features and then use a lower and an upper threshold for each feature.

An even more general approach, which we call the *feature plot* method, is to plot the feature values and to recognize clusters in the feature plot to identify the voxels belonging to the structure. To be more specific, suppose we associate two features f_1 and f_2 with each voxel and we plot the values of these features for every voxel in the scene (see Figure 8). If our choice of features is good, we expect the points representing feature values of voxels belonging to the structure of interest to form a cluster in this plot. To identify the voxels in the structure, then, we first create such a plot, construct a line separating the cluster representing the structure from the rest of the plot, and then find out voxel by voxel on which side of the line its feature values (plotted as a point) fall. In general it may not be possible to isolate the cluster by a line, and we may have to determine a curve that separates the cluster optimally.

An example of a situation where two-feature cluster partitioning may be useful[14] is illustrated in Figure 8(c). Suppose T_1, T_2, and T_3 represent three different tissue regions in a slice such that the mean voxel densities t_1, t_2, t_3 in these regions are in the order $t_1 > t_2 > t_3$. Since voxels in the boundary between T_1 and T_3 are likely to contain both types of tissues T_1 and T_3, their density lies in between t_1 and t_3 (a phenomenon known as *partial volume artifact*) and is thus very close to t_2. Hence, if we try to extract T_2 by setting an upper threshold $t_2 + \Delta t$ and a lower threshold $t_2 - \Delta t$, where Δt is chosen such that we cover the range of variation of the density within T_2, then in addition to the voxels within T_2 those in the boundary between T_1 and T_3 are also classified as belonging to T_2. No matter how we choose Δt it is not possible to extract T_2 without including such boundary voxels. If we include a second feature, in addition to density, namely the gradient magnitude, computed at each voxel (say, using Equation 2), we can differentiate between voxels within T_2 and those in the boundary between T_1 and T_3 since the latter voxels have a much higher gradient value than those within T_2. A feature plot in this situation would have three clusters as illustrated in Figure 8(d), corresponding to the three tissue regions. All clusters are situated close to the f_1 axis since the gradient value inside a region is very small. Since the boundary voxels have a high gradient value, they are represented by the arch between T_1 and T_3 clusters in this figure. (There are similar but small arches between the clusters of T_3 and T_2, and T_2 and T_1 corresponding to the boundary voxels between the respective regions.) To extract the T_2 region, we simply identify all those voxels whose feature values fall into the T_2 cluster. A medical imaging example [Figure 8(c)] of this situation is the MR imaging of an infarcted myocardium where T_1 represents the myocardium, T_2 the infarcted region, and T_3 the region outside the myocardium.

Two-feature cluster partitioning is becoming popular in MR imaging applications.[15-17] The features selected are from among the measured MR properties of tissues such as T_1 and T_2 relaxation times and proton density of tissues or even those derived from the measured properties. Cluster partitioning tools are useful in experimentally determining which measured and/or derived features are effective in identifying various types of tissues.

The above partitioning method can be generalized to more than two features. However, interactive processing such as pointing to a cluster of interest on a display of the feature plot or interactive isolation of the cluster would become difficult or impossible for more than two features. There is a well-developed body of theory and techniques[18] for automatically classifying feature vectors (in our case, voxels) based on statistical variations of feature values for each class (in our case, tissue type). Such techniques do not use feature plot displays to do the classification. The use of such techniques in 3D image segmentation remains largely unexplored.

A few other types of region-based segmentation techniques for 3D imaging have been reported.[19-21] Some of the recent ones are in the beginning stages of their clinical evaluation and hence their effectiveness is not known at this time.

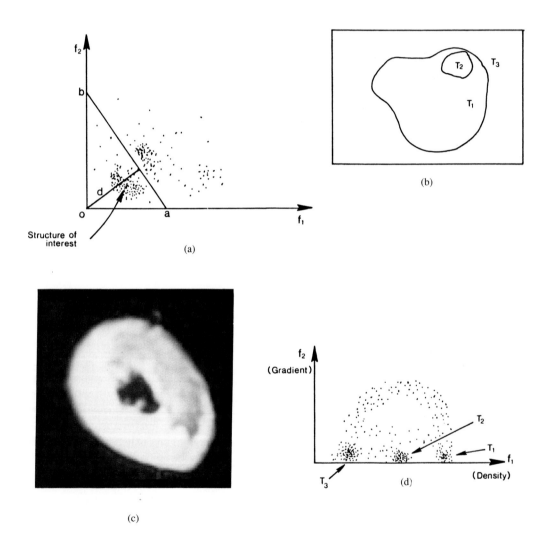

FIGURE 8. Illustration of feature cluster partitioning. (a) Two-feature case. A point (f_1, f_2) in the feature plot corresponds to a voxel with feature values f_1 and f_2. A dense cluster signals that a number of voxels in the scene have similar feature values. The segmentation rule for the example shown is given by: assign density 1 to the voxel if $f_1/a + f_2/b < d$, d being the distance of line ab from the origin; otherwise assign density 0 to the voxel. (b and c) An example of the two-feature case. Here f_1 is the voxel density and f_2 is the rate of change of density (gradient) at the voxel. The regions T_1, T_2, T_3 have mean densities t_1, t_2, t_3 such that $t_1 > t_2 > t_3$. The region T_2 is extracted by assigning the density 1 to those voxels that map into the middle cluster in (d).

E. MASKING

This is an operation that takes a scene and a binary scene as input and produces a binary scene as output. The domain for all scenes should be identical.

The purpose of masking is to complement segmentation. Quite often automatic segmentation fails and the only alternative is to have the medical expert draw on a display of the slices of the input scene to indicate where the structure of interest lies. In the worst case he may have to draw precisely the 2D boundary of the structure on all slices of the scene. Usually, however, it is enough to specify a region called the *mask region* such that automatic segmentation confined to this region can extract the structure of interest.[22] This implies that precise tracing of the boundary may be needed, if at all, at only small sections of the boundary, and often the mask region may not change over a few slices. As an example, suppose we are interested in the 3D imaging of the articular surfaces of a joint. Because of the partial volume phenomenon mentioned earlier, it is often impossible to automatically segment the bony components of the joint. Using the masking operation the user has to draw the boundary as best as he can where the boundaries of adjacent components appear to fuse in the slice image. Where the boundaries are far apart he just has to make sure that the component under consideration is inside the mask region (see Figure 9).

It is readily seen that the set of all mask regions specified on all slices (of course, some of them may not have any mask region) can be represented as a binary scene wherein a voxel has density 1 if it is on or inside the curve traced by the user (clearly, for "inside" to be well defined the curve should be "closed"); otherwise the voxel has density 0. The binary scene is easily generated from the traced curves by sorting the voxels in each row of the scene in an ascending order of their x-coordinates [see Figure 9(b)] and then generating voxels (with density 1) between every successive pair of voxels of the curve in the sorted list. Once the binary scene representing the mask region is generated, the output binary scene is easily computed by assigning the density 1 to those voxels of it which have a density 1 in both the mask and the input binary scene. All other voxels are assigned the value 0. (Note that the masking operation as we have described is suited for supporting region-based segmentation. It is possible to modify boundary-based techniques so as to accommodate masking.)

If the masking operation is indispensable in an application, we strongly recommend the use of shape-based interpolation described earlier. The reason is that if another scene interpolation scheme is used instead, masking will have to be done on a much larger number of slices. As an example, suppose we wish to segment a 4D scene representing time-varying cardiac MR images consisting of 10 slices for each of 8 time instances, with (cuboid) voxels of uniform size 0.7 mm $\times$ 0.7 mm $\times$ 5 mm. Assuming we create an interpolated scene with cubic voxels of size 0.7 mm $\times$ 0.7 mm $\times$ 0.7 mm, the interpolated scene will have close to 560 slices. If we wish to create more, say 16, time instances, also through interpolation (often desirable for providing a smooth depiction of dynamics), the resulting 4D scene will have more than 1100 slices! The masking operation quickly becomes an impossible task. If we use shape-based interpolation, the masking operation is required on only the 80 (original) slices (often less, since the mask can be duplicated by the computer for a set of contiguous slices), irrespective of how much interpolation is needed or desired.

In the above description of the scene processing operations, intentionally we have not indicated the order in which these operations should be done, because, in some sense, they are independent. However, the outcome may not be the same for two different sequences of processing operations. For example, the result of filtering followed by scene interpolation may not be the same as that of interpolation followed by filtering. We have already explained why segmentation followed by binary scene (shape-based) interpolation is generally superior to scene interpolation followed by segmentation. A sequence that we highly recommend is filtering (median)–segmentation (and masking)–shape-based interpolation. In an actual computer implementation, for reasons of efficiency, it is not necessary to carry out an operation on the

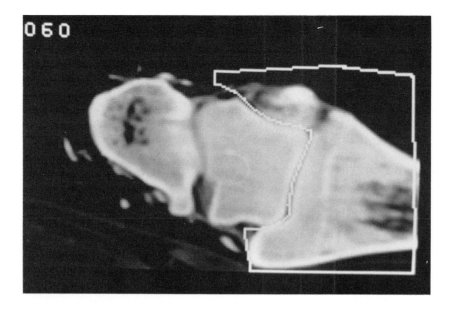

(a)

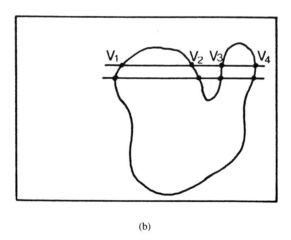

(b)

FIGURE 9.   Illustration of masking. (a) Masking is needed in disarticulation of joints. Here we are interested in viewing the tibia and the fibula independently from the rest of the bones at the ankle using CT data. (b) Illustration of how a binary scene corresponding to the mask can be generated.

whole scene prior to applying the next operation; rather the operations can be applied in combination. For example, in boundary-based segmentation based on filtered interpolated scenes, it is not necessary to carry out filtering and interpolation on the entire scene. It is often more efficient to determine the filtered and interpolated density of only those voxels for which we need the density value during boundary detection (see Figure 7). This implies that we do filtering and interpolation only along the path traced out by the detection process.

This concludes our discussion of the commonly used preprocessing operations. In the following three sections we will describe the commonly used techniques for visualizing 3D structures.

III. SLICE IMAGING

The premise in slice imaging is to break up the 3D scene into 2D subscenes in a variety of ways and to visualize the structure of interest through visualization of the subscenes. Slice imaging has two aspects to it. The "slice" part refers to computing the subscenes, which we call a "reslicing" operation, and the "imaging" part refers to how the subscenes are displayed. The former operation represents transformations from the scene space into the scene space (see Figure 1), and the latter represents transformation from the scene space into the view space.

A. RESLICING

To facilitate our description, we refer to the subscenes resulting from reslicing as *R-slices*. The simplest reslicing operation is to get one of the slices of the given scene. This, of course, does not involve any computation, and in this case an R-slice is simply a K-th slice of the scene.

A somewhat more complex operation is to compute R-slices orthogonal to the xy-plane of the scene space [see Figure 10(a)]. If we assume that the slices of the scene represent transaxial cross-sections of the human body lying prone or supine on a plane parallel to the xz-plane, then reslicing parallel to the yz-plane yields sagittal R-slices and that to the xz-plane yields coronal R-slices. If the voxels of the given scene are of the same size as that of the R-slice and if both are cubic, then this operation consists of simply picking up the voxel densities corresponding to the R-slice from the given scene. Usually the voxel densities of the scene are stored in a computer disk file as a sequential list in a row-by-row fashion starting from the 1st slice and going up to the last (Nth) slice. Hence the order of the voxels for storing their density is $(1,1,1)$, $(2,1,1),\ldots,(L,1,1)$, $(1,2,1)$, $(2,2,1),\ldots,(L,2,1),\ldots,(L,M,1)$, $(1,1,2)$, $(2,1,2),\ldots,(L,M,N)$. For the $3 \times 3 \times 3$ scene shown in Figure 10(b), this order is: 11, 21, 43, 27, 22, 33, 42, 56, 57, 15, 17, 40, 20, 25, 30, 44, 50, 61, 16, 19, 48, 32, 29, 35, 51, 63, 65. To compute, say, a sagittal R-slice that goes through the second column of the scene, we have to pick up densities in the following order of the voxels: $(2,1,1)$, $(2,1,2)$, $(2,1,3),\ldots,(2,1,N)$, $(2,2,1)$,. $(2,2,2),\ldots,(2,2,N),\ldots,(2,M,N)$. In the example, the sequence of densities will be: 21, 17, 19, 22, 25, 29, 56, 50, 63. These values should, of course, be interpreted as the densities of the R-slice arranged in a row-by-row fashion. Constructing an R-slice in this manner is faster if the whole scene can be stored in the main memory (not the disk) of the computer than when only a part of the scene can be stored. The main reason is that values can be accessed much faster from the main memory than from the disk. A $512 \times 512 \times 64$ scene for example has over 16 million voxels. Typically each voxel density requires 2 bytes of storage space (a byte is the space required to store one character), and hence the storage required for the scene is over 32 MB (MB is an abbreviation for mega (million) bytes). Though there are sophisticated workstations with this size of memory, it overwhelms most CT scanner computers by a factor of roughly 256. Note that the resulting R-slice itself needs to be stored somewhere, maybe for later display. Though in this simple reslicing operation there is no computation involved, when the memory is insufficient, it is not a trivial task to devise a strategy of accessing data from the disk so as to minimize the total number of accesses and to optimize speed.

Now suppose the voxels of the R-slice are not of the same size as that of the given scene and that their cross-section within the plane of the R-slice is a square (we are usually not interested in rectangular cross-sections since the pixels of the display screen are always squares; otherwise we will have trouble viewing the slices). The R-slice now may not nicely coincide with a layer of voxels of the scene as in the previous situation even though we assume the plane of the R-slice to be parallel to the yz- or the xz-plane. This implies that we have to compute the R-slice through some form of interpolation. The situation is the same even when we consider the R-slice to be oriented obliquely as in Figure 10(c). Since the location, orientation, and size of the rectangular block defining the scene domain of the R-slice are completely known (because the user has specified these parameters), we can start building voxels from one corner of this block. For each voxel v' in the R-slice, hence, we know the coordinates of its center C'. As described under "Interpolation", we may use one of a number of techniques to assign a density to v'. The simplest rule is to assign to v' the density of the voxel in the input scene whose center is closest to C'. A more sophisticated technique is to use trilinear interpolation as illustrated in Figure 6(c).

The above technique can be readily extended to the computation of curved R-slices. All we have to do is compute the center C' of each new voxel in the curved slice.

The location and orientation of the R-slices in all cases of orthogonal, oblique, and curved

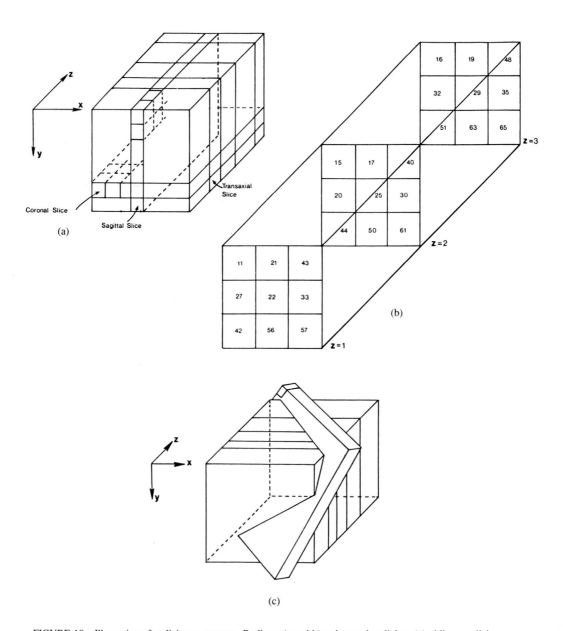

FIGURE 10. Illustration of reslicing to compute R-slices: (a and b) orthogonal reslicing, (c) oblique reslicing.

R-slices can be specified in a variety of ways. The simplest is to have the user specify a line or a curve on the display of a slice of the scene to indicate a plane or a curved surface perpendicular to the plane of the slice. Alternatively, we may compute an R-slice, say a sagittal slice, and then indicate the line or the curve on this slice. A third possibility is to display a cuboid in proportion to the scene region together with a mathematical plane or curved surface which the user can interactively alter.

Slice imaging is the earliest form of 3D imaging attempted. References 23 to 26 provide some examples of early publications reporting slice imaging. Commonly called multiplanar reconstruction by scanner manufacturers, such tools are currently available on most CT and MRI scanner display consoles.

B. DISPLAY

The pixels in the screen of a display device are all of identical size and the brightness of each pixel can be independently controlled through a computer program. Since the voxels in a slice (or an R-slice) are uniformly spaced in the plane of the slice, a simple way of visualizing the density distribution in the slice is to assign a brightness to each pixel proportional to the density of the corresponding voxel in the slice. If we have a 512×512 slice, we need a 512×512 pixel area in the screen to make this one-to-one assignment. This is the simplest form of transformation from the scene space to the view space. As we will see later on under "Volume Rendering", there are other complex transformations when a one-to-one correspondence cannot be established.

Suppose the lowest and the highest densities in the slice are G_L and G_H, and the brightness levels, called the grey values of the display device, range from 0 to H (the range of grey values is called the *grey scale* of the device). Then the above simple assignment of G_L to 0, G_H to H, and all in-between densities to in-between grey values in a proportional fashion [see Figure 11(a)] is not always useful in examining a particular type of tissue in detail. A technique commonly used to overcome this difficulty is called *windowing* wherein only the densities within a "window" are considered for proportional (linear) assignment, and the densities below and above the window are assigned grey values 0 and H, respectively [see Figure 11(b)].

It is not always necessary to make a linear mapping of density to grey values. In fact, often it is preferable to assign grey values in a nonuniform fashion.[27] It is known that the human eye is more sensitive to differences in grey values at the lower end of the grey scale than to those at the higher end, the sensitivity being dependent on both the observer and the device. Hence, to make optimum use of the grey values, it is better to assign more grey values to the lower end of the density window [see Figure 11(c)] than to the higher end. This type of mapping can be worked out for any given display device through observer experiments. Assuming that the perceptual variation between those observers involved in the experiment and users of the device is negligible, once the mapping is determined, it can be performed automatically through a computer program.

Another form of nonlinear density to grey value assignment is called *histogram equalization*.[28] Suppose a 100×100 slice (or an R-slice) of a scene consists of four regions each with uniform density as shown in Figure 12(a). The plot of the density versus the number of voxels with each density, called a *histogram,* of the slice is as shown in Figure 12(b). Suppose also that the grey scale of the display device ranges from 0 to 150 and that we wish to visualize the entire density distribution from 0 to 3000 in a single display. If we do a linear assignment [Figure 11(a)], then the displayed image will have pixel values as shown in Figure 12(c) and its histogram will be as in Figure 12(d). Though the boundaries BO and DO would be clearly discernible in the image, AO and CO would be barely visible because of the low contrast between the regions separated by them. We could not have used windowing since we wished to visualize the entire density range, and the nonlinear mapping of Figure 11(c) would perhaps enhance these boundaries a little but not much. Suppose we use the following assignment: 0 to 0, 100 to 50, 2900 to 100, 3000 to 150; then the image and its histogram would be seen as in Figures 12(e) and (f). Clearly, the contrast between the regions is now optimum and all boundaries are clearly (and more or less equally well) discernible. Although the histogram of the slice is extremely nonuniform, the histogram of the image resulting from the above assignment has become uniform or *equalized* (hence the name histogram equilization). What we have done essentially is to assign grey values based on how much each density value is made use of in the slice. If a certain segment of the density range is used heavily (meaning the histogram has high values in that segment), we assign a large number of grey values to that segment, or else only a few grey values are allotted. Of course, the effect of this nonuniform assignment is to equalize the histogram of the image. A somewhat less trivial example is given in Figures 12(g), (h), and (i). Here we made a slight modification to the density distribution: we assume that the density to the left of DB varies from 0 to 100 from the first to the last row, and to the right of DB it varies from 2900 to 3000 from the first to the last row. Clearly if we do a linear assignment, we cannot discern the variation in density within either the left half or the right half of the image. However, if we assign grey values to densities as in Figure (h), not only the resulting histogram will be equalized [Figure (i)], but the contrast within each half region will be enhanced.

The enhancement achieved through histogram equalization is, in some sense, global. The operation is not very sensitive to regional properties of the slice. An improvement called adaptive histogram equalization[29] compensates for this by adapting the equalization operation to regional histograms. The slice is partitioned into a number of smaller rectangular regions and the assignment of grey values to densities is determined separately for each region by equalizing its histogram as described earlier. (Artificial boundaries may appear in the image between adjacent regions since assignment is done independently in each region. A commonly used solution is to do the assignment within a smaller rectangular region enclosed within each rectangular region and subsequently to interpolate the grey values in the gaps remaining between the small regions.) Improved variants of the adaptive method have also been studied.[30] The effects of these enhancement operations on observer diagnostic performance have been evaluated through observer experiments.[31] Though no significant improvement has been observed over, say, the conventionally used windowing operation, the main advantages of the equalization methods are the convenience of not having to change the window setting and the resulting relatively user-independent standard mode of observing the image. Such operations are currently available on many scanner display consoles and 3D imaging workstations.

One final form of slice display we wish to mention is the so-called *cine display*. The slices are displayed very rapidly (10 or more per second) so that the display creates the illusion of "dialing through" the 3D scene. This enables perceiving in a smooth and continuous fashion how structures change from slice to slice. This form of display is necessary in order to visualize the dynamics of a moving organ such as a heart. Unfortunately, this option is not available on most CT scanner display consoles and some 3D imaging workstations.

This concludes our discussion of slice imaging methods. The strongest aspect of these methods is that all information pertaining to the structure as well as everything else is presented, though in a piecemeal fashion, with no obscuration, unlike in projective imaging (surface and volume rendering) wherein some form of information reduction and/or obscuration is inevitable. Unfortunately, this is also the cause for its weakest point, namely the difficulty of perceiving static, and even more so dynamic, structures in their entirety. As long as highly accurate and reproducible methods of automatically identifying structures are not available, the medical end user will have to rely on slice imaging and measurable density distribution to answer diagnostic and therapeutic questions.

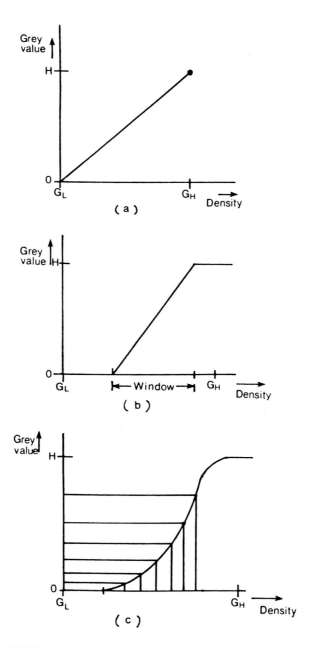

FIGURE 11. Density to grey value assignment for slice display. (a) Proportional (linear) assignment. (b) Windowing: linear assignment is only for the range of the window. (c) Nonlinear assignment.

IV. SURFACE RENDERING

Under this class of approaches we consider all those that explicitly form surfaces of the structure of interest and then somehow depict the surface on the screen of a display device. These approaches provide operations to transform information from the scene space to the view space via the object and image space (see Figure 1) as we will demonstrate in this section.

There are two aspects to surface rendering: *forming* the *surface* of the structure and *rendering* or depicting the surface on a 2D screen.

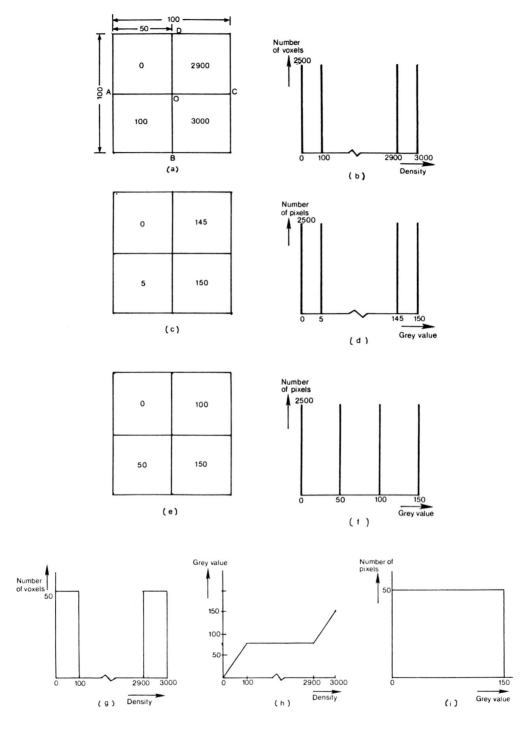

FIGURE 12.   Histogram equalization. (a) The slice in our example has four uniform regions and (b) its histogram. (c) The displayed image of the slice, and (d) the histogram of the image. (e) The displayed image after an equalizing density to grey value assignment. (f) The equalized histogram. (g) The histogram of a slice similar to that in (a) but with uniformly varying density across the rows 0 to 100 in the left half and 2900 to 3000 in the right half. (h) The equalizing density to grey value assignment. (i) The equalized histogram.

A. SURFACE FORMATION

These operations may be thought of as transformations from scene space to object space since their purpose is to determine a description of the structure of interest. We denote the coordinate system of the object space by $x^{\circ}y^{\circ}z^{\circ}$.

There are three most commonly used approaches to surface formation which we call *contouring, patching,* and *surface tracking.* In contouring the idea is to compute the structure boundary in each individual slice as contour lines and to treat the 3D boundary of the structure as a stack of these contour lines. In patching, the 3D boundary is estimated as a set of patching elements (usually triangles) by patching up the contour lines of two successive slices with such elements. In surface tracking, the 3D boundary is determined as a set of those faces of voxels that bound the structure.

1. Contouring

Since most imaging scanners acquire scene information in a slice-by-slice fashion, it is natural to ask why not approximate the surface from the boundaries of the structure in the slices. We have already studied one such method of boundary determination under segmentation. A more often used technique, however, is to track boundaries in slices of binary scenes. The assumption here is that the binary scene is already computed using a region-based method of segmentation.

To begin with, consider as an example a slice of a binary scene as shown in Figure 13(a). Our aim is to determine the boundaries separating 1-voxels from 0-voxels (the 0-voxels have not been labeled explicitly; all blank voxels are 0-voxels), such as those labeled C_1, C_2, and C_3. We may consider each boundary as a *ribbon* of faces (see Figure 13(b) which shows C_1

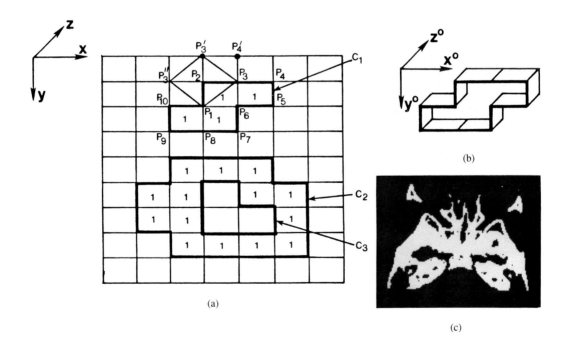

(a)

(b)

(c)

FIGURE 13. Illustration of contouring. (a) A slice of a binary scene. Only the 1-voxels are labeled; all the rest are 0-voxels. There are three boundaries C_1, C_2, C_3 in the slice, all represented as contours. For example, C_1 consists of the sequence of points P_1, P_2,...,P_{10}. (b) The contour C_1 is represented as a ribbon. Note that the slice is defined in the scene space (xyz system) whereas the boundaries such as the ribbon are defined in the object space ($x^{\circ}y^{\circ}z^{\circ}$ system). (c) A binary slice obtained from CT.

in the ribbon form) as we did previously under "Segmentation". Instead, we may consider the boundary as a sequence of points (for C_1 the sequence is P_1, P_2,....,P_{10}), which we call a *contour*. Note that these two descriptions are equivalent since we know the extent of the slice. The notion of a contour is useful in describing surface formation while that of a ribbon is more appropriate in describing rendering.

For the time being assume that we are given a pair of adjacent points, such as (P_1, P_2) in Figure 13(a), on the contour that we wish to determine (later on we shall explain how these can be found automatically). To detect all points on the contour that contains these two points, we start from one of these, say P_2, and determine which of the neighboring points should be included next. To decide this, we move around P_2 in counterclockwise direction starting from P_1 along the periphery of the polygon P_1, P_3, P_3', P_3'', P_1 that links the horizontal and vertical neighbors of P_2. One of these points at which the first transition from a 1-voxel to a 0-voxel along the path takes place is taken to be the next point on the contour. In our example P_3 qualifies as such a point. Note that a point such as P_1 which is already on the contour will not be selected by this method. We next move to the point P_3 just selected and repeat the procedure: starting from P_2 move along P_2, P_6, P_4, P_4', P_2 and detect the first transition at P_4, and hence select P_4. This procedure is repeated until the next point selected is the first point P_1 of the initially given pair of points. The set of all points P_1, P_2,....,P_{10} selected during the tracking procedure constitutes the contour we wanted to determine. Note that instead of starting the selection process from P_2, if we had started from P_1 (but searched peripheral points in the clockwise order), the contour would have been traversed in the reverse order P_1, P_{10}, P_9,....,P_3, P_2.

Several variations of this algorithm exist.[12,32,33] An import mathematical result[34] underlying these algorithms is that they terminate (meaning we always come back to the starting point) and that the contour they determine indeed represents a closed curve with a well-defined "inside" and "outside".

Coming back to how the starting pair of points (P_1, P_2) can be determined automatically; note that a transition from a 0-voxel to a 1-voxel in this slice represents the presence of a face of the ribbon, which in turn indicates the presence of an edge such as P_1, P_2. Hence to determine an initial pair of points, we have to determine where a transition occurs. In general, a slice contains many boundaries [see Figure 13(c)]. We may be interested in just one boundary, in which case it may be specified by pointing to a location in its proximity on a display of the slice. We may also wish to determine all boundaries automatically, in which case it is not necessary to specify any starting point. We can search the slice systematically for a transition to determine the starting points. The previous algorithm can be modified[33] so that soon after a contour is determined it is removed from the slice through a subtraction operation such that further transitions are detected only if other boundaries remain in the subtracted slice.

Clearly, the entire boundary of the structure of interest can be formed by applying one of the above procedures to all the slices of the binary scene.

2. Patching

The problem of patching is to determine a surface that envelops a given set of contours as a set of surface patches.

The contours determined by the above algorithms are discrete in nature and hence contain a large number of points (a sufficiently complex contour may consist of the order of a thousand points). The patching methods require that the contours be smooth and contain as few points as possible. Hence, discrete contours are usually preprocessed prior to patching to reduce the number of points.

One such method[35] is to select only certain points in the discrete contour. For illustration, let us consider the contour shown in Figure 14(a). We always assume the first point P_1 to be

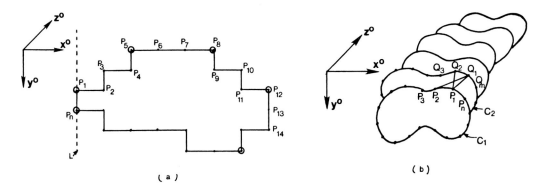

FIGURE 14. (a) Smoothing of discrete contours. The contour $P_1, P_2,...,P_n$ is smoothed to the contour defined by the circled points by the algorithm described in the text. (b) Tiling of smoothed contours. The surface enveloping all the contours is determined by building triangles between each pair of successive contours (see text).

on the smoothed contour. Then we try to approximate the curve $P_1, P_2,...,P_n$ by a single line P_1P_n. We consider this possible if the distance of the farthest point of all points between P_1 and P_n from the line L defined by P_1, P_n is less than a predetermined threshold, say of value 1 (we assume that the sides of the voxels are all equal to unity). Clearly, P_{12}, P_{13}, P_{14} are all at the same maximum distance of 7 from L. Next we choose one of these points (the idea is to choose a point farthest from L), say P_{12}, and see if we can approximate the curve $P_1, P_2,...,P_{12}$ by the line P_1, P_{12} using the maximum distance criterion. We immediately see that P_5, P_6, P_7, P_8 all are at a distance of 2 from the line P_1, P_{12}, and hence our criterion is still not met. If we choose P_5, we readily see that the line P_1, P_5 satisfies the approximation criterion, and hence, P_5 qualifies as a point on the smoothed curve. We repeat the entire procedure starting from P_5 and continue in this manner until the end of the contour is reached. The points selected by this algorithm when it terminates are encircled in the figure. Clearly, the result is sensitive to the threshold distance used in the algorithm. The greater the threshold value, the greater is the smoothing effect. Many variations of this algorithm are possible, and many other forms of contour smoothing algorithms are also available.[36]

The simplest patching element, which is also the most commonly used, is the triangle. Given a set of smoothed contours covering the entire structure, our task then is to "tile"[37] the contours using triangular tiles so as to form a surface [see Figure 14(b)]. We break this task into a number of smaller tasks of determining the tiles between contours from two adjoining slices. We make a further simplification by assuming that tiling is to be done between exactly two contours C_1 and C_2, one coming from each slice, as in Figure 14(b).

Let $P_1, P_2,...,P_n$ and $Q_1, Q_2,...,Q_m$ be the points representing contours C_1 and C_2, respectively. Suppose that we know an initial side, often called a "span", of some triangle. (There are many ways of determining a "good" starting span; one simple way is to pick that pair of points between which the distance is minimum compared to all possible distances between points in C_1 and C_2). Let P_1Q_1 be such a span. (This is not fortuitous; if the starting span is something else, say P_5Q_{10}, we simply relabel all points, starting with P_5 as P_1 and Q_{10} as Q_1.) Given P_1Q_1, there are only two possible ways we can select the next span to form a triangle: P_1Q_2 and P_2Q_1. If we choose P_1Q_2, the resulting triangle is $P_1Q_2Q_1$, and in the other case we get $P_2Q_1P_1$. To decide the choice we may use a variety of criteria;[37-40] for example, we may choose the shorter of the two spans.[39] Suppose P_1Q_2 satisfies this criterion. The next choice to be made then is between P_1Q_3 and P_2Q_2. We continue selecting spans in this fashion until P_1Q_1 again becomes a part of a triangle.

Several other tiling methods that use criteria such as maximizing the volume enclosed by the surface,[37] minimizing the surface area,[38] and maximizing the similarity of span orientation[40] are available. The method described above is a part of a widely distributed computer graphics software package called MOVIE.BYU.[41]

In spite of the appeal of tiling techniques and years of hardware and software development in the computer graphics industry towards the formation and rendering of tiled surfaces, their usefulness in 3D imaging remains questionable. There are two main reasons for this. First, even when automatic segmentation of the structure is possible, meaning that we can get the smoothed contours bounding the structure automatically, automatic tiling cannot be guaranteed. The main difficulty is that the number of contours in two adjoining slices is not always the same because the structure may split or merge from one slice to another. And even when this is not the case, it is not always possible to determine automatically how the contours should be paired so as to perform tiling between them. Although there have been several attempts to automate the process of establishing correspondence between contours,[39,42-45] their repeatability and fool-proof behavior have not been established. Hence, often an operator has to indicate on the computer screen how the contours should be matched.[43] Second, even when automatic tiling is possible, there is no guarantee that the resulting arrangement of triangles would "look right" when the surface is displayed, in the sense that the surface may appear warped or skewed. Because of these reasons, many, including some commercial vendors of 3D imaging products who started with tiling techniques for surface rendering, switched over to voxel-based surface and volume-rendering methods. At their current state of development, the patching methods do not seem to be usable routinely in a clinical 3D imaging set up. Their strongest asset, however, is the availability of a variety of hardware and software products that have incorporated tiling-based rendering.

Patching based on techniques other than contour tiling as well as nontriangular patch elements have been investigated.[22,46-48] Some of these do not have some of the drawbacks mentioned above.

3. Surface Tracking

The purpose of surface tracking is to produce a surface of the structure as a set of faces of voxels. Unlike patching methods, the surface tracking schemes start with a precise mathematical definition of the surface of a structure and then output precisely such a surface. The starting point in these techniques is a scene or a binary scene. When the input is a scene, the location of the boundary is not known unequivocally, and hence, in addition to tracking the surface (i.e., collecting all boundary faces) the boundary has to be also "detected". This, as we discussed earlier, also represents a segmentation operation. The detectability of where the boundary lies depends on a number of variables (the imaging modality (CT, MRI, etc.), the scanning parameters (keV and mA settings, field strength, pulse sequence used, method of image reconstruction), the scanning geometry (the patient orientation, voxel dimensions), and the organ system imaged and is less tractable mathematically. The tracking strategy of how to go about collecting faces, on the other hand, depends only on the mathematical properties of objects and their surfaces. Our aim in this section is to concentrate on these latter aspects. We shall, nevertheless, point out how tracking strategies can be combined with certain detection strategies (such as those described earlier under segmentation) to yield a complete surface detection system.

Roughly speaking, an object is a connected entity. In the 2D example of Figure 15(a), there are two objects O_1 and O_2. A surface of an object is the interface between the object and a connected component in the background. In the example, the background has two connected components: B_1, which is everything outside O_1 and O_2, and B_2, which is the hole within O_1. Obviously O_1 has two surfaces: one between O_1 and B_1 and the other between O_1 and B_2. Of course, O_2 has only one boundary. Our aim is to track the surface between a specified object and a specified connected component of the background. First we need to translate these concepts to the voxel environment.

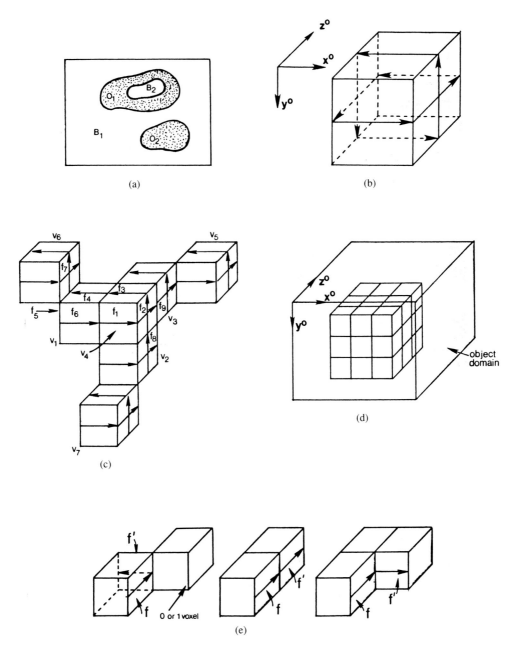

FIGURE 15. (a) Intuitive definition of objects and boundaries using a 2D example. (b) A voxel with two circuits wrapped around it. This model is used in the definition of the object and its surface. (c) The definition of an object. The given set S consists of seven voxels v_1, v_2,...,v_7. There are four objects in S per our definition, the first consisting of v_1, v_2, v_3, v_4, and the second, third, and fourth each consisting of exactly one voxel, viz., v_5, v_6, and v_7 respectively. Note that just v_1, v_4, and v_3 together do not constitute an object since they do not form a maximally connected set although they are connected. This is because v_2 (and no more) can be added to this set since it is connected to v_4. (The results of this section are valid for many other types of connection, for example connection by edges parallel to the y^o and z^o axis. If we use this connection, there are two objects in S, the first consisting of v_1, v_2,...,v_6 and the second consisting of only v_7.) (d) Definition of object domain and the background of an object (see text). (e) Definition of next face. In the figure f is the current face and f' is the next face in the direction of the circuit shown. Only the 1-voxels sharing the edge e are shown. For type-2 faces (which f is), a next face is defined in the direction of both circuits through the face. To see how these cases should be used, the face f_2 in (c) has f_3 and f_9 as its next faces. The former corresponds to the first case and the latter to the second case.

Consider a single voxel around which two sequences of arrows (called *circuits*) are assigned as shown in Figure 15(b). The actual direction of these circuits (i.e., whether clockwise or counterclockwise) does not really matter. (Note also that there are two other possible assignments for which the results of this section apply: the circuits going around directions parallel to the x^o and the y^o axis, and the x^o and the z^o axis.) Our intention is to first specify ways in which the voxels may be connected to each other to form an object and then to examine how their faces may be connected to form a surface.

There are a variety of ways in which voxels may be connected to form an object, for example, by faces, edges, vertices, or by only certain edges, or using combinations of these. We consider one particular method of connection — joining voxels by faces. An *object* in a given set S of voxels then is simply a maximally connected set, that is a connected component, of voxels. The maximality of connection implies that every voxel in S which is connected to some voxel in the object is in the object (see Figure 15(c) for some examples).

In order to define a surface of the object we need to cover the object all around with voxels that do not belong to it. We do this through the notion of an *object domain,* which is simply a cuboid region [see Figure 15(d)] inside which the object is situated. We think of the object domain as an array of voxels. Obviously, a subset of these voxels belongs to the object. We denote the object by O and the set of all those voxels in the object domain that do not belong to O (called the *background* of O) by $\bar{O}$. In the example in Figure 15(d), the object domain is a $5 \times 5 \times 5$ array of voxels, and O consists of all voxels in a $3 \times 3 \times 3$ block of voxels (situated centrally within the object domain) except the voxel in the center of the block. Clearly $\bar{O}$ consists of the central voxel as well as a one-voxel thick layer around and outside the block. We define a connected component of $\bar{O}$ to be a maximally connected set of voxels in $\bar{O}$. The method of connection of voxels in $\bar{O}$ is by the edges through which a circuit passes (i.e., by the edges parallel to the y^o- and z^o-axis). (There are many other methods of connection of voxels in $\bar{O}$ as well as of those in O for which the results of this section are valid.) In the example of Figure 15(d), $\bar{O}$ has two connected components: the first consisting of just the central voxel and the second made up of the remaining voxels of $\bar{O}$. Finally, the *surface* of the object O with a connected component B of its background is simply the set of all voxel faces that are common to the voxels in both O and B. In our example, the object has two surfaces: the first made up of the 6 faces comprising the hole in the center and the other made up of the 54 outer faces of the block.

Having defined what objects and surfaces are, the surface tracking problem may be stated as: *given* a set S of voxels, *find* the surface of a specified object O in S with a specified connected component B of the background of O. Note that the problem as formulated in Figure 15(a) has not changed. All that we have done is to make various intuitive notions precise. When we are given a binary scene, the set S corresponds to the set of 1-voxels in it. One way of specifying O and B is to specify a face f_0 that is common to two voxels v and v' such that v is in O and v' is in B. Then O and B are the maximally connected sets that contain v and v', respectively. In practice, f_0 can be specified by indicating a location (a voxel) near the surface of interest on a display of a slice of the binary scene. The computer program can then search in the vicinity of this voxel to find a transition from a 1- to a 0-voxel or vice versa.

Now we come to the solution of the problem, for which we need the notion of how to connect faces. Coming back to the example in Figure 15(c), note that, within any given surface (for example, the surface of the object made up of v_1, v_2, v_3, v_4), if f and f' are any two faces, then f' can be reached starting from f following the arrows and crossing only the faces of the surface, and vice versa. For example, f_3 can be reached starting from f_1 via f_2, and f_1 can be reached from f_3 via f_4, f_5 (the invisible face of v_1 on the side) and f_6. At edges such as that between v_1 and v_6 connection should be defined carefully so that the next face is still on the surface. This will not be the case if we choose f_7 to be connected to f_4 since v_6 is not a part of the object. The three situations shown in Figure 15(e) completely define how the connection should be done at any edge for the above property to hold good. For edges of other orientations, these illustrations have to be properly rotated. However, the principle is simple: to choose the next face f′ that is a part of the current surface at an edge e, we should check the (0,1) values assigned to the voxels that share e and determine f′ using Figure 15(e).

Note that some of the faces have two circuits going through them (we call these type-2 faces), while the rest (called type-1 faces) have just one circuit. Clearly, every type-2 face can be reached from two faces (for example, f_2 can be reached from f_1 and f_8) and every type-1 face is reachable from one other face. To track a surface which is specified by an initial face f_0 as mentioned earlier (we assume that f_0 is of type-2; if it is not, the following algorithm has to be slightly modified), we proceed from f_0 in the two directions indicated by the arrows assigned to it. We use Figure 15(e) to determine the next face in each of these directions. Each next face may be a type-1 or a type-2 face. From each of these faces we further discover its next faces and so on. When a face is discovered we put a label o on it to indicate that it represents an open end meaning that search for the next face(s) should continue from it. When its next face(s) have been discovered we change its label to c to indicate it is no longer an open end. Our aim is to keep discovering new faces as long as there are open ends. When no open ends are left we will have discovered all faces of the surface specified by f_0. More precisely, the procedure consists of the following steps:

Step 1. Put label o on f_0.
Step 2. If there are no faces with label o, stop.
Step 3. Else, take a face with label o, output it, change its label to c and find its next face(s).
Step 4. For each next face found in Step 3, if it has no label, mark it with o; otherwise leave it alone.
Step 5. Go back to Step 2.

The main conjecture in the above description (which has not yet been mathematically proved) is that a surface as defined earlier is a maximally connected set of faces (this guarantees that the above algorithm terminates). This description is based on a recently reported method of surface tracking.[49] Several earlier publications have addressed the mathematical and algorithmic aspects of surface tracking.[50-56]

Extending the above algorithm to 3D grey scales is complicated mainly because the set O (or the set S) of voxels and hence the set B is not specified to us unequivocally as in a binary scene. As we noted in "Segmentation", the boundary may have to be modified as it is tracked if the segmentation operation signals that the tracked boundary deviates from a "desirable" boundary. It is possible to incorporate modifications to the above algorithm so as to accommodate a boundary correction step. Another important difference would be in the way in which the next face is determined. We have to determine which one of the situations of Figure 15(e) holds good by applying a segmentation operation. The details of such procedures[12] are beyond the scope of this tutorial.

B. RENDERING

Given a surface of the structure of interest, the purpose of the rendering operation is to create a depiction of the structure on a computer display screen, usually upon making certain geometric transformations such as magnification/minification, translation, and rotation of the surface. This operation hence represents a transformation of object information from the object space to the view space via the image space. Our aim is to make the image as closely as possible resemble the perceived or a photographic image of the structure.

The nature of the rendered image depends, in addition to the rendering method, on the assumed viewing conditions. We assume the following scenario for all rendering methods (see Figure 16). The direction of viewing is along the z^i axis of the $x^i y^i z^i$ image space system (The image space coordinate system can be thought of as fixed to the display device, as in the figure, and the $x^o y^o z^o$ system to be attached to the surface. The coordinate system of the screen (the view space) then is $x^i y^i$.) The structure is illuminated with a source emitting parallel light rays. The direction of the light rays coincides with the viewing direction. This obviously avoids shadowing not just of the whole structure but also of the structure on itself. Some techniques use other illuminating situations including moving light sources. These ignore the shadowing effect for computational simplicity. This introduces some artificiality and consequently the full effect of the sophisticated lighting conditions does not come through. We assume no perspective effect in viewing. Consequently, aspects of the structure of identical size situated close to and far away from the viewpoint appear the same size. (Note that if we are viewing a tall building from a short distance, the building appears to taper off toward its top. The magnitude of this perspective effect depends very much on the size of the structure and the viewing distance. In 3D imaging, this effect is usually ignored because of the resulting computational simplifications.) With this assumption, to create an "image" of something on the screen, say of a voxel face, we simply do an "orthogonal projection" of the face onto the screen. This means the $x^i y^i$ coordinates of the vertices of the face in the screen are the same as their x^i, y^i coordinates in the image space. Thus the projection operation reduces a point $x^i y^i z^i$ in the 3D image space to a point $x^i y^i$ in the screen. (This dimensionality reduction operation is common to and an essential part of all (surface as well as volume) rendering techniques. That is why we called them collectively projective imaging techniques, Figure 2).

To make up for dimensionality reduction, the rendering techniques have to somehow create an illusion of three dimensionality in the rendered images. A number of techniques are available for this purpose, most important among these being suppression of parts of the surface not visible from the viewpoint, shading visible parts, dynamic rotation of the surface, and stereo projection. Although these principles have been borrowed from computer graphics,[57] a number of techniques to implement these principles have originated in 3D imaging,

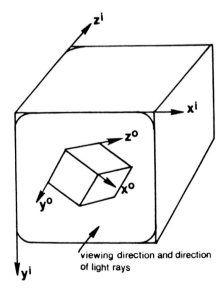

FIGURE 16.   Description of the viewing geometry used in all rendering methods. $x^o y^o z^o$ is the object space coordinate system, and $x^i y^i z^i$ and $x^i y^i$ are the image and view space systems, respectively.

mainly due to the inadequacy of previously existing computer graphics techniques. In discussing these aspects, we shall concentrate on techniques original to 3D imaging.

Prior to creating an illusion of three dimensionality, the structure should be properly scaled (magnified/minified) and rotated. These geometric operations are common to all (surface and volume) rendering techniques. Given a point $x^o y^o z^o$ in the object space (scene space, in the case of volume rendering as we shall see in the next section), this transformation amounts to finding the location $x^i y^i z^i$ of the same point in the image space after the transformation. The point may represent, for example, the vertex of a polygon. To compute the new coordinates $x^i y^i z^i$ of the point, the vector whose elements are the x^o, y^o, and z^o coordinates of the point should be multiplied by a matrix. The details of how to compute these matrices are given in any standard computer graphics text.[57] For our discussion, it is enough to know that given a set of points and the nature of the geometric transformation, we can compute the new coordinates of the points in image space.

Given a surface representation, a typical sequence of operations required to create a rendered image consists of geometric transformation, projection, hidden part removal, and shading.

1. Hidden-Part Removal

We will discuss this operation for ribbon (contour) form and for polygonal representation separately. Given a surface in image space, our aim here is to determine for every pixel of the screen which aspect (point) of the surface should be associated with it, so that an intensity (shading value) can be assigned to the pixel (as described under shading) depending on the light reflected from that aspect of the surface toward the view point. (Of course, many pixels may not correspond to any aspect of the surface if these pixels are outside the region of projection of the surface.)

Ribbons — Consider the ribbon $f_1, f_2, \ldots, f_n$ shown in Figure 17(a). The orientation of the ribbon as shown is achieved by rotating the object about the x^i axis 90° counterclockwise (equivalent to making the patient stand up from an initial lying position with his long axis along the z^i axis). The slice plane that contains the ribbon is shown for convenience. Now imagine that the inside of each face in the ribbon (i.e., the side which is inside the object) is painted white and its outside painted black. If we were to determine "visibility" for this orientation of the ribbon under the viewing conditions described earlier, our task is to find out for each pixel in the screen which face among $f_1, f_2, \ldots, f_n$ is first interested by the line drawn from the center of the pixel in the direction of the z^i axis.

The faces in a ribbon can be divided into four types based on the direction in which their black side faces. We call a face a x^o_+ face if its black side faces the $+ x^o$ direction. Other types — x^o_-, y^o_+, and y^o_- — are similarly defined. Note that for the orientation chosen in Figure 17(a), or for that matter for any orientation resulting from rotating the ribbon in the slice plane starting from the orientation shown in the figure, only the black sides of the faces are potentially visible ("visibility" should always be interpreted in the way described earlier, "potentially visible" means the face stands a chance to appear in the image). Further, since the faces of a particular type all have the same orientation, they as a group are either potentially visible or invisible. For example, in Figure 17(a) the y^o_+ faces are invisible since their black side faces away from the viewing direction, but the y^o_- faces are potentially visible. For any given rotation of the type described above, it is easily seen that at least two types of faces are invisible (in the example y^o_+, x^o_+, and x^o_- are invisible). Given the degree of rotation, it is a simple matter to determine which of the four types are definitely invisible.

Given a potentially visible set of faces to determine final visibility, however, is not as straightforward. A face, though potentially visible, may not eventually appear in the image because other potentially visible face(s) may obscure it. For example, in Figure 17(a), though both f_5 and f_1 are potentially visible, f_1 completely obscures f_5. It is possible that a face may

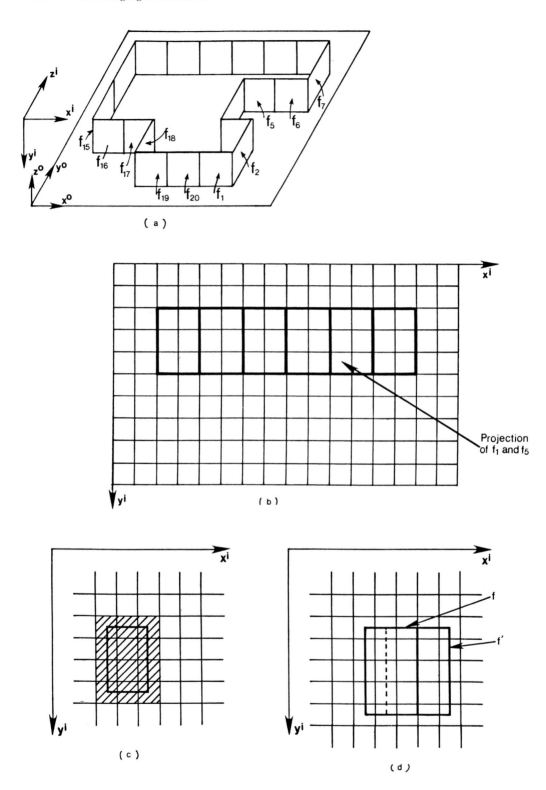

FIGURE 17. Hidden-part removal for ribbons. (a) A ribbon rotated 90° counterclockwise about the x^i axis. The different views of the surface are obtained by rotating the surface (hence the ribbon) about the y^i axis. (b) A projection of the ribbon in (a) onto the screen. Note that faces f_1 and f_5 project to the same screen area. (c) The situation where the projection of a face covers some pixels only partially. (d) The situation where the projection of faces f and f' partially overlaps.

only partially obscure another face. This is why we have to investigate for each pixel in the image which is the face "seen" by the pixel. A commonly used procedure to determine final visibility in this fashion is called the *z-buffer technique*, which may be explained as follows. Suppose we determine the projection of each of the potentially visible faces onto the screen as shown in Figure 17(b) (we assume for now that no pixel is partially covered in the projection). Clearly, except for f_1 and f_5, every face projects onto a distinct region in the screen. Suppose we associate two numbers with every pixel irrespective of whether or not faces overlap in the projection. The first number represents the z^i coordinate of the face most recently projected onto the pixel and the second represents the shade associated with that face (shading will be described in the next section). When a new face is to be processed, all pixels covered by its projection are first determined. For each such pixel, the z^i coordinate value associated with it is compared to that of the current face. If the latter is smaller than the former (meaning the face is closer to the viewer than the previous closest face that covered that pixel), then the pixel's shade value is changed to that of the current face. (Of course, its z^i coordinate value is also upgraded to the z^i coordinate of the face to reflect the fact that the current face is the closest to the viewpoint as seen from that pixel.) Otherwise, the values associated with the pixel are not modified. If, before commencing the processing of faces, we assign a fixed z^i coordinate value to all pixels that is greater than the z^i coordinate of the farthest face from the viewpoint and a fixed minimum shade value (say 0), then when all faces in the ribbon have been processed as described above, the set of shade values associated with all pixels represents the rendered image of the ribbon. To get the image of the entire surface, we simply have to repeat the above procedure for all ribbons constituting the surface.

In general, unlike our assumption in 17(b), the projection of a face may cover some pixels only partially as in Figure 17(c). One way of ensuring that no gaps occur in such a situation between adjacent faces in the rendered image is to consider all pixels — those covered entirely as well as partially — to be in the projection. Of course, some pixels may be considered as associated with more than one face. This does not cause any problem as can be seen from the above procedure. It is also possible that two faces may not overlap exactly (as did f_1 and f_5), as demonstrated in Figure 17(d) (for example, when we rotate the ribbon in Figure 17(a) in the slice plane by a few degrees). This also is taken care of automatically by the z-buffer algorithm.

A number of tricks can be employed to refine the basic technique described above, and as such a variety of forms of ribbon-based rendering techniques are available.[33,36,58-61] A major drawback of these techniques is the restriction on the mode of rotation, namely that the viewing direction should always be parallel to the slice plane. To create other views, these techniques often reslice the given scene along planes parallel to the desired viewing direction, recreate ribbons, and then render them.

Surfaces — A commonly used polygonal representation in medical 3D imaging is surface approximation via voxel faces as described earlier. Other forms, such as triangular tile approximations, are used extensively in computer graphics. Since a variety of techniques for hidden-part removal are described for the latter in many text and tutorial books[57] on computer graphics, we shall concentrate here only on the former type.

Using the black/white-side description of faces used for ribbons we can easily see that there are six types of faces in a surface, which we call the x_+^o, x_-^o, y_+^o, y_-^o, z_+^o, and z_-^o *faces*. The definition of these types is identical to that used for ribbons except that we now have two additional types, z_+^o, z_-^o faces whose black sides face the $+z^o$ and the $-z^o$ direction, respectively. Similar to the case of ribbons, at most three of these six types of faces are potentially visible for any orientation of the surface.[62] This is easily seen for the surface of a single voxel. During surface formation the faces can be easily sorted into six lists each containing the faces of a given type. For any given orientation of the surface it is simple to figure out which of these six lists can be ignored because the faces in them are not potentially visible. The final visibility among potentially visible faces can be then determined using the z-buffer technique described earlier. The only difference here is that, since there is no restriction in the orientation of the surface as in ribbons, the projection of a face onto the screen for a general surface orientation is a parallelogram [see Figure 18(a)]. Several variants of this technique are available.[63-65] They are all based on the following common mathematical property[63] of surfaces made up of faces of cubic voxels: if f_1 and f_2 are any two faces in the surface [see Figure 18(b)] on the same side of a viewpoint O such that the distance to their centers OC_1 and OC_2 obeys $OC_1 < OC_2$, then $OP_1 < OP_2$ for any points P_1 and P_2 that are colinear with O in f_1 and f_2, respectively. This property is what makes the application of the z-buffer technique to such surfaces computationally efficient. Scene interpolation so that the voxels in the scene are cube-shaped ensures the validity of this property.

2. Shading

In our natural vision, for an object to be visible it should be properly illuminated and the light it reflects should reach the eye. Obviously, the hidden parts of an object are invisible since light reflected by them does not reach the eye. Among those aspects of the object that are visible, not all appear the same because the nature of the light they reflect towards the eye is not the same. The purpose of shading is to somehow determine this entity.

The intensity itself is governed mainly by two factors: the conditions of illumination and of viewing, and the optical properties of the object, such as its color, its light transmission property (opaqueness/transparency), and the smoothness of its surface (glossiness). When the object is opaque it is enough to consider the optical properties of its surface. When this is not the case, the properties of the entire object may have to be considered. But in natural vision, the latter situation is encountered far less frequently than the former. The nature of most of these properties for a given organ is known only in a general sense. It may vary from patient to patient, it may also vary within a given organ (for example, the color of an organ may vary over its surface). Such information is not available in the scene data representing the organ. Hence, for the purpose of shading, the question to be first settled is what is an appropriate

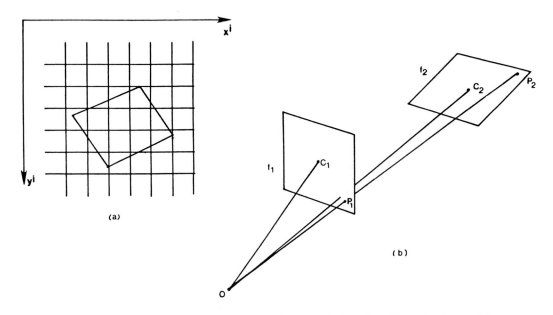

FIGURE 18.   (a) The projection of a face onto the screen for a general orientation of the surface is a parallelogram. Hence it is computationally more difficult to determine the pixels in the projection than for the case of ribbons. (b) Illustration of a common mathematical property of surfaces made up of faces of cubic voxels (see text).

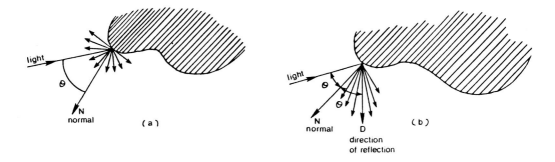

FIGURE 19. Light reflection properties of surfaces used in shading. (a) Diffuse reflection. The reflected light intensity is the same in all directions and depends on the angle θ. (b) Specular reflection. In a perfect mirror, all reflected light is in the direction D of reflection. For a glossy surface it is more realistic to assume that light intensity is maximum along D but dies away on either side of it.

optical model of the object surface that will enable portrayal of most medically useful information in the rendered image. Though the general nature of the models used can be described, their relative effectiveness in portraying medically useful information is largely unknown at present. They can be characterized by various combinations of the following set of assumed properties, given the illumination and viewing conditions described previously.

P1. Distance-only: The intensity of light reflected from a point on the surface that reaches the eye decreases with the distance of the point from the viewing plane $x^i y^i$.

P2. Diffuse reflection: The intensity of light reflected from a point on the surface is the same in all directions (hence independent of the viewing direction) but varies as some function of the cosine of the angle θ between the direction of the light rays and the normal to the surface at that point [see Figure 19(a)].

P3. Specular reflection: The intensity of the reflected light is maximum in the direction D of reflection which is at an angle θ to the normal **N** on the other side of **N** but dies off on either side of D [see Figure 19(b)]. This property says that the surface behaves somewhat like a mirror (in that the incident angle equals the angle of reflection), but the reflected light diminishes as we move away from D according to some rule, usually as the cosine of the angle α from D. Obviously, since we are interested in the light reflected towards the viewer and since the viewing direction is the same as the direction of light rays, $α = 2θ$.

P4. Color: The color (hue) of the light reflected from different points of a surface is the same and is identical to the color of the surface.

P5. Transparency: A fraction of the light incident at a point on a surface is transmitted through the object with unit refractive index (i.e., no bending of light rays at the surface) and is the same for all points of the surface.

P6. Stereo: The views of an object as seen by the left and the right eye are rotated with respect to each other by a small angle (4° to 6°). This creates a strong perception of depth and a sense of where different aspects of the surface are disposed in space irrespective of whether or not they appear with similar brightness.

Clearly, the above properties are independent in terms of the perceived effect they generate. Some of them individually are not valid or may be unrealistic. (For example, P5 may not be valid because most organ surfaces are not transparent. P4, on the other hand, is somewhat unrealistic because the color of an organ is likely to vary somewhat over its surface.) The premise in rendering is that these properties in combination may, nevertheless, yield renderings that carry medically useful information.

The simplest property to transform into a shading rule is that given by P1. If Z represents the distance from the $x^i y^i$ plane of the farthest point on a surface, then by assigning to each pixel an intensity proportional to $(Z-z^i)$ where z^i represents the distance of that aspect of the surface that projected onto the pixel, we will have created a *distance-only shaded* image. The main drawback of this technique is that subtle as well as prominent features including edges, ridges, sutures, etc. are portrayed poorly in the image (see Figure 20), since for most viewing angles the points on either side of the features are approximately at the same distance from the viewpoint. Nonetheless, this technique has been used[66] extensively in the study and management of patients with craniofacial anomalies.

Among the remaining properties, the perceived shape and details of a surface are determined mainly by P2 and P3. As is clear from the description of these properties, their effect on perceived shape depends on the normal to the surface at various points. Hence, for faithful portrayal of the shape of a surface, it is just as important to determine these normals accurately as it is to determine the surface itself as precisely as possible. Of course, neither the real organ surface nor the normals to it are known to us. The surface formation techniques described earlier only *estimate* the location of the surface. Similarly, in order to make use of properties P2 and P3, we need to somehow estimate normals to the surface.

The normal estimation techniques may be classified as scene-space, object-space, or view-space techniques depending on the space in which the normals are computed, as described below. Given the normal at a point $P = (x^i, y^i, z^i)$ on the surface, a general shading formula[67] that incorporates properties P1 to P3 can be given by

$$SHADE(P) = k(Z - z^i) \left[k_d \left(\cos \frac{\theta}{n_d} \right)^{p_d} + k_s \left(\cos \frac{\alpha}{n_s} \right)^{p_s} \right] + A \qquad (6)$$

where k, k_d, k_s, p_d, p_s, n_d, n_s, and A are constants. The values assigned to these parameters may be left for modification under user control in a program or may be empirically determined for optimum effect and fixed in the program once and for all.

Scene-space shading — In these techniques the surface normals are estimated independently of how the surface is formed. The basic premise here[68] is that if the surface represents

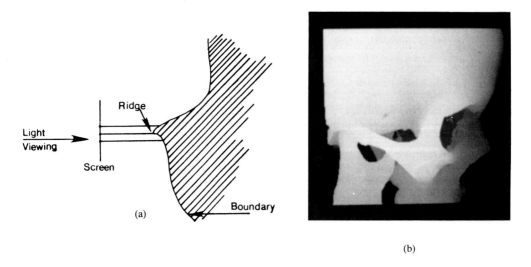

FIGURE 20. Distance-only shading uses only property P1. (a) Natural edges and other subtle features are lost in this technique because there is no significant change in the distance to points in the vicinity of such features from the viewing plane. Figure shows a 2D example. (b) Surface rendering of a patient skull. The scene data were obtained from a GE 9800 CT scanner. The size of the scene data after a VOI and a linear interpolation operation was 256 × 256 × 240.

the set of points at which the rate of change of density (i.e., the gradient) is maximum, then the direction in which this rate is maximum should be perpendicular to the surface and should hence represent the direction of the normal. Suppose P is a point on a surface (such as the center of a face of a voxel or a vertex of a triangular tile) that has already been determined using one of the methods described earlier. To estimate the normal at P, we first locate P in the original 3D scene from which the surface was derived and then apply a gradient operator at P. The result of this operation is a vector $\mathbf{N} = (n_1, n_2, n_3)$ in scene space whose direction indicates the direction of the normal at P. Many definitions of this operator are possible. The simplest of these consists of first determining the center of a voxel v closest to P and then computing $n_1 = |g(v_1) - g(v_3)|$, $N_2 = |g(v_4)|, - g(v_2)|$, $n_3 = |g(v_6) - g(v_5)|$, where $v_1, v_2,...,v_6$ are the six neighbors of v as in Figure 5(b). Alternatively, one could determine densities $g(P_1)$, $g(P_2),...,g(P_6)$ through interpolation at the six points $P_1, P_2,...,P_6$ that are equidistant from P along the six principal directions and determine $\mathbf{N}$ using the above formula replacing v_1, $v_2,...,v^6$ with $P_1, P_2,...,P_6$. While this technique, now commonly used in 3D imaging,[48,68,70] produces normals that are quite sensitive to subtle features of the surface, it also retains sensitivity of normals to artifacts such as those due to sampling. A variety of more sophisticated methods have been proposed recently[71] which seem to suppress sensitivity to noise yet retain sensitivity to actual features. Some rendering examples from one of these latter methods is shown in Figure 21.

Object-space shading — In these methods the normal estimation is based on the geometry of the estimated surface itself. As an example,[72] consider the discrete surface produced by the surface tracking algorithm described in "Surface Formation". Every voxel face f in such a surface has four adjoining faces that share an edge with f (see Figure 22). If $\mathbf{N}_0, \mathbf{N}_1,...,\mathbf{N}_4$ represent the normals to the plane of the face f and its four neighbors as in Figure 22(a), then we may approximate the normal to the organ surface at the center of f to be an appropriate weighted sum of these vectors. A rendering created using such a method is shown in Figure 22(b). Similar techniques that use geometric information in a small neighborhood of a point on the surface that consider both 3D neighborhoods[40,73] as well as neighborhoods in the vicinity of the boundary in the slice plane[58,36] are also available. In general, the object-space techniques do not seem to be able to capture details comparable to those provided by scene-space techniques. Whether this makes any difference in the outcome of a clinical task is not known.

View-space shading — These methods[74,75] estimate the surface normal only for visible parts of the surface from a projection of the surface in view space. Suppose we create a projection of the surface using the z-buffer technique described previously. We know that the z^i-coordinate values associated with each pixel P on the screen at the end of this process represent the distance of that visible aspect of the surface that projected onto P from the $x^i y^i$ plane. If a spatially contiguous part of the surface projected onto the pixel and its immediate vertical and horizontal neighbors, then from the z^i-values associated with these pixels we can infer the orientation of that part of the surface and hence its normal at the point that projected onto P [see Figure 23(a)]. The normal to be associated with P can be given by the vector $(z_1^i - z_2^i, z_3^i - z_4^i, -1)$ [see Figure 23(b)], where z_1^i, z_2^i, z_3^i, and z_4^i are the z^i-values associated with the four neighbors of P.

A drawback of this method is that a spatially contiguous part of the surface does not always project onto a group of neighboring pixels [see Figure 23(c)]. As a consequence, around the edges and ridges of the surface, normal estimation is erroneous in some views. This error manifests in the form of dark rings in such regions. A number of tricks can be employed[74,76] to minimize the rings. Nevertheless, one of the main shortcomings of view-space shading techniques is the lack of view to view coherence which becomes apparent when a contiguous sequence of views is visualized in a cine mode. An example of the rendering by this method[76] is shown in Figure 23(d).

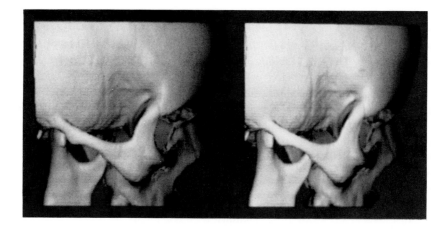

(a)

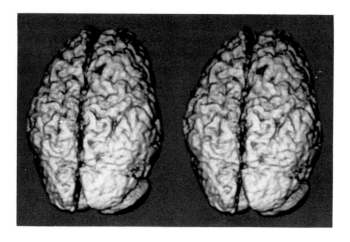

(b)

FIGURE 21. (a) The patient skull data of Figure 20 rendered using a scene-space shading technique.[71] Note especially how the sutures and other fine textures are depicted. (b) An MRI brain data rendered using the same technique. Both examples show a stereo pair of images.

The use of the remaining properties, namely P4, P5, and P6, for 3D imaging is studied somewhat lesser than those aspects that we have already discussed. The use of color in surface rendering has been mainly for the purpose of distinguishing between different surfaces in a single display, although some have attempted[73] to map regional properties of the surface onto the surface display using color. Although the computer processing involved in the use of stereo is rather trivial (we need to create a pair of views, instead of a single view, that are rotated by an appropriate angle, see Figure 21), the display of the images to create the proper stereo effect is not trivial.[77] Stereo projection is not routinely used in 3D imaging mainly because of the specialized display hardware required and due to the lack of hardware and software tools that allow interaction with surface images (such as pointing to features/locations on surfaces). However, since such tools are now beginning to appear as an integral part of graphics workstations, medical imaging workstations in the near future may come with these powerful capabilities. The transparency property is useful in the simultaneous display of multiple overlying surfaces. A simple mechanism of creating this effect[78] when two surfaces — one transparent and one opaque — are to be displayed is to render the surfaces separately as described previously. The two resulting images can then be combined into a single image knowing the z^i-value and the shade value associated with every pixel in the two images and an assumed transparency coefficient of the first surface. However, perhaps there is no point in trying to create the effect of transparency with utmost care using all laws that govern the transmission of light through opaque bodies, since, in natural vision, we are not accustomed to viewing complex transparent and opaque surfaces in combination and in various colors. However, to the extent transparency can convey a sense of overall context and spatial relationship between multiple surfaces it will be useful in many 3D imaging applications. We close the discussion on surface rendering with an example of the use of transparency, shown in Figure 24.

V. VOLUME RENDERING

Under this class of approaches we consider all those techniques that produce a 3D rendition directly from (grey-level or binary) scene data. These approaches can be hence characterized by transformations from scene space directly to view space often via the image space (see Figure 1) as we will describe in this section.

There are two aspects to volume rendering: *preprocessing* the volume and then *rendering* the volume. Our following discussion of these aspects in two separate sections should not be construed as implying that computations should always be done in two steps. This division is only conceptual, and often it is more efficaceous to do preprocessing during rendering.

A. PREPROCESSING

In simple terms, volume rendering can be described as a process of projective imaging

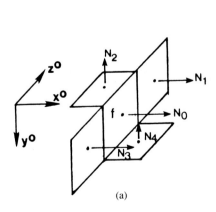

(a)

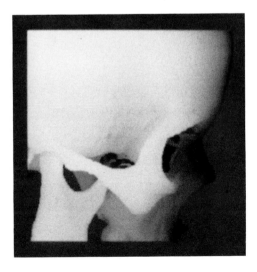

(b)

FIGURE 22.   Object-space shading. (a) Surface normal is estimated from surface geometry in object space. The effective normal at the center of f is assumed to be a weighted sum of the vectors N_0, N_1,...,N_4. (b) The patient skull data of Figure 20 rendered using this method.[72]

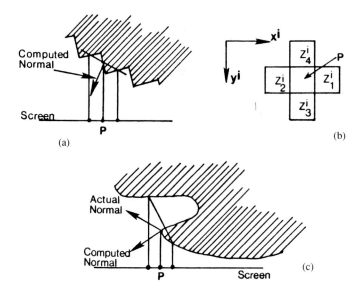

(a)

(b)

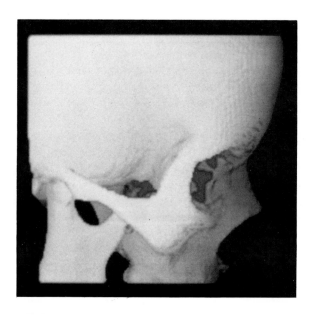

(d)

FIGURE 23.   View-space shading. (a) Surface normal is estimated from the z^i-values associated with pixels P_1 and P_2 in the vicinity of P to compute the shading to be assigned to P. This figure shows a 2D boundary. (b) In 3D, the z^i-values of all four neighbors of P are considered. (c) The neighboring pixels may correspond to distinct parts of the surface and hence normal estimation may be meaningless. (d) The patient skull data of Figure 20 rendered using this technique.[76]

somewhat similar to but more complex than radiographic imaging. There is no real irradiation of any sort in the former process which is purely computational having no physical analog. While in the latter case the tissue property at every point in the path of projection is not known, in the former process, since the property value associated with the tissue within every voxel is known, it is possible to create the projection in a variety of sophisticated ways so as to emphasize or de-emphasize selected aspects of the scene.[75,79] Further, certain assumed optical behavior can be imposed on various tissue interfaces in the projection path to create shading effects so as to portray shape as well.[70,80-82] In summary, the conceptual model used in volume rendering is such that the scene is considered to represent a block of translucent colored jelly whose color and opacity (degree of translucency) are different for different tissue regions. The goal of rendering is to compute images that depict the appearance of this block from various angles simulating the transmission of light through the block as well as its reflection at tissue interfaces, through careful selection of color and opacity. Since it is necessary to specify which aspects should be emphasized (low degree of translucency) and which should be de-emphasized (high degree of translucency) as well as their color, it is imperative that these aspects are identified in a voxel-by-voxel fashion through some form of scene processing, including the VOI operation, masking, filtering, interpolation, and segmentation. The techniques for realizing these operations described under "Scene Processing" are applicable and have been used in volume rendering as well. In addition, segmentation, also called classification, is often done somewhat differently, as we now describe.

The purpose of segmentation here is to assign to every voxel in the scene an opacity and a color. Of course, if we assign the same color to all voxels and an opacity of 100% to all those voxels that belong to the object and an opacity of 0% to all the rest, then the result is essentially a binary scene. In the more general situation, we may consider multiple objects, a different color for each object, and an opacity between 0 and 100% based on the mixture of object material contained in each voxel.

The simplest method is to assign the same color to all voxels but varying opacity based on voxel density[79] [Figure 25(a)]. A fixed opacity between 0 and 100% is assigned to each fixed range of voxel density with the premise that the type of the tissue within each voxel can be identified based on the interval in which its density falls.

A more sophisticated method[80] is to assign opacity and color to a voxel based on an estimated tissue mixture in the voxel. Suppose we know how the property representing the voxel density (e.g., CT number in the case of X-ray CT) varies for each type of tissue that forms our object of interest [see Figure 25(b)]. Further, if the tissues surround each other within the body in the fashion depicted in Figure 25(c), (i.e., TISSUE1 neighbors TISSUE2, TISSUE2 neighbors TISSUE1 and TISSUE3, etc.), then we can guess the tissue mixture (percentage) within each voxel knowing the voxel's density and the density distribution for the various tissues. In fact, we can compute a graph of this percentage versus density as

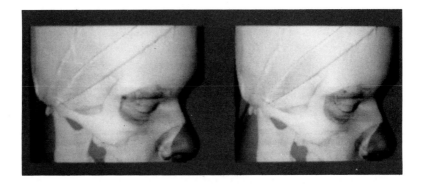

FIGURE 24. The skin surface and the skull surface rendered using a scene-space shading technique[71] to illustrate the use of transparency. The original scene data are the same as those used in Figure 20. The figure shows a stereo pair of images.

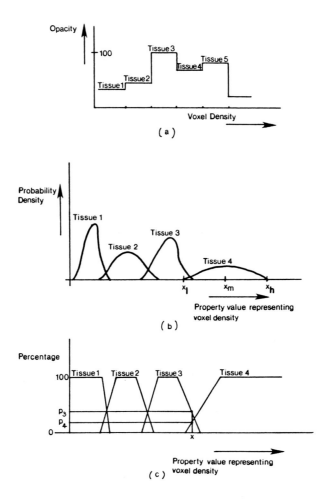

FIGURE 25.   Segmentation for volume rendering. (a) Simple assignment of opacity based on voxel density interval. (b) These graphs, assumed to be known, represent the distribution of the property value for each type of tissue. For example, if the property value corresponds to CT number and TISSUE4 is bone, then its graph tells us that CT number for bone varies between x_l and x_h and that x_m is the most typical CT number for bone. (c) These graphs, based on those in (b), indicate the material mixture for a given property value. Again using CT for example, if a voxel's density (CT number) is x, then the graph tells us that the voxel contains p_3% of TISSUE3 and p_4% of TISSUE4. Note that for this estimation to be valid, the adjacency of tissues should be as indicated; for example TISSUE3 is surrounded only by TISSUE2 and TISSUE4, etc. (d) This graph is based on assumed typical values for TISSUE1,...,TISSUE4 and assumed maximum opacities $\alpha_1,...,\alpha_4$ for various tissues at maximum gradient magnitude. The opacity is assumed to vary linearly with gradient as well as between typical tissue property values. If a voxel has density x and gradient magnitude y, then its opacity α is read off from the graph as shown. In actual computation this graph can be represented as a precalculated table of x, y, and α values.

illustrated in Figure 25(c). Now suppose we assume a certain fixed opacity and color for each type of tissue based on the type of display we wish to create (e.g., opaque bone within semi-transparent (say 25% opacity) skin). Then, knowing the tissue percentage mixture within each voxel, the effective color and opacity of the voxel can be calculated. For illustration, suppose a voxel contains two types of tissues with opacities α_1 and α_2 and colors $C_1 = (R_1, G_1, B_1)$ and $C_2 = (R_2, G_2, B_2)$. (It is customary in computer graphics to express color as a vector whose components indicate the intensities of red, green, and blue, since these are the primary colors used in display devices; there are of course other ways of expressing color.[83]) If the percentages of these tissues in the voxel are p_1 and p_2, respectively, then the effective color associated

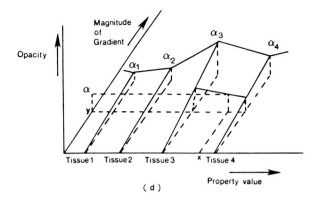

FIGURE 25 (continued)

with the voxel is $[(p_1\alpha_1R_1 + p_2\alpha_2R_2), (p_1\alpha_1G_1 + p_2\alpha_2G_2), (p_1\alpha_1B_1 + p_2\alpha_2B_2)]$ and its effective opacity is $(p_1\alpha_1 + p_2\alpha_2)$. If there is strong evidence of a surface passing through the voxel, then it is possible to keep the opacity and color associated with each tissue within the voxel separate for the purpose of determining the contribution of this voxel to the final rendition,[80] as we will describe later on.

An alternative method of segmentation[70] is motivated by the idea of volume rendering only the tissue interfaces rather than the entire tissue volumes. Suppose we use the magnitude of a gradient evaluated at every voxel in the scene to signal the strength of evidence of an interface and assign an opacity proportional to the gradient magnitude. Then interfaces with high contrast will have high opacity and those with low contrast will have low opacity. Further, we may assign different maximum opacities to different tissue interfaces for displaying multiple surfaces, where each tissue region is specified by a typical property value as illustrated in Figure 25(d). The color of a voxel is specified by its diffuse and specular reflective characteristics for each (R,G,B) component of the incident light. Although this method[70] does not describe how these characteristics should be chosen, the typical values used for indicating tissue regions and interfaces can also be used to specify these characteristics (assuming they are fixed) for each tissue interface.

Another preprocessing operation that often becomes necessary in volume rendering is masking. Often, voxels constituting structures which we do not really intend to include in the display get assigned high opacities (and/or inappropriate color) by the automatic segmentation process with the consequence that they appear predominantly in the rendering. Such voxels can be suppressed by assigning a low opacity (essentially zero) value to them using the masking operation. Masking is also needed if we wish to treat a part of the same structure differently from the rest. For example, we may wish to display the skin surface on one half of the face with high opacity and the other half with low opacity so as to visualize the underlying osseous structures. Volume rendering, hence, is no exception to the tedious requirement of the (often slice-by-slice) specification of a subregion of the scene to which segmentation should be restricted. Recall that these restrictions apply equally well to surface rendering techniques. There is nothing inherent in either of these techniques that grants a significant waiver to the limitations of segmentation techniques. We will get back to this discussion of surface vis-a-vis volume techniques after describing the rendering strategies.

B. RENDERING

The computational goal of volume rendering is to determine the color to be assigned to every pixel in the view space (the screen) for any given orientation of the scene in the image space. A fundamental computational step towards this goal is to determine the set of voxels that contribute to each pixel. Two techniques are commonly used for this purpose: *voxel projection* and *ray casting*.

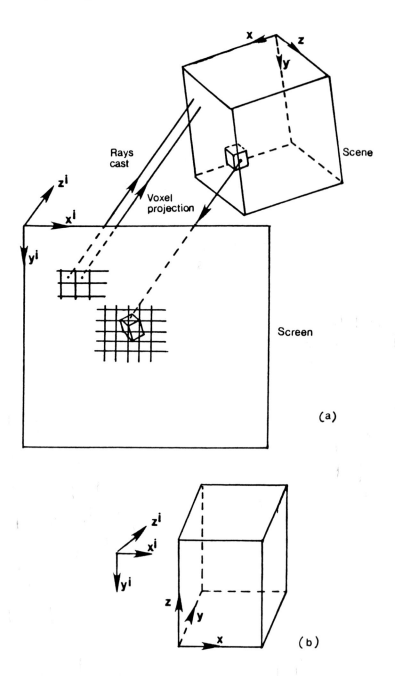

FIGURE 26. (a) Illustration of voxel projection and ray casting. In voxel projection, voxels are projected in a certain order according to their distance from the screen, either from farthest to closest or vice versa. In ray casting, a ray is sent from each pixel in the screen and voxels in the vicinity of the ray are identified. (b) For the orientation of the scene shown, to produce a farthest-to-closest sequence we should project the voxels in the order: from the first to the last slice, and from the last to the first row within each slice. The order of the voxels in each row does not really matter.

In the projection method,[84-86] voxels are projected onto the screen in a certain order of their distance to the screen — either farthest to the closest or vice versa. There are many ways of projecting voxels one-by-one which ensure that the voxels which project onto any given pixel in the screen are in one of these orders. For example, consider the rotated scene shown in Figure 26(a). Suppose we store this scene as a 3D array in a row-by-row and slice-by-slice order. It is easily seen that if we project the voxels exactly in this order, starting from the first slice and ending with the last slice, then the voxels projecting onto any given pixel are visited in the farthest-to-closest order. Clearly, projecting voxels in the reverse order, that is starting from the last slice and ending with the first, corresponds to a closest-to-farthest order. Note that, in some situations, we may have to reverse the order of the rows (i.e., last row to first row) and even the order of the voxels within a row [i.e., last voxel to first voxel, see Figure 26(b)] to guarantee an orderly projection. In summary, for any given orientation of the scene in image space, it is possible to determine the order of the slices, of the rows within a slice, and of the voxels within a row in which the voxels should be projected so as to guarantee that the criterion of the selected order of their distance from the screen is satisfied.

In ray casting,[68-70,75,79,81,82] the voxels that influence the color of a pixel are identified by sending a ray from the pixel into the scene [see Figure 26(a)] and noting the voxels in the vicinity of the ray as it passes through the scene. The vicinity may be defined in several different ways. The simplest method is to consider the voxels that are intercepted by the ray. A more sophisticated method is to take a number of equally spaced points between the point of entry and the point of departure of the ray through the scene region and to estimate opacity and color at each of these points via interpolation. Most of the density interpolation techniques described under the section "Scene Processing" can be used for this purpose (in interpolating color each of the R,G,B components should be estimated separately).

In the following discussion of how the pixel colors are actually determined we will use both the ray-casting and voxel projection paradigm depending on whichever results in a simpler description. In almost every instance these strategies can be used interchangeably although they have subtle computational and subjective image quality differences.

1. Binary Volumes

If we assign a single opacity value (100%) and color to all voxels containing the tissue (object) to be rendered, we can disregard both these entities and essentially have a binary scene. Several elegant techniques are available for rendering such binary volumes.

The most straightforward of these[84,87] uses a *rectangular array* representation (i.e., a row-by-row, slice-by-slice order of storing voxel data) for the binary volume. If we project only the 1-voxels in the volume onto the screen in the farthest-to-closest order, identify the pixels covered by the projection of every voxel, assign a shading value to each voxel, and always overwrite the pixel values by the shading of the voxel, then when all voxels have been projected we will have created a rendition of the projected volume which very much resembles a surface rendition of the object. Any of the shading techniques described earlier under "Surface Rendering" can be used to determine the shading value for the projected voxels. Note that in this method much computation is wasted on the voxels that are behind the voxel closest to the viewpoint since the shading value of such voxels always gets overwritten by that of the closest voxel. An alternative method is to use a closest-to-farthest projection of 1-voxels[85] and paint only those pixels that have not yet been painted. This method is somewhat more efficient than the former since we can stop computing the shading value as well as the process of overwriting, once we find that all pixels covering the projection of a voxel have already been painted. A still more efficient technique[88] is to retain only the boundary voxels (i.e., those voxels at least one of whose six neighbors sharing a face with it is a 0-voxel) in the binary volume, project only those among such voxels that are in the "front half" of the object, and hence do projection, shading, and pixel overwriting only for a minimal number of voxels. Figure 27 is an example of a rendering created by such a method.

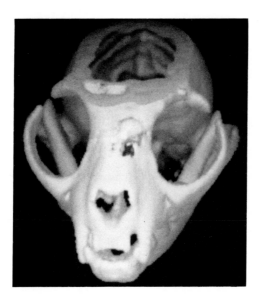

FIGURE 27.   A volume-rendering example for binary
volumes. A scene-space shading technique was used to
assign shading values to voxels. The CT data pertain to
the skull of a cat, a $256 \times 256 \times 112$ volume of cubic
voxels.

An alternative class of techniques[89-91] uses, instead of rectangular arrays, a representation
known as *octrees*. An octree is a hierarchical representation, applicable to both binary and grey
volumes, derived by repeated subdivision of the scene into subscenes. The idea is illustrated
in Figure 28 with an example of a binary scene. The scene region is divided into eight equal
subregions which are identified by the numbers associated with them as in Figure 28(b). If a
subregion is either empty (meaning it contains *no* 1-voxels) or full (meaning it contains *only*
1-voxels), then the subdivision of that subregion stops. Otherwise it is further divided into
eight subregions. The process of subdivision stops when there are no partial subregions left.
We represent the process of subdivision by an octal tree (meaning a tree in which each node
has eight branches) in the following way. In our example, since the scene region is neither full
nor empty (i.e., it is partial) we represent it by a partial node at the top of the tree [Figure 28(c)]
and subdivide it into eight subregions. Each of these subregions is in turn represented by a
node in the tree diagram as either empty, partial, or full. Since subregions numbered 4 and 8
are partial we further subdivide them. We continue in this fashion until all terminal nodes of
the tree are either empty or full.

A useful property of the octree organization is that for any rotation of the scene we can
determine, with some simple calculations, the order in which the subregions should be
considered to get a farthest-to-closest (or vice-versa) ordering (a property very similar to that
of rectangular arrays described earlier). For example, for the viewing conditions assumed
earlier (Figure 16), the farthest-to-closest ordering of the subregions of the unrotated scene of
Figure 28(a), is 5, 6, 7, 8, 1, 2, 3, 4. The same ordering applies to all subregions of each of
these subregions. Our aim is to project only those subregions represented by full nodes in the
tree onto the screen in this order. Since many such subregions are big blocks containing many
voxels, we can compute the projection of the entire block without requiring calculation of the
projection of the individual voxels. This is an advantage of the octree method. Using this rule
then the subregions in our example are projected in the following order of their nodes: 85, 86,

87, 88, 81, 83, 84, 48, 438. The shading value associated with the subregion should overwrite the pixel values in its projection. (The object-space and view-space shading techniques described earlier are applicable and have been used in octree-based volume rendering, but the scene-space techniques, though applicable, significantly mitigate the computational advantages gained in projection.)

2. Grey Volumes

The simplest technique to render grey volumes is to assume that all voxels (tissues) have

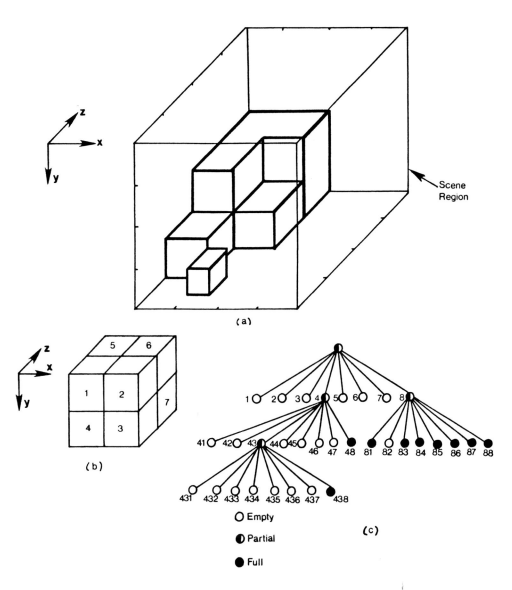

FIGURE 28. Octree representation of scenes. Here a binary scene is shown in (a) where the heavy lines indicate blocks containing only 1-voxels. The rest of the space is filled with 0-voxels. The scene region is repeatedly divided into eight subregions. These subregions are numbered as shown in (b) (the eighth subregion is behind the fourth). The octree representation of the binary scene in (a) is shown in (c). Each subregion is represented by an empty, partially filled, or completely filled circle and a number which indicates which of the eight subregions of the bigger region it represents.

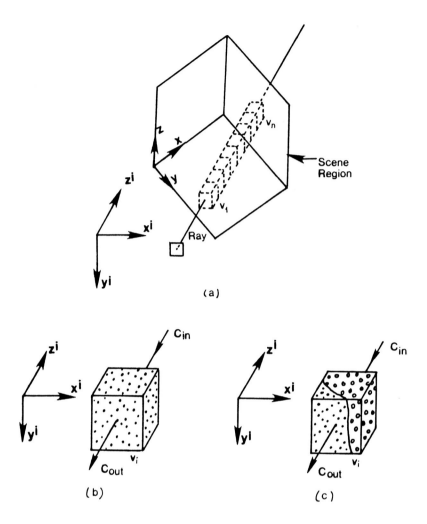

FIGURE 29.   Volume rendering using material mixtures. (a) The resampled voxels $v_1, v_2, \ldots, v_n$ along a ray from a pixel. (b) When the material within a voxel is homogeneous no boundary passes through the voxel and it only emits light because of its luminence. The color of the light coming out, C_{out}, depends on that coming in, C_{in}, the voxel's emission, and its absorption due to its opacity. (c) When a surface does pass through the voxel, C_{out} depends on C_{in}, the voxel's emission, and absorption, as well as the reflection from the surface.

the same color, that their opacities are determined based on voxel density intervals as in Figure 25(a), and then to simulate a radiographic reprojection using the assigned opacities to control the relative radio-opacity of each voxel.[79] Suppose $d_1, d_2, \ldots, d_n$ represent densities estimated at a number of uniformly sampled voxels along a ray cast from a pixel into the scene [see Figure 29(a)]. If $\alpha_1, \alpha_2, \ldots, \alpha_n$ are the opacities corresponding to their densities as determined from an assumed opacity assignment scheme as in Figure 25(a), then the pixel is assigned a grey value (or a fixed color of intensity)

$$\left(\sum_{i=1}^{n} \alpha_i d_i \right) \Big/ \left(\sum_{i=1}^{n} \alpha_i \right)$$

By assigning high opacities to the density intervals of interest and low opacities to the rest we can simulate "tissue dissolution" and enhancement, still retaining the context of the less interesting tissue regions. An example of the type of rendition created by this method can be found in the chapter by Hoffman[92] (Figure 4) in this book.

A more sophisticated method[80] (from the viewpoint of the complexity of the created renditions) is to impose a model of optical behavior on tissue interfaces in the scene and to assign a color to a pixel based on the light reflected toward the pixel. Although the original method[80] uses voxel projection we use ray casting to simplify the description of the main idea. Suppose we sample equally spaced points along a ray cast from a pixel. This is equivalent to determining a sequence of abutting voxels $v_1, v_2, \ldots, v_n$ along the ray with the sampled points as their centers [see Figure 29(a)]. Let $\alpha_1, \alpha_2, \ldots, \alpha_n$ and $C_1, C_2, \ldots, C_n$ be the opacities and colors assigned to these voxels as estimated from the opacities and colors associated with the voxels of the scene via interpolation. We assume that the voxels are luminous and that they emit light whose nature depends on the color of the voxels. This accounts for how volumes are seen. Since we also wish to see surfaces, we use an external light source so that the light reflected from the surfaces can depict their shape. Our aim is to estimate the color of the light reaching the pixel as a result of a combination of these optical behaviors.

If a voxel v_i along the ray does not have any tissue interface (i.e., surface) passing through it [Figure 29(b)], then the tissue it contains is more or less homogeneous. The color C_{out} of the light coming out of it may then be given as

$$C_{out} = C_i + (1 - \alpha_i)C_{in} \tag{7}$$

The first term in the sum represents the color emitted by v_i and the second term the part of C_{in} that v_i transmits. (We have assumed that the opacities are the same for the R,G,B components of color. However, this restriction is not a necessity; we can write a similar equation for each component to make opacity color-component-dependent.) To determine whether or not a surface passes through v_i, we can estimate a gradient vector $\mathbf{N}$ (see "Scene-space shading") at the center of v_i. The magnitude $|\mathbf{N}|$ of this vector indicates the "strength" of the surface and its direction indicates a normal to the surface within the voxel. In practice, $|\mathbf{N}|$ is rarely exactly zero even for voxels within a homogeneous region, but at tissue interfaces it has a large value. When a surface passes through a voxel [Figure 29(c)], C_{out} is determined by the color C_i^B of the tissue behind the surface, the color C_i^F of the tissue in front of the surface, and the color C_i^S of the light reflected from the surface. If all of these entities are known we can apply Equation 7 in three steps: first to calculate the color C_{out}^B of the light coming out of the tissue behind the surface, then to calculate the color C_{out}^S of the light coming out of the surface, and finally to compute the color C_{out} of the light coming out of the tissue in the front.

$$C_{out}^B = C_i^B + (1 - \alpha_i^B)C_{in} \tag{8}$$

$$C_{out}^S = C_i^S + (1 - \alpha_i^S)C_{out}^B \tag{9}$$

$$C_{out} = C_i^F + (1 - \alpha_i^F)C_{out}^S \tag{10}$$

Using the material mixture model described earlier [Figure 25(b) and (c)], we know C_i^B, C_i^F, α_i^B and α_i^F. We can assume the opacity α_i^S of the surface to be identical to the opacity α_i^B of the tissue in the back. (We can determine which of the types of tissues in the mixture is in the back and which is in the front by checking the sign of the z^i-component of $\mathbf{N}$. If this is positive, then the higher density tissue is in the back; otherwise it is the tissue with lower density.)

The only unknown in the above equations, C_i^S, depends on the assumed reflective properties of the tissue interface. Any of the previously described models of surface reflection can be used to compute C_i^S. In particular, Equation 6 can be used with appropriate modifications to account for the color and strength of the surface to influence what the surface reflects. The general formula can be given in the following form:

$$C_i^S = \left[f_d(\mathbf{N}, \mathbf{L}) C_d + f_s(\mathbf{N}, \mathbf{L}) C_1 \right] |\mathbf{N}| + A \tag{11}$$

where C_d is the color of the diffusely reflected light which we can assume to be identical to C_i^B, C_l is the color of specular reflection assumed to be identical to the color of the light source, L is a vector indicating the direction of light rays (we have assumed that the direction of L and viewing are identical), and f_d and f_s are appropriate functions describing diffuse and specular reflection.

We have so far described how to compute C_{out} knowning C_{in} for a particular voxel v_i. To determine the color to be assigned to the pixel in Figure 29(a), we start from v_n, assume C_{in} = (0,0,0) (i.e., black) for v_n, and compute its C_{out}, which of course is C_{in} for $v_n - 1$. We work towards v_1 in this fashion and finally compute C_{out} for v_1 which is the color assigned to the pixel under consideration. The entire process described so far should be repeated for all pixels in the screen to create one complete rendition. Examples of the rendered images created by this method can be found in other chapters[93-95] in this book.

If we leave out emission and consider only reflection and absorption for every voxel, we can depict only tissue interfaces (surfaces) through volume rendering.[70] Suppose then that we assign opacities to voxels based on both their density and gradient magnitude as in Figure 25(d). Then, clearly, only voxels at the interfaces will have any significant opacity and those within homogeneous regions will have negligible opacity. C_{out} and C_{in} for each voxel along the ray are now related as follows:

$$C_{out} = \alpha_i^S C_i^S + (1 - \alpha_i^S) C_{in} \tag{12}$$

where α_i^S, the opacity of the surface through v_i, is assumed to be equal to the opacity assigned to v_i. C_i^S, as before, is the color of the light reflected from the surface through v_i.

$$C_i^S = F_d(\mathbf{N}, \mathbf{L}) + F_s(\mathbf{N}, \mathbf{L}) + A \tag{13}$$

F_d and F_s are functions that determine the color reflected from the surface through diffuse and specular reflection, respectively. (This method[70] does not specify the color dependency of these functions.) The color to be assigned to the pixel under consideration is determined, as in the previous method, starting from v_n assuming C_{in} = (r,g,b) — some background color — and working towards v_1 taking C_{out} of v_i to be C_{in} for $v_i - 1$. Examples of the rendered images created by this method can be found in the chapter by Rosenman[96] in this book.

VI. MANIPULATION AND ANALYSIS

Our discussion so far has concentrated on how to visualize structures in scenes. While visualization is essential because our decision process is heavily influenced by what we see, it represents only a qualitative aspect of the total analysis of the data. Quantitative analysis that can more concretely describe biological forms and their variations strongly complements visualization and often alone can fulfill a 3D imaging objective. However, research efforts on this aspect so far have been rather minimal compared to those on visualization, and consequently standard methodologies for this aspect have yet to evolve.

A. INTERACTIVE MANIPULATION

For truly exploring the data, either for the purpose of visualization or analysis, interactive manipulability becomes necessary. Often visualization itself is exploratory. What we want to eventually examine closely is not predeterminable and preliminary visualization is used in combination with manipulation to reach the end result. Since the objective of manipulation is to alter structures (in the above context this may be for removing an obscuring structure to create a nonoverlapping display),[22] it may be represented by a transformation from object space into object space (Figure 1).

The more exciting avenue for interactive manipulation is in surgical planning. The idea here is to indicate osteotomies directly on the rendition of the structure using a pointing device (such as a mouse) as a knife, to separate segments, and to reposition them for structure normalcy. The questions to be addressed here are: how to specify osteotomies with a natural action, what are appropriate object representation schemes and rendering techniques, how to separate segments for repositioning, and what are the repositioning strategies. The answers to these questions together with some of the mensuration tools discussed in the next section determine the effectiveness of interactive manipulation in surgical planning.

The few published techniques[87,88,97-102] on interactive manipulation use either surface rendering or binary volume rendering for the 3D display of the structures being manipulated. The osteotomies are usually specified by drawing a curve on the surface display, with the assumption that the osteotomy goes perpendicular to the $x^i y^i$ plane in the image space to a specified depth. This is a rather limited and somewhat unnatural way of indicating a cut on the object surface. Yet quite a few useful operations can be specified in this fashion by appropriately selecting the object orientation for drawing the osteotomy curve.

From the point of view of algorithm development the most challenging issues arise from the speed requirement. Clearly, since interactive response is paramount, fast rendering and fast manipulation (isolation of osteotomy segments, their repositioning, etc.) become essential. Unfortunately, there are some contradictory requirements for efficient rendering and manipulation. For example, since our aim in rendering is to get a depiction of the surface, our rendering technique should do little or no computation for voxels interior to the object, preferably even on some of those voxels in the vicinity of the boundary that are not visible in the chosen orientation. Hence the object representation should be boundary oriented. In manipulation, on the other hand, since voxels interior to the object can potentially fall in the vicinity of the boundary of the manipulated object or of the separated segment, they have to be retained in the object representation. If we use separate representations for rendering and manipulation, some time is wasted in converting the representation for manipulation to that of rendering before the modified object and the segments are displayed. The actual algorithm used for separating the segments based on the specified osteotomy depends very much on the representation scheme.

One simple method of segment repositioning is to use the pointing device to point to a segment, to drag it to a desired location, and to rotate it to an appropriate orientation so that the display combined with some of the mensuration techniques described below can indicate whether or not structural normalcy has been achieved. The value of such procedures lies in the fact that many aspects of the surgical procedure — where is the osteotomy optimally placed, how much the segments should be moved, what is the bone graft requirement, how a tumor should be approached for biopsy or removal with minimum damage to critical structures — can be quantified prior to surgery. In addition, alternative strategies can be experimented with. It is also possible to automatically determine how a segment should be moved for optimal results based on a computer representation of what is normal, once an osteotomy is specified and the segment is separated.[103] Figure 30 shows some of the manipulative operations that are currently possible. CT data of a dry skull are used for illustration.

Most of the manipulative operations considered above are currently possible with binary volume and some surface rendering techniques. Some of these operations are currently not possible using grey volume rendering techniques.

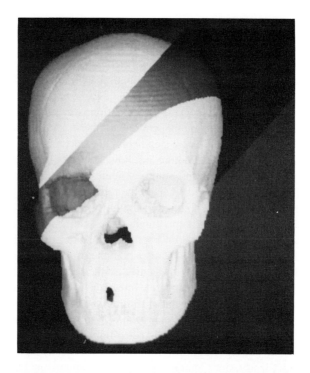

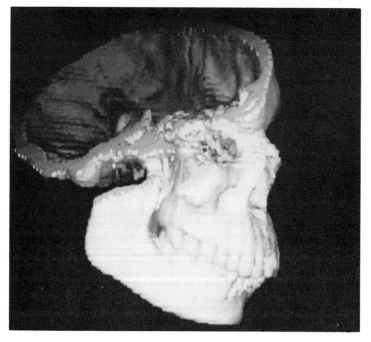

(a)

FIGURE 30.   Interactive manipulation for surgical planning. Though the procedures illustrated are hypothetical (since the data pertain to a normal skull), they demonstrate the potential of the manipulative tools. (a) Cut away to reveal the interior. The cutting plane can be specified interactively and the cutting operation can be carried out in a couple of seconds. (b) Mirror reflection. In cases of unilateral deformity, the normal structure can be reflected about any user-specified plane, the reflected and the normal structures can be displayed simultaneously, and they can be measured with respect to each other. (c) Complex osteotomy. Osteotomy curves can be drawn on the surface for separating a segment and for subsequently repositioning it. The separated segment can be freely moved and rotated in space. All of these operations including the interactive display can be done in a couple of seconds on a SUN workstation. These illustrations are produced using a recently reported method.[88]

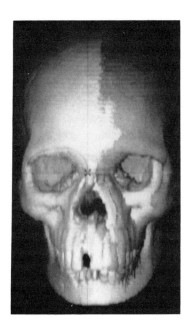

FIGURE 30(b)

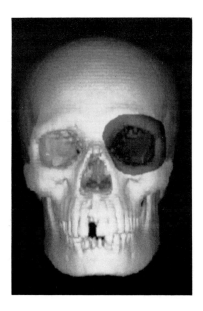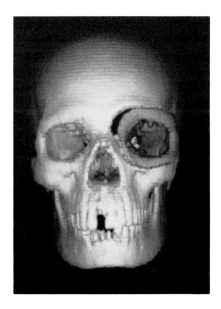

FIGURE 30(c)

B. ANALYSIS

The purpose of analysis in 3D imaging is to acquire quantitative information about the structure in the scene. Usually the end result is the value of a parameter such as the volume of a cavity, a tumor or a bone graft, the diameter of a vessel, the mean density of a 3D region, or the instantaneous velocity of a certain aspect of a dynamic organ. It may also be the distribution of a parameter in space such as the magnitude of the velocity at various points within the myocardium or the distribution of multiple parameters such as the magnitude and direction of velocity as well as density at various points in the myocardium. Hence, most analysis operations can be considered as transformations from scene space to parameter space usually via the view space (Figure 1).

The simplest techniques of analysis are based on measuring directly on slice display. Distances, areas, and angles can all be measured in 2D using a pointing device to indicate what is to be measured. Statistics of density distribution (mean, standard deviation, median, etc.) can be measured within closed 2D regions indicated by the user. Such capabilities are common to most software packages available on CT and MRI scanners. Though all of these measurements can be made in 3D, it is awkward or often impossible to indicate on slices the 3D entity to be measured. The ability to measure on 3D display, however, is not so commonly available, although the associated techniques[104] are rather simple. Note that if we use a z-buffer technique, since we know the x^i, y^i, z^i coordinates of all visible points on the surface, the 3D distance between any two of them can be easily calculated. This capability is sufficient for computing the angle (in 3D) subtended at a point by any other two points. Of course it is required that all points be visible but not necessarily in the same view. The distance measurement can be extended to the measurement of the length of a curve on the surface between two points by indicating a sequence of points between them on the curve. Volume is computed by calculating the volume enclosed by the modeled surface.[105-108] Just counting the voxels satisfying a segmentation criterion is generally not enough; some form of connectivity and/or anatomic membership also needs to be considered. There are no satisfactory techniques available for computing surface area. Though the triangulation techniques[37-45,48] discussed under "Surface Formation" are perhaps more appropriate for computing surface area (by summing the areas of polygons forming the surface) than techniques based on voxel faces,[49-52] the former do not always guarantee a foolproof, automatic solution to surface formation. Other less commonly used measurements are center of mass, moment of inertia,[109] and surface curvature.[110]

There are analysis techniques beyond direct mensurations described above. One of the more common techniques is the composition of multiple scenes obtained for the same organ system from different imaging modalities.[111-114] The main purpose of such analysis is to combine complementary information obtained from different modalities, for example, bone information from CT and soft-tissue details from MRI, anatomic information from CT and/ or MRI and regional metabolic information from positron emission tomograms, and anatomy from CT and external body geometric information from light scanners. The method consists of representing the multiple scenes in the same scene by optimally matching some known aspects of a structure that are present in all scenes. The same technique has also been used in the comparative analysis of a single structure some aspects of which may change with time to analyze just the change.[115,116] For example, consider the longitudinal follow up of the bone grafts implanted within a patient's craniofacial structure that has been surgically corrected. Since grafts fuse with the native bone after some time, it becomes impossible to isolate them for the purpose of measuring resorption. If we can identify a set of landmarks that can be accurately located in renditions of the cranial structure, pre- as well as postsurgically, then we can register the scenes, subtract postsurgical from presurgical scenes and obtain just the grafts, provided of course the landmark locations remain unaltered postsurgically.

Another type of analysis is the study of the mechanical functionality of a structure by investigating the stress and strain distribution through the structure. Such analysis is done mainly for the purpose of designing prostheses.[117]

The 3D analysis techniques described above are realizable via surface- and binary volume-rendering techniques. Grey volume-rendering methods do not have the notion of a structure and hence cannot provide these measurements at present.

VII. HARDWARE AND SOFTWARE ASPECTS

Both hardware and software approaches have been taken for implementing 3D imaging methodologies.

The hardware approach is necessitated mainly from speed considerations. Often, a computationally significant step may have to be repeated a large number of times to accomplish a specific process. For example, in binary volume rendering using the farthest-to-closest projection method the following sequence of steps is repeated millions of times (once for every 1-voxel in the binary scene): rotation of the voxel (corresponding to the given rotation of the scene), computing its projection and shading, and overwriting pixel values. Sometimes, for practical reasons, a process may have to be realized in real time (usually means in a small fraction (1/20 to 1/10) of a second) or at least at interactive speeds (close to a second or two). An example is the simulation of osteotomies described earlier (see also Figure 30). Clearly, if the computation of the segment specified by an osteotomy takes more than a couple of seconds or if the segment cannot be repositioned with the movement of a mouse or a similar device, such tools are of no practical use. The premise in hardware approaches is to achieve speed by designing specialized electronic processing components that realize the computationally intensive steps in interactive processes. Several approaches[89,118-122] have been pursued for implementing mainly visualization techniques. Some[118] are based on rendering polygonal surface representations, and others[89,119-122] realize binary volume-rendering techniques. While real-time visualization is possible using some of these, for which prototypes have already been built, interactive manipulation and analysis of the type described in the previous section have not been considered in these implementations. In fact, this points to one of the major drawbacks of most of the hardware approaches, namely their specialized design to implement a specialized algorithm or methodology. This often implies that they cannot be programmed to do other general tasks or even other processing operations without extensive research and/ or training on the specialized hardware. It is not at all clear how a design that was specialized for real-time visualization can readily cater to the interactive needs of other applications such as surgical planning or comparative and composite analysis of multimodality images. This extensibility issue combined with the fact that, due to rapid advances in workstation technology, many, including some manipulative, operations can be realized at close to interactive speeds on general-purpose workstations makes the practical utility of the hardware approaches questionable, although most of them are very exciting from purely a research point of view. From the extensibility and general utility points of view, a device that incorporates in hardware a set of basic computationally intensive operations common to most imaging methodologies and that makes these operations available in a programmable fashion forms an ideal 3D imaging machine. Maybe such a machine will eventually be developed from research on hardware aspects of 3D imaging.

Software approaches are, by far, the most commonly used in both research and commercial distribution related to the 3D imaging technology. The technology is currently available to the end user in three forms: on-the-scanner software, imaging workstations, and research software.

Perhaps, there is no (CT) scanner manufacturer at present who does not yet have a 3D imaging software product that runs on the manufacturer's scanner display console. (The details of commercially available product options can be found in the chapter by Hemmy[123] in this book.) The processing options available in these software products are confined mostly to visualization. These include slice imaging (reslicing, windowing, sometimes histogram equalization), either surface rendering or binary volume rendering with masking, and some analysis tools (region-of-interest pixel density statistics, 2D distances on slices, and 3D distances on surfaces). The total processing time depends on the hardware options available on the display console. Many offer hardware enhancements along with the software (which, of course, has a price tag) which bring the processing time enviably close to those of some of the imaging workstations. The processing time with and without such enhancements can range between a few minutes for generating a couple of dozen views at a few seconds per view to a couple of hours at a few minutes per view. The main deficiencies of on-the-scanner software are the lack of manipulative and analysis tools and cine display. Since the technology of display consoles is rather outdated, they cannot provide the many facilities that the modern workstation can offer such as drop-down menus, color options, multiple mini-screens, and a variety of natural mechanisms of communicating with the machine via graphic input devices.

Imaging workstations, in general, have more power and options than on-the-scanner software packages. In addition to the imaging options provided by the latter, they often support curved reslicing, filtering, cine display, volume rendering, and more advanced application-specific analysis tools such as additional display capabilities for radiation therapy, object modeling for prostheses design, and interactive segmentation. Their processing speed ranges from real-time performance to close to an hour for visualization tasks of the type mentioned above. Some of their current deficiencies are the lack of manipulative and advanced analysis tools, the lack of extensibility and programmability for high-performance systems, and the specificity of the methodologies implemented. We feel that since a clear superiority is not indicated at present among major methodologies or even among approaches within a methodology, providing flexibility and alternatives should be beneficial to the end user. This is best done via software at the current state of development of 3D imaging. Workstation technology is better poised for making this provision than the scanner display consoles. Unfortunately, such choices are typically not provided by commercial 3D imaging workstations.

The software developed at universities and research laboratories under federal programs is the final category we wish to discuss. Although this activity is quite significant in magnitude, the software available in the public domain for 3D imaging in the form of usable packages is very scanty. While some general image processing packages are available, the only 3D imaging package currently available in source code is rather outdated.[124] A package that can provide the basic platform necessary for 3D imaging-related development can contribute significantly to 3D imaging research by eliminating repeated, application-independent, common developmental efforts.

VIII. HISTORIC NOTES

The first reported work on biomedical 3D display was done by Greenleaf et al.[125] The earliest 3D rendering of anatomic structures from CT images was by Ledley and his co-workers.[126,127] The 3D imaging work at the Medical Image Processing Group started with the Ph.D. thesis work of Liu[8] under the direction of G.T. Herman.[23,63] Some brisk activity followed this initial investigation by this group, and the subsequent work of Herman, Udupa and co-workers till the early eighties[1,22,50-52,64,105] elicited a burst of development elsewhere. Early investigations were pursued at about the same time by Fuchs and Pizer,[27,38] Cook et al.,[40] and Rhodes.[128] Though many perceive volume rendering to be a recent development, the main idea underlying this method was first developed and applied to cardiopulmonary imaging by Harris et al.[79]

Although clinical studies related to 3D imaging started in the late seventies,[129,130] serious application began about 1983.[131] Craniomaxillofacial applications were actively pursued by Hemmy et al.[129] and Marsh et al.[131] Early studies of orthopedic applications[132-134] used surface rendering. Recently Fishman et al.[95] used volume rendering to study extensively this application. Though the use of interactive manipulation for surgical planning was studied in principle and from the point of view of algorithm development earlier,[97,99] Cutting[98,103] and his co-workers started the study of the clinical issues and their solutions as related to this problem.

Early development related to radiation therapy planning is due to McShan et al.[135] (see also the chapter by Rosenman[96] in this book). Cardiovascular applications of 3D imaging have long been studied at the Mayo Clinic since their earliest reported work,[125] and especially after the installation of an early version of DISPLAY82[124] in 1980, and recently by Hoffman[92] after the development of the Analyze package[136] by Robb and his co-workers.

On-the-scanner and commercial software development started with the implementation of the DISPLAY82 package on the display console of the General Electric 8800 CT scanner in 1980 by the author of this article. By about 1985, most of the CT scanner manufacturers had a 3D software option on their scanners. The earliest 3D imaging workstation development is due to Phoenix Data Systems.[89,109]

IX. CONCLUDING REMARKS

Considering visualization, interactive manipulation and analysis together to constitute 3D imaging, a bulk of the work done during the past decade addresses issues relating mostly to visualization. Though rapid progress has been made in visualization, many old issues still remain and new ones also have surfaced. Manipulation and analysis have remained largely unexplored areas, posing some technically very challenging issues.

A. ISSUES RELATED TO VISUALIZATION
1. Segmentation
This is perhaps the most crucial issue. Whatever the imaging task — visualization, manipulation, or analysis — it is imperative that the structure we are interested in is somehow *identified*. If this cannot be done, the most useful visualization one can create is either via slice display or via the display of radiograph-like images (in projective imaging) wherein the voxel densities along a ray are simply added to compute a pixel value. Identification of the structure of interest is made difficult by a variety of blurring effects that are introduced by the imaging scanner or inherent in the body (e.g., overlapping distribution of the property imaged of adjoining tissues). It appears that higher level anatomic knowledge is necessary in complex situations to resolve voxel-level ambiguities. To convert this knowledge to voxel-level information or vice-versa is quite challenging. Sophisticated strategies for automatic segmentation are being pursued by several research groups[12,21,137,138] and the coming few years will perhaps bring some tangible solutions to this problem.

2. Surface Versus Volume Rendering
A controversial subject in visualization is related to the relative merits and demerits of volume rendering vis-a-vis that of surface rendering. Many claims have been made on this subject in recent talks, interview reports, and even in technical papers, without substantiating evidence. We will not delve into a detailed discussion of this subject, which deserves a full paper on its own, but will attempt to clarify some often-raised issues.

(i) "(Gray) volume rendering does not require object segmentation". From our description under "Preprocessing" in "Volume Rendering" it is clear that this is not true. The only situation where this is the case is when projective imaging is done without any knowledge of the object and the result is a radiographic image. This obviously does not fulfill the 3D imaging objective.

(ii) "(Voxel-based) surface rendering cannot capture fine details such as sutures of the skull and gyrations on the brain surface and it introduces jaggedness". Referring the reader to Figure 21 we only wish to state that inability to capture details is not a weakness of surface rendering (or for that matter of volume rendering). It may be a weakness of a particular technique but it is independent of the methodology. The geometry describing the surface (say as a set of voxel faces) may not actually "contain" a suture or a fine fracture. Nevertheless, by appropriately calculating the normal in the vicinity of where a fine structure may be present, it is possible to depict the structure in the rendering. To that extent the precise geometry resulting from segmentation is not very crucial. As to jaggedness, we wish to point out that the renderings in Figure 21 were created without doing any form of image smoothing (commonly known as *antialiasing*). We wish to emphasize that with appropriate combination of scene interpolation and normal estimation techniques[7,71] it is possible to create remarkably smooth and natural renditions of discrete surfaces.

(iii) "(Gray) volume rendering is less prone to false positives and false negatives". The false-positive and -negative rates depend on a number of factors: the observer, the clinical task, the object identification method, the rendering technique, the display technique, etc., irrespective of whether the surface or the volume method is used. In general, both surface and volume renderings show a good deal of both false positives and false negatives. Whether these matter in a clinical application cannot be settled through theoretical arguments. For no spurious objects we should display the slices; for no missing objects the best display (in the projective imaging sense) is the radiographic image.

(iv) "In surface rendering the original density information is lost". Clearly, this applies to volume rendered images as well. It is not possible to infer the original voxel density information from volume displays or for that matter even from a radiographic image. Obviously, that information is not lost (because one can retain the original slices and, if desired, map such information onto substructures in surface renditions[135,139] and even in volume renditions) nor is it readily (visually) measurable from the display irrespective of the 3D mode of display.

(v) It is sometimes believed that thresholding is synonymous with surface rendering. Thresholding is a particular technique for segmentation which can be used in conjunction with both surface and volume methods. There are nonthresholding techniques (see under "Segmentation") that are more sophisticated than thresholding that can also be used in both surface and volume rendering.

(vi) Finally, contrary to popular belief, volume rendering is not a recent development. To the best of our knowledge, the main idea of volume rendering via reprojection through the volume was developed and applied by Harris et al.[79] more than ten years ago, although the recent developments are undoubtedly more sophisticated.

In summary, the strength of volume rendering is in providing a "bird's eye view" of selected contents of the scene, especially when the structure is diffused, like a cloud, rather than discrete. For example, a volume renderer, in principle, is a better candidate for the 3D imaging of a diffused tumor than a surface renderer. Of course, the effectiveness of the display, which depends on how effectively opacities (and color) can be assigned to the tumorous tissue in a perceptibly different manner relative to other tissues of interest, is a different matter altogether. The strength of surface rendering is in its ability to provide structure-oriented interactive manipulation capabilities, which are currently not possible using (grey) volume rendering. An added strength (our personal opinion which requires further study) is that keeping common all processing operations that can be kept common, a surface renderer (of the type used in Figure 21) can create sharper depiction of fine details such as sutures and natural texture than a volume renderer. On the other hand, the fuzziness afforded by volume rendering appears to somewhat make up for the deficiencies in object identification, but it is not at all clear that the accuracy of portrayal of useful information is necessarily better in volume rendering than in surface rendering or vice versa.

3. Multidimensional Multiparametric Data

As computer processing methods and/or imaging modalities become more sophisticated it would increasingly become possible to compute and/or gather data pertaining to multiple properties associated with structures (in a pointwise fashion) such as regional motion/growth, perfusion, contractibility, etc. The composite display of such information poses some problems worth pursuit. There is a question of whether this can be done at all on a 2D screen in a useful manner. Of course, color and texture give additional display dimensions. However, if the methodology is not devised carefully, the display can be more confusing than helpful.

Finally, there are some interesting theoretical questions as to how to do scene processing and visualization in a unified, scene-dimensionality independent way.

B. ISSUES RELATED TO MANIPULATION AND ANALYSIS

The problem of interactively manipulating structures has been studied so far only sparsely. Unlike visualization, systems for the practical use of manipulation are not available. Much further work is needed toward developing natural ways of providing interaction. The two degrees of freedom available in commonly available pointing devices (like the mouse) greatly limit the type of operations that can be specified. The technology of 3D pointing devices is now becoming available. There are some very challenging hardware, software, and algorithmic issues relating to how such devices can be used (possibly in conjunction with stereo display) for indicating various user actions for osteotomy specification, marking the surface, moving segments, overlaying/inlaying bone grafts, draping soft tissues, etc. The provision of constant visual and/or kinesthetic feedback, which become necessary for naturalness of some of the actions, poses new technological challenges.

Some of the complex operations cannot be realized at present in real time or even at interactive speeds on general-purpose workstations. Investigation of algorithms (including parallel) to achieve acceptable speeds is hence an important activity. As computerized surgical planning advances, the need to generate and maintain large data sets of normal patients increases. Such data have already been gathered in craniofacial surgery for landmark data based on cephalograms.[103] The creation of normative structural data based on tomographic images raises questions as to how 3D form should be characterized, how the characterization should be stored in the computer so that it can be conveniently exchanged between research centers, and how such data can be used to optimize surgical plans.

Finally, though many exciting 3D imaging techniques have been developed over the years, the effort toward *validation* of results and objective *evaluation* of methods has been rather abysmal (see chapters by Herman[140] and Gillespie[141] for a detailed account). By validation we mean the assessment of the fidelity with which a qualitative or a quantitative measure that is sought after is produced by a 3D imaging technique. By evaluation we mean a comparative assessment of the performance of two competing techniques. This is an area to which the clinician end users of the 3D imaging technology can make significant contributions.

ACKNOWLEDGMENTS

The research of the author is supported by NIH Grants HL28438 and CA50851.

The author is grateful to Mary A. Blue for typing the manuscript, to Marilyn Kirsch for drawing the illustrations, and to Steve Strommer for photography.

Some of the surface renderings illustrated in this article are the outcome of the author's joint work with his colleagues Mr. Dewey Odhner, Dr. Keh-Shih Chuang, Dr. Sai Raya, and Ms. Hsiu-Mei Hung.

REFERENCES

1. Udupa, J.K. Display of 3D information in discrete 3D scenes produced by computerized tomography, in *Proc. IEEE* 71:420–431 (1983).

2. Harris, L.D. Display of multidimensional biomedical image information, in *Three Dimensional Biomedical Imaging*, R.A. Robb (Ed.) (Boca Raton, FL: CRC Press, 1985, v. 2, pp. 125–139.

3. Udupa, J.K. 3D imaging in medicine, in *Proceedings of the Eighth Annual Conference and Exposition of the National Computer Graphics Association* (Fairfax, VA: NCGA, 1987), v. I, pp. 73–104.

4. Barillot, C., and B. Gibaud. Computer graphics in medicine: A survey, *CRC Crit. Rev. Bioengr.* 15:269–307 (1988).

5. Farrell, E.J., and R.A. Zappulla. Three-dimensional data visualization and biomedical applications, *CRC Crit. Rev. Bioengr.* 16:323–363 (1989).

6. Udupa, J.K., G.T. Herman, P.S. Margasahayam, L.S. Chen, and C.R. Meyer. 3D98: A turnkey system for the display and analysis of 3D medical objects, in *Proceedings of the Society of Photo-Optical Instrumentation Engineers* (Bellingham, WA: SPIE, 1986), v. 67, pp. 154–168.

7. Raya, S.P., and J.K. Udupa. Shape-based interpolation of multidimensional objects, *IEEE Trans. Med. Imag.*, to appear (1989).

8. Liu, H.K. Two- and three-dimensional boundary detection, *Comput. Graph. Image Proc.* 6:123–134 (1977).

9. Zucker, S.W., and R.A. Hummel. An optimal three-dimensional edge operator, in *Proceedings of IEEE Computer Science Conference on Pattern Recognition and Image Processing* (Piscataway, NJ: IEEE, 1978), pp. 162–168.

10. Morgenthaler, D., and A. Rosenfeld. Multidimensional edge detection by hypersurface fitting, *IEEE Trans. Patt. Anal. Mach. Intel.* PAMI-3:482–486 (1981).

11. Herman, G.T., and H.K. Liu. Dynamic boundary surface detection, *Comput. Graph. Image Proc.* 7:130–138 (1978).

12. Chuang, K.S., and J.K. Udupa. Boundary detection in grey level scenes, in *Proceedings of the Tenth Annual Conference and Exposition of the National Computer Graphics Association* (Fairfax, VA: NCGA, 1989), v. I, pp. 112–117.

13. Prewitt, J.M.S., and M.L. Mendelsohn. The analysis of cell images, *Ann. N.Y. Acad. Sci.* 128:1035–1053 (1966).

14. Trivedi, S.S., G.T. Herman, and J.K. Udupa. Segmentation into three classes using gradients, *IEEE Trans. Med. Imag.* MI-5:116–119 (1986).

15. Vannier, M.W., R.L. Butterfield, D.L. Rickman, D.M. Jordon, W.A. Murphy, and P.R. Biondetti. Multispectral magnetic resonance image analysis, *CRC Crit. Rev. Bioengr.* 15:117–144 (1989).

16. Herman, G.T., M.I. Kohn, and R.E. Gur. Computerized three-dimensional volume analysis from magnetic resonance images for the characterization of brain disorders, in *Proceedings of a Special Symposium on Maturing Technologies and Emerging Horizons in Biomedical Engineering* (Piscataway, NJ: IEEE, 1988), pp. 65–67.

17. Kennedy, D.N., P.A. Filipek, and V.S. Caviness, Jr. Anatomic segmentation and volumetric calculations in nuclear magnetic resonance imaging, *IEEE Trans. Med Imag.* MI-8:1–7 (1989).

18. Fukunaga, K. *Introduction to Statistical Pattern Recognition* (New York, NY: Academic Press, 1972).

19. Hwang, J.J., E.L. Hall, C.C. Lee, W.J. Nalesnik, and C.R. Archer. Global local edge coincidence segmentation for computer tomography images, in *Proceedings of the Sixth Conference on Computer Applications in Radiology and Computer-Aided Analysis of Radiological Images* (Piscataway, NJ: IEEE, 1979), pp. 312–320.

20. Lemke, H.J., H.S. Stiehl, H. Scharnweber, and D. Jackel. Applications of picture processing, image analysis and computer graphics techniques to cranial CT scans, in *Proceedings of IEEE/ACR Conference on Computer Aided Analysis of Radiological Images* (Piscataway, NJ: IEEE, 1979), pp. 341–354.

21. Pizer, S.M., J.M. Gauch, and L.M. Lifshitz. Interactive 2D and 3D Object Definition in Medical Images Based on Multiresolution Image Descriptions, Technical Report 88-005, Department of Computer Science, University of North Carolina at Chapel Hill (1988).

22. Udupa, J.K. Interactive segmentation and boundary surface formation for 3D digital images, *Comput. Graph. Image Proc.* 18:213–235 (1982).

23. Herman, G.T., and H.K. Liu. Display of three-dimensional information in computed tomography, *J. Comput. Asst. Tomogr.* 1:155–160 (1977).

24. Rhodes, M.L., W.V. Glenn, Jr., and V.M. Azzawi. Extracting oblique planes from serial CT sections, *J. Comput. Asst. Tomogr.* 4:649–657 (1980).

25. Maravilla, K.R. Computer reconstructed sagittal and coronal computed tomography head scans: Clinical applications, *J. Comput. Asst. Tomogr.* 2:120–123 (1978).

26. Rothman, S.L.G., G.D. Dobben, M.L. Rhodes, W.V. Glenn, and Y-M. Azzawi. Computerized tomography of the spine: Curved coronal reformation from serial images, *Radiology.* 150:185–180 (1984).

27. Pizer, S.M. Intensity mappings to linearize display devices, *Comput. Graph. Image Proc.* 17:262–268 (1981).

28. Hummel, R. Histogram modification techniques, *Comput. Graph. Image Proc.* 4:209–224 (1975).

29. Pizer, S.M., J.D. Austin, J.R. Perry, H.D. Safrit, and J.B. Zimmerman. Adaptive histogram equalization for automatic contrast enhancement of medical images, in *Proceedings of the Society of Photo-Optical Instrumentation Engineers* (Bellingham, WA: SPIE, 1986), v. 626, pp. 242–250.

30. Pizer, S.M., E.P. Amburn, J.D. Cromartie, R.A. Geselowitz, T. Green, B.H. Romeny, J.B. Zimmerman, and K. Zuiderveld. Adaptive histogram equalization and its variations, *Comput. Vis. Graph. Image Proc.* 39:355–368 (1987).

31. Zimmerman, J.B., S.B. Cousins, M.E. Frisse, K.M. Hartzell, and M.G. Kahn. A psychophysical comparison of two methods for adaptive histogram equalization, in *Proceedings of the Society of Photo-Optical Instrumentation Engineers* (Bellingham, WA: SPIE, 1989), v. 1091, pp. 154–161.

32. Rosenfeld, A., and A.C. Kak. *Digital Picture Processing,* v. 2 (New York, NY: Academic Press, 1982).

33. Tuy, H.K., and J.K. Udupa. Representation, display and manipulation of 3D discrete scenes, in *Proceedings of the 16th Hawaii International Conference on System Sciences* (North Hollywood, CA: Western Periodicals, 1983), v. II, pp. 397–406.

34. Rosenfeld, A. *Picture Languages,* (New York, NY: Academic Press, 1979).

35. Duda, R.O., and P.E. Hart. *Pattern Classification and Scene Analysis* (New York, NY: John Wiley & Sons, 1973), p. 338.

36. Heffernan, P.B., and R.A. Robb. A new method for shaded surface display of biological and medical images, *IEEE Trans. Med. Imag.* MI-4:26–38 (1985).

37. Keppel, E. Approximating complex surfaces by triangulation of contour lines, *IBM J. Res. Develop.* 19:2–11 (1975).

38. Fuchs, H., Z.M. Kedem, and S.P. Uselton. Optimal surface reconstruction for planar contours, *Comm. ACM* 20:693–702 (1977).

39. Christiansen, H.N., and T.W. Sederberg. Conversion of complex contour line definitions into polygonal element mosaics, *Comput. Graph.* 12:187–192 (1978).

40. Cook, L.T., S.J. Dwyer, III, S. Batnitzky, and K.R. Lee. A three-dimensional display system for diagnostic imaging applications, *IEEE Comput. Graph. Appl.* 3:13–19 (1983).

41. MOVIE.BYU: A General Purpose Computer Graphics System, Department of Civil Engineering, Brigham Young University, (Provo, UT: Brigham Young University, 1981).

42. Shantz, M. Surface definition for branching contour-defined objects, *Comput. Graph.* 15:242–259 (1981).

43. Mills, P.H., H. Fuchs, S.M. Pizer, and J.G. Rosenman. IMEX: A tool for image display and contour management in windowing environment, in *Proceedings of the Society of Photo-Optical Instrumentation Engineers* (Bellingham, WA: SPIE, 1989) v. 1091, pp. 132–142.

44. Toennies, K.D. Automatic surface approximation from contours in serial sections, in *Proceedings of COMPINT'87,* (Piscataway, NJ: IEEE, 1987), pp. 93–96.

45. Shaw, A., and E.L. Schwartz. Construction of polyhedral surfaces from serial sections: Exact and heuristic solutions, in *Proceedings of the Society of Photo-Optical Instrumentation Engineers* (Bellingham, WA: SPIE, 1989) v. 1091, pp. 221–233.

46. Mazziotta, J.C., and K.H. Huang. THREAD (three-dimensional reconstruction and display) with biomedical applications in neuron ultrastructure and computerized tomography, *Amer. Fed. Infor. Proc. Soc.* 45:241–250 (1976).

47. Sunguroff, A., and D. Greenberg. Computer generated images for medical applications, *Comput. Graph.* 12:196–202 (1978).

48. Cline, H.E., W.E. Lorensen, S. Ludke, C.R. Crawford, and B.C. Teeter. Two algorithms for the three-dimensional reconstruction of tomograms, *Med. Phys.* 15:320–327 (1987).

49. Gordon, D., and J.K. Udupa. Fast surface tracking in three-dimensional binary images, *Comput. Vis. Graph. Image Proc.* 45:196–214 (1989).

50. Artzy, E., G. Frieder, and G.T. Herman. The theory, design, implementation and evaluation of a three-dimensional surface detection algorithm, *Comput. Graph. Image Proc.* 15:1–24 (1981).

51. Udupa, J.K., S.N. Srihari, and G.T. Herman. Boundary detection in multidimensions, *IEEE Trans. Patt. Anal. Mach. Intel.* PAMI-4:41–50 (1982).

52. Herman, G.T., and D. Webster. A topological proof of a surface tracking algorithm, *Comput. Vis. Graph. Image Proc.* 23:162–177 (1983).

53. Frieder, G., G.T. Herman, C.R. Meyer, and J.K. Udupa. Large software problems for small computers: An example from medical imaging, *IEEE Software* 2:37–47 (1985).

54. Morgenthaler, D.G., and A. Rosenfeld. Surfaces in three-dimensional digital images, *Infor. Cont.* 51:227–247 (1981).

55. Lorensen, W.E., and H.E. Cline. Marching cubes: A high resolution 3D surface construction algorithm, *Comput. Graph.* 21:171–179 (1987).

56. Udupa, J.K. A unified theory of objects and their boundaries in multi-dimensional digital images, in *Proceedings of Computer Assisted Radiology CAR'87* (Berlin, West Germany: Springer-Verlag, 1987), pp. 779–784.

57. Foley, J.D., and A. Van Dam. *Fundamentals of Interactive Computer Graphics* (Reading, MA: Addison-Wesley, 1982).

58. Dev, P., S.L. Wood, J.P. Duncan, and D.N. White. An interactive graphics system for planning reconstructive surgery, in *Proceedings of the 4th Annual Conference and Exposition of the National Computer Graphics Association* (Fairfax, VA: NCGA, 1983), pp. 130–135.

59. Kliegis, V.G., W. Schwerig, and H. Weigel. An integrated system for the automatic manufacturing of organ models, in *Proceedings of the Ninth Annual Conference and Exposition of the National Computer Graphics Association* (Fairfax, VA: NCGA, 1988), v. III, pp. 128–142.

60. Dekel, D. A fast, high-quality, medical 3D software system, in *Proceedings of the Society of Photo-Optical Instrumentation Engineers* (Bellingham, WA: SPIE, 1987), v. 767, pp. 505–508.

61. Udupa, J.K., H.M. Hung, and L.S. Chen. Interactive display of 3D medical objects, in *Pictorial Information Systems in Medicine,* K.H. Höhne (Ed.) (Berlin, West Germany: Springer-Verlag, 1986), pp. 445–457.

62. Artzy, E. Display of three-dimensional information in computed tomography, *Comput. Graph. Image Proc.* 9:145–198 (1979).

63. Herman, G.T., and H.K. Liu. Three-dimensional display of human organs from computed tomograms, *Comput. Graph. Image Proc.* 9:1–29 (1979).

64. Herman, G.T., R.A. Reyolds, and J.K. Udupa. Computer techniques for the representation of three-dimensional data in a two-dimensional display, in *Proceedings of the Society of Photo-Optical Instrumentation Engineers* (Bellingham, WA: SPIE, 1982), v. 367, pp. 3–14.

65. Reynolds, R.A. Fast Methods for 3D Display of Medical Objects, Ph.D. thesis, Department of Computer and Information Science, University of Pennsylvania, Philadelphia, (1985).

66. Vannier, M. W., J.L. Marsh, and J.O. Warren. Three-dimensional computer graphics for craniofacial surgical planning and evaluation, *Comput. Graph.* 17:263–274 (1983).

67. Phong, B.T. Illumination for computer generated images, *Comm. ACM* 18:311–317 (1975).

68. Höhne, K.H., and R. Bernstein. Shading 3D images from CT using gray-level gradients, *IEEE Trans. Med. Imag.* MI-5:45–47 (1986).

69. Schlusselberg, D.S., W. Smith, D.J. Woodward, and R.W. Parkey. Use of computed tomography for a three-dimensional treatment planning system, *Comput. Med. Imag. Graph.* 12:25–32 (1988).

70. Levoy, M. Display of surfaces from volume data, *IEEE Comput. Graph. Appl.* 8:29–37 (1988).

71. Chuang, K.S., J.K. Udupa, and S.P. Raya. High-Quality Rendering of Discrete Three-Dimensional Surfaces, Technical Report MIPG130, Medical Image Processing Group, Department of Radiology, University of Pennsylvania, Philadelphia (1988).

72. Chen, L.S., G.T. Herman, R.A. Reynolds, and J.K. Udupa. Surface rendering in the cuberille environment, *IEEE Comput. Graph. Appl.* 5:33–43 (1985).

73. Heffernan, P.B., and R.A. Robb. A new procedure for combined display of 3-D cardiac anatomic surfaces and regional functions, in *Proceedings of Computers in Cardiology* (Piscataway, NJ: IEEE, 1984), pp. 111–114.

74. Gordon, D., and R.A. Reynolds. Image-space shading of three-dimensional objects, *Comput. Vis. Graph. Image Proc.* 29:361–376 (1985).

75. Lenz, R. Processing and presentation of 3D images, in *Proceedings of International Symposium on Medical Images and Icons* (Piscataway, NJ: IEEE, 1984), pp. 298–303.

76. Barrett, W.A., S.P. Raya, and J.K. Udupa. A low-cost PC-based image workstation for dynamic interactive display of three-dimensional anatomy, in *Proceedings of the Society of Photo-Optical Instrumentation Engineers* (Bellingham, WA: SPIE, 1989), v. 1091, pp. 346–355.

77. Herman, G.T., W.F. Vose, J.M. Gomori, and W.G. Gefter. Stereoscopic computed three-dimensional surface displays, *RadioGraphics* 5:825–852 (1985).

78. Herman, G.T., J.K. Udupa, D.M. Kramer, P.C. Lauterbur, A.M. Rudin, and J.S. Schneider. The three-dimensional display of nuclear magnetic resonance images, *Opt. Engr.* 21:923–926 (1982).

79. Harris, L.D., R.A. Robb, T.S. Yuen, and E.L. Ritman. Non-invasive numerical dissection and display of anatomic structure using computerized X-ray tomography, in *Proceedings of the Society of Photo-Optical Instrumentation Engineers* (Bellingham, WA: SPIE, 1978), v. 152, pp. 10–18.

80. Drebin, R.A., L. Carpenter, and P. Hanrahan. Volume rendering, *Comput. Graph.* 22:65–74 (1988).

81. Upson, C., and M. Keeler. V-buffer: Visible volume rendering, *Comput. Graph.* 22:5–964 (1988).

82. Sabella, P. A rendering algorithm for visualizing 3D scalar fields, *Comput. Graph.* 23:51–58 (1988).

83. Meyer, G.W., and D.P. Greenberg. Perceptual color spaces for computer graphics, *Comput. Graph.* 14:254–261 (1980).

84. Frieder, G., D. Gordon, and R.A. Reynolds. Back-to-front display of voxel-based objects, *IEEE Comput. Graph. Appl.* 5:52–60 (1985).

85. Reynolds, R.A., D. Gordon, and L.S. Chen. A dynamic screen technique for shaded graphics display of slice-represented objects, *Comput. Vis. Graph. Image Proc.* 38:275–298 (1987).

86. Farrel, E.J., Yang, W.C., and R.A. Zapulla. Animated 3D CT imaging, *IEEE Comput. Graph. Appl.* 5:26–32 (1985).

87. Trivedi, S.S. Interactive manipulation of three-dimensional binary scenes, *Vis. Comput.* 2:209–218 (1986).

88. Udupa, J.K., and D. Odhner. Display of medical objects and their interactive manipulation, *Proceedings of the 15th Canadian Conference on Computer Graphics and Computer Vision* (Palo Alto, CA: Morgan Kauffman, 1989), pp. 40–46.

89. Meagher, D. Geometric modeling using octree encoding, *Comput. Graph. Image Proc.* 19:129–147 (1982).

90. Gargentini, I. Linear octrees for fast processing of three-dimensional objects, *Comput. Graph. Image Proc.* 20:365–374 (1982).

91. Gargentini, I., T.R. Walsh, and O.L. Wu. Viewing transformations of voxel-based objects via linear octrees, *IEEE Comput. Graph. Appl.* 6:12–21 (1986).

92. Hoffman, E.A. A historical perspective of heart and lung 3D imaging, in *3D Imaging in Medicine,* J.K. Udupa and G.T. Herman (Eds.) (Chelsea, MI: Lewis Publishers, 1989).

93. Vannier, M.W., C.F. Hildebolt, D.E. Gayou, and J.L. Marsh. Introduction to 3D imaging, in *3D Imaging in Medicine,* J.K. Udupa and G.T. Herman (Eds.) (Chelsea, MI: Lewis Publishers, 1989).

94. Levin, D.N., X. Hu, K.K. Tan, S. Galhotra, A. Hermann, C.A. Pelizzari, G.T.Y. Chen, R.N. Beck, C.T. Chen, and M.D. Cooper. Integrated 3-D display of MR, CT, and PET images of the brain, in *3D Imaging in Medicine,* J.K. Udupa and G.T. Herman (Eds.) (Chelsea, MI: Lewis Publishers, 1989).

95. Fishman, E.K., D. Magid, D.R. Ney, and J.E. Kuhlman. Three-dimensional imaging: Orthopedic applications, in *3D Imaging in Medicine,* J.K. Udupa and G.T. Herman (Eds.) (Chelsea, MI: Lewis Publishers, 1989).

96. Rosenman, J. 3D imaging in radiotherapy treatment planning, in *3D Imaging in Medicine,* J.K. Udupa and G.T. Herman (Eds.) (Chelsea, MI: Lewis Publishers, 1989).

97. Brewster, L.J., S.S. Trivedi, H.K. Tuy, and J.K. Udupa. Interactive surgical planning, *IEEE Comput. Graph. Appl.* 4:31–40 (1984).

98. Cutting, C., F. Bookstein, B. Grayson, L. Fellingham, and J.G. McCarthy. Three-dimensional computer-aided design of craniofacial surgical procedures, in *Proceedings of the First International Congress of the International Society of Cranio-Maxillo-Facial Surgery* (Berlin, West Germany: Springer-Verlag, 1987), pp. 17–18.

99. Udupa, J.K. Computerized surgical planning: Current capabilities and medical needs, *Proceedings of the Society of Photo-Optical Instrumentation Engineers* (Bellingham, WA: SPIE, 1986), v. 626, pp. 474–482.

100. Yokoi, S., T. Yasuda, Y. Hashimoto, J. Toriwaki, M. Fujoka, and H. Nakajima. A craniofacial surgical planning system, in *Proceedings of the Eighth Annual Conference and Exposition of the National Computer Graphics Association* (Fairfax, VA: NCGA, 1987), v. III, pp. 152–161.

101. Dev. P., L.L. Fellingham, A. Vassiliadis, S.T. Woolson, D.N. White, and S.L. Young. 3D graphics for interactive surgical simulation and implant design, in *Proceedings of the Society of Photo-Optical Instrumentation Engineers* (Bellingham, WA: SPIE, 1984), v. 507, pp. 52–57.

102. Herman, G.T., S.S. Trivedi, and J.K. Udupa. Manipulation of 3D imagery, in *Progress in Medical Imaging,* V.L. Newhouse (Ed.) (New York, NY: Springer-Verlag, 1988), v. I, pp. 123–157.

103. Cutting, C.C., B. Grayson, F. Bookstein, L. Fellingham, and J.G. McCarthy. Computer-aided planning and evaluation of facial and orthognathic surgery, *Comput. Plast. Surg.* 13:449–461 (1986).

104. Trivedi, S.S., G.T. Herman, J.K. Udupa, L.S. Chen, and P. Margasahayam. Measurements on 3D surface displays in the clinical environment, in *Proceedings of the Seventh Annual Conference and Exposition of the National Computer Graphics Association* (Fairfax, VA: NCGA, 1986), v. III, pp. 93–110.

105. Udupa, J.K. Determination of 3D shape parameters from boundary information, *Comput. Graph. Image Proc.* 17:52–59 (1981).

106. Cook, L.T., P.N. Cook, K.R. Lee, S. Batnitzky, B.Y.S. Wong, S.L. Feritz, J. Ophir, S.J. Dwyer, III, L.R. Bigongiari, and A.W. Templeton. An algorithm for volume estimation based on polyhedral approximation, *IEEE Trans. Biomed. Engr.* BME-27:493–500 (1980).

107. Walser, R.L., and L.V. Ackerman. Determination of volume from computerized tomograms, *J. Comput. Asst. Tomogr.* 1:117–130 (1977).

108. Bentley, M.D., and R.A. Karwoski. Estimation of tissue volume from serial tomographic sections: A statistical random marking method, *Invest. Radiol.* 23:742–747 (1988).

109. Meagher, D.J. Interactive solids processing for medical analysis and planning, in *Proceedings of the Fifth Annual Conference and Exposition of the National Computer Graphics Association* (Fairfax, VA: NCGA, 1984), v. II, pp. 96–106.

110. Bookstein, F.L., and C.C. Cutting. A Proposal for the apprehension of curving craniofacial form in three dimensions, in *Craniofacial Morphogenesis and Dysmorphogenesis,* K. Vig and A. Burdi (Eds.) (Ann Arbor, MI: Center for Human Growth and Development, 1988), pp. 127.

111. Bajcsy, R., R. Lieberson, and M. Reivich. A computerized system for the elastic matching of deformed radiographic images to idealized atlas images, *J. Comput. Asst. Tomogr.* 7:618–625 (1983).

112. Gamboa-Aldeco, A., L. Fellingham, and G.T.Y. Chen. Correlation of 3D surfaces from multiple modalities in medical imaging, in *Proceedings of the Society of Photo-Optical Instrumentation Engineers* (Bellingham, WA: SPIE, 1986), v. 626, pp. 467–473.

113. Pelizzari, C.A., G.T.Y. Chen, D.R. Spelbring, R.R. Weichselbaum, and C.T. Chen. Accurate three-dimensional registration of CT, PET, and MR images of the brain, *J. Comput. Asst. Tomogr.* 13:20–26 (1989).

114. Apicella, A., J.H. Nagel, and R. Duara. Fast multimodality image matching, in *Proceedings of the Annual International Conference of the IEEE Engineering in Medicine and Biology Society* (Piscataway, NJ: IEEE, 1988), v. 10, pp. 414–415.

115. Llewellyn, J.A., R.S. Belsole, M.M. Dale, D.R. Hilbelink, and S.A. Stenzler. Quantitative Carpal geometry from CT scans without implanted markers, in *Proceedings of the Eighth Annual Conference and Exposition of the National Computer Graphics Association* (Fairfax, VA: NCGA, 1987), v. III, pp. 99–103.

116. Toennies, K.D., J.K. Udupa, and G.T. Herman. Segmentation of implanted bone grafts using anatomical landmarks, in *Proceedings of the Tenth Annual Conference and Exposition of the National Computer Graphics Association* (Fairfax, VA: NCGA, 1989), v. I, pp. 207–214.

117. Robertson, D.D. Three-dimensional modelling and the design of hip and knee prostheses, in *3D Imaging in Medicine,* J.K. Udupa and G.T. Herman (Eds.) (Chelsea, MI: Lewis Publishers, 1989).

118. Fuchs, H., J. Poulton, A. Paeth, and A. Bell. Developing pixel-planes: A smart memory-based raster graphics system, in *Proceedings of the 1982 Conference on Advanced Research on VLSI* (Cambridge, MA: M.I.T., 1982), pp. 137–166.

119. Goldwasser, S.M., R.A. Reynolds, T. Bapty, D. Baraff, J. Summers, D.A. Talton, and E. Walsh. Physician's workstation with real-time performance, *IEEE Comput. Graph. Appl.* 5:44–57 (1985).

120. Kaufman, A. The CUBE workstation — A 3D voxel-based graphics environment, *Vis. Comput.* 4:210–221 (1988).

121. Jackel, D. The graphics PARCUM system: A 3D memory based computer architecture for processing and display of solid objects, *Comput. Graph. Forum* 4:21–22 (1985).

122. Cook, R.L., L. Carpenter, and E. Catmull. The Reyes image rendering architecture, *Comput. Graph.* 21:95–102 (1987).

123. Hemmy, D.C., and P.M. Brigman. A comparison of modalities, in *3D Imaging in Medicine,* J.K. Udupa and G.T. Herman (Eds.) (Chelsea, MI: Lewis Publishers, 1989).

124. Udupa, J.K. DISPLAY82 — A System of Programs for the Display of Three-Dimensional Information in CT Data, Technical Report MIPG67, Medical Image Processing Group, Department of Radiology, University of Pennsylvania, Philadelphia (1983).

125. Greenleaf, J.F., T.S. Tu, and E.H. Wood. Computer-generated three-dimensional oscilloscopic images and associated techniques for display and study of the spatial distribution of pulmonary blood flow, *IEEE Trans. Nucl. Sci.* NS-17:353–359 (1970).

126. Huang, H.K., and R.S. Ledley. Three-dimensional image construction from *in vivo* consecutive transverse axial sections, *Comput. Biol. Med.* 5:165–170 (1975).

127. Ledley, R.S., and C.M. Park. Molded picture representation of whole body organs generated from CT scan sequences, in *Proceedings of the First Annual Symposium of Computer Applications in Medical Care* (Piscataway, NJ: IEEE), pp. 363–367.

128. Rhodes, M.L. An algorithmic approach to controlling search in three-dimensional image data, in *Proceedings of SIGGRAPH79* (New York: NY, Association for Computing Machinery, 1979), pp. 134–142.

129. Hemmy, D.C., G.T. Herman, E.A. Miller, and V.M. Haughton. Three-dimensional reconstruction of the spine and calvarium in children utilizing computed tomography, in *Proceedings of the Second International Child Neurology Congress* (Sydney, Australia, 1979), p. 8.

130. Coin, C.G., and G.T. Herman. The use of three-dimensional computer display in the study of disc disease, *J. Comput. Asst. Tomogr.* 4:564–567 (1980).

131. Marsh, J.L., M.W. Vannier, W.G. Stevens, J.O. Warren, D. Gayou, and D.M. Dye. Computerized imaging for soft tissue and osseous reconstruction in the head and neck, *Clinics Plast. Surg.* 12:279–291 (1985).

132. Burk, D.L., D.C. Mears, L.A. Cooperstein, G.T. Herman, and J.K. Udupa. Acetabular fractures: Three-dimensional computed tomographic imaging and interactive surgical planning, *CT: J. Comput. Tomogr.* 10:1–10 (1986).

133. Sartoris, D., D. Resnick, D. Gershuni, D. Bielecki, and M. Meyers. Computed tomography with multiplanar reformation and 3-dimensional image analysis in the preoperative evaluation of ischemic necrosis of the femoral head, *J. Rheum.* 13:153–163 (1986).

134. Roberts, D., J. Pettigrew, J.K. Udupa, and C. Ram. Three-dimensional imaging and display of the temporo-mandibular joint, *Oral Surg. Oral Med. Oral Patho.* 58:461–474 (1984).

135. McShan, D.L., A. Silverman, D.M. Lanza, L.E. Reinstein, and A.S. Glicksman. A computerized three-dimensional treatment planning system using interactive color graphics, *Brit. J. Radiol.* 52:478–481 (1979).

136. Robb, R.A., P.B. Heffernan, J.J. Camp, and D.P. Hanson. A workstation for multidimensional display and analysis of biomedical images, *Comput. Meth. Prog. Biomed.* 25:169–184 (1987).

137. Raya, S.P. Rule-based segmentation of multidimensional medical images, in *Proceedings of the Tenth Annual Conference and Exposition of the National Computer Graphics Association* (Fairfax, VA: NCGA, 1989), v. I, pp. 193–197.

138. Sander, P.T., and S.W. Zucker. Inferring Differential Structure from 3-D Images: Smooth Cross Sections of Fibre Bundles, Technical Report TR-CIM-88-6, Computer Vision and Robotics Laboratory, McGill Research Center for Intelligent Machines, McGill University, Montreal, Quebec, Canada, (1988).

139. Höhne, K.H., R.L. Delapz, R. Bernstein, and R.C. Taylor. Combined surface display and reformatting for the three-dimensional analysis of tomographic data, *Invest. Radiol.* 22:658–663 (1987).

140. Herman, G.T. Quantitation using 3-D images, in *3D Imaging in Medicine,* J.K. Udupa and G.T. Herman (Eds.) (Chelsea, MI: Lewis Publishers, 1989).

141. Gillespie, J.E. Three-dimensional computed tomographic reformations: Assessment of clinical efficacy, in *3D Imaging in Medicine,* J.K. Udupa and G.T. Herman (Eds.) (Chelsea, MI: Lewis Publishers, 1989).

Chapter 2

INTRODUCTION TO 3D IMAGING

Michael W. Vannier, Charles F. Hildebolt, Donald E. Gayou, and Jeffrey L. Marsh

TABLE OF CONTENTS

I. ABSTRACT

Computer graphics, especially reconstruction from contiguous high-resolution computed tomography (CT) scans have found application in the management of patients with a variety of disorders of the head, spine, pelvis, and other sites in the musculoskeletal system. Magnetic resonance imaging (MRI) offers additional high-quality soft tissue information from the central nervous system, heart, and other soft tissue structures. The sources of medical images, methods for modeling complex morphology of many organs, clinical applications, and problems encountered with these methods are considered. The design and evaluation of surgical procedures based on surface and volumetric images of hard and soft tissue interfaces found in CT and MRI scans has been achieved.

II. INTRODUCTION

Many modern medical imaging instruments are based on digital mini- and microcomputers, including CT,[42] MR, ultrasound (US), digital vascular imaging (DVI), nuclear medicine, and positron emission tomography (PET).[2,34] Typically, each unit contains a sensor system with associated signal conditioning electronics, an analog-to-digital converter, a digital computer, a frame buffer with CRT display, and in some cases an array processor (Figure 1).

The purpose of these instruments is the production of diagnostic, morphologic, and functional information in the least amount of time with the greatest possible margin of safety for the patient.

Morphologic diagnosis is often achieved with these instruments as the result of viewing a large number of images, most often a serial set of tomographic "slices".[39] For the study of a single organ or body region, 50 to 100 or more slice images may be available. These images are reviewed by a radiologist using transparent film hardcopy as they are assembled in sequence on a set of X-ray viewboxes.

CT imaging is widely used for diagnostic evaluation of the head (Figure 2), neck, and skeleton. CT imaging is capable of interrogating the full volume of an organ or body part and is intrinsically three dimensional (3D), representing body regions as a stack of transaxial slices. It is also relatively insensitive to body and organ geometry and thus can be used to inspect organs with complex shapes such as found in the skull and skeleton.[34]

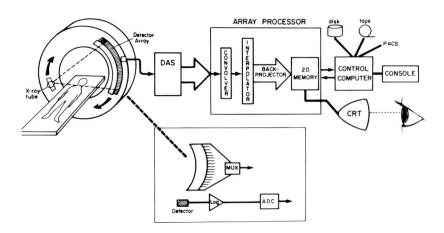

FIGURE 1. Block diagram of a CT scanner.

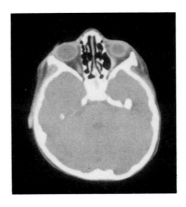

FIGURE 2. CT scan slice of the head.

III. COMPUTED TOMOGRAPHY

Since the early 1980s, we have concentrated on acquisition, quantitative evaluation, and 3D display of high-resolution, contiguous, thin-slice CT scan data in patients with major craniofacial deformities, either congenital or acquired.[23] Pelvic,[48] calcaneal, wrist, and spine fractures,[60] bone tumors, and selective cases of advanced arthritis[9] have been examined this way.

CT scans are taken contiguously through an area of interest to reveal their internal structure as a series of slices. If the slices are sufficiently thin (2 mm thickness is typical) and the object being inspected is sufficiently large (e.g. a skull or pelvis), a very large number of slices may be produced (75 or more for a single study). The analysis of these CT scan data sets presents a formidable obstacle to efficient interpretation and diagnosis.

IV. SEGMENTATION AND DISPLAY

The separation of a scene into its components parts is termed image segmentation.[14] A head CT scan slice contains air, soft tissue, and bone in the simplest model. The measured values for each of these components is significantly different, and the detection of edges by simple thresholding is possible.[30,36,47,52]

Level slicing at fixed thresholding has been found empirically to be sufficient to accurately delineate the surface boundaries (Figure 3) found at air-soft tissue interfaces (threshold = −100 HU) and soft tissue-bone interfaces (threshold = +175 HU) (Figure 4). Surprisingly, these values have been used with a large number of cases from several CT scanners operated over a period of several years, and the reconstructed results have been consistently satisfactory (Figure 5).

More complex edge detectors are required to identify and track edges found at other than air-soft tissue-bone interfaces in CT scan slices.

In the segmentation scheme that we use, each pixel is examined without knowledge of its neighbors. If the pixel lies above a given threshold, its index (x, y location) is saved in an output array. As pixels are evaluated in columns from top to bottom, or rows from right to left or vice versa, the first instance of a threshold transition is saved in the output array and the evaluation of the current vector ceases (painter's priority algorithm[13]), resulting in opaque surface generation.

The index of threshold transitions is saved in an output vector. 256 × 256 resolution is sufficient for this application, because the facial structures in most patients can be included in a scanned area of 180 × 180 mm (0.7-mm pixel edge length). With 2 mm slice thickness, the volume sampling elements (voxels) have 3 times higher spatial resolution in the plane of section (0.7 mm) than along the direction of table motion (2 mm), so larger matrix sizes (256 × 256 to 512 × 512 or 1024 × 1024) increase the data handling requirements without a significant improvement in the final surface reconstruction images. With 256 × 256 input scans, only 8 bits are required to store the indices found at threshold transitions. If output data files are created which have the same format as ordinary CT scans, these 8 bits may be scaled to 11 or 12 bits to make the display window settings more convenient for the operator.

CT scanners routinely obtain slices oriented in a transverse plane. As a maximum, most

O	O	O	70	100	100
O	O	2	82	100	100
O	O	25	100	100	100
O	O	35	100	100	100
O	O	70	100	100	100
O	8	92	100	100	100
O	20	100	100	100	100

SAMPLED DATA

FIGURE 3. Sampled data from a typical two-dimensional CT scan slice is shown. Once a threshold is set, the surface in this CT scan slice can be approximated with the curved lined shown in the illustration. The quality of these interpolated results can significantly influence the quality of the displayed image.

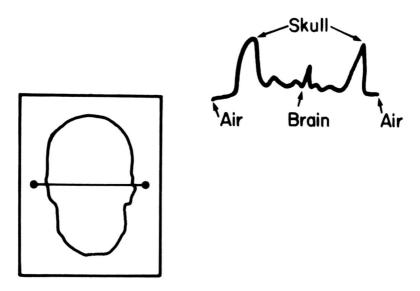

FIGURE 4. The horizontal profile tracing shown in a CT scan slice includes intensity variations due to air, skull, brain, and followed by air.

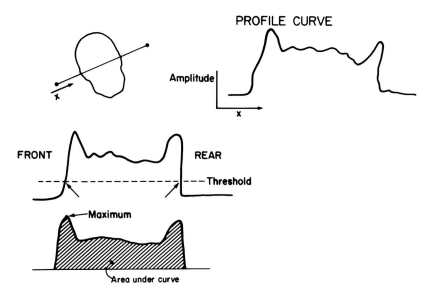

FIGURE 5. If an oblique profile is displayed through a head CT scan, the amplitude will vary with position. If this is done from the front to the rear direction and a threshold is set, several measures can be extracted such as the location of initial and final threshold crossings, the maximum value encountered in the scan profile, the area under the scan profile curve, or others. Each of these parameters can be mapped to the display to form an image.

modern CT scanners can tilt the orientation of this plane by 20 degrees to the long axis of the body. In MR and US imaging, the orientation of section images is less restricted.

In selected cases, it has been possible and useful to reformat the serial sections in a two dimensional form. The simplest reformatting is the construction of images that show slices oriented orthogonal to the original plane of section. By convention, we call these planar reconstructions: sagittal if they pass through the midline and are oriented front to rear, coronal if they pass through the midline and are oriented left to right, parasagittal or paracoronal if they are parallel to the sagittal or coronal plane and do not pass through the midline, and paraxial if they are obliquely oriented.

Planar reconstructions from serial sections may be obtained with orientations that are not perpendicular to the plane of section. The software that performs this "multiplanar reconstruction" is available on most modern CT scanners.[45] This facility has been used in selected cases with some success.

Standard sectional anatomy texts and atlases divide the body into transverse, sagittal, and coronal planes. By producing planar reconstructions parallel to these reference planes, diagnostic medical images are easier for physicians to understand. Images oriented at arbitrary planes may be obtainable from serial scans, but their usefulness is often limited because they are much more difficult to interpret.

Regardless of the availability of reformatted images, the radiologist without exception studies the original transverse slice images. These transverse slices are the most familiar format to radiologists, and the role of additional reconstructed images is adjunctive.

Surface reconstruction images have found application in the study of craniofacial,[18] orthopedic,[11,28] wrist, and intracranial disorders.[57] Increasing the availability of these techniques in a clinical environment will expand their range of application. Below, we discuss these methods as applied to imaging of the head.

V. CRANIOFACIAL SURGICAL PLANNING

Regional abnormalities in facial morphology are the hallmark of many craniofacial syndromes.[37,38,56,58] The clinical utility of employing CT scans and 3D reconstructions as discriminators of abnormal form is predicated upon establishing a normal range of variation for these images. Although it is often assumed that the normal skull is symmetric and homogeneous in form, recent studies suggest that there is significant heterogeneity in the pattern of skull structure both by section and segment (within any given sectional level).

Complex craniofacial surgical procedures have been planned using manual methods based on measurements taken from cephalometric skull X-rays in the frontal (anteroposterior or AP) and lateral projections. The drawbacks of skull X-rays for surgical planning include geometric distortion due predominantly to magnification, lack of soft tissue contrast, overlap of structures (front and rear on the AP view, right and left on the lateral view), misregistration and positioning errors on serial examinations, and failure to reveal the intracranial or orbital contents.[35]

Attempts to obtain accurate information about the spatial variability of sectional and segmental skull structure have been hindered by technical limitations in the field of craniofacial imaging. These limitations include the use of nontomographic imaging techniques (cephalometric skull radiography) and the inability to obtain multilevel, high-resolution, transaxial, spatially registered images in patients. Moreover, the aforementioned techniques are dependent on fixed geometric assumptions to determine facial structure, volumes, or mass. While many of these mathematical constructs may be reasonable when applied to the normal skull, they are questionable when the skull is misshapen secondary to congenital or posttraumatic changes.[54]

In the late 1970s, high-resolution CT scanning became available permitting the acquisition of thinner slices (1 or 2 mm or less instead of 1 or 1.5 cm) with greater spatial resolution (submillimeter pixel size) in the plane of section. High soft tissue contrast, excellent geometrical characteristics, and consistent reproducibility of CT scan examinations encouraged their application to evaluation of many disorders of the head and neck.

The use of serial CT scans for planning and evaluating surgical procedures[3,4] is complicated by the difficulty found in associating the location and relationship of transaxial slice findings with the patient's skull and its contents in the operating room. Adjunctive images including scan projection radiographs acquired by the CT scanner to assist localization and sagittal/coronal/oblique planar reconstructions have been used to assist scan interpretation. The limitations of scan projection radiographs are much the same as ordinary skull X-rays, except for the substitution of superior soft tissue contrast for decreased spatial resolution. The limitations of reformatted planar images are much the same as the transaxial slices from which they are derived — both are abstract, flat pictures extracted from a complicated 3D object.

Even a skilled observer experiences difficulty in visualizing a patient's appearance when presented with a sequence of CT scans containing the facial structures.

A new skull imaging modality, 3D-CT reconstruction has recently become available.[7,19,20] This CT scanning method permits the acquisition of high-resolution, transaxial images from the mandible through the calvarium. Recent studies have demonstrated that skull CT can precisely determine quantitative geometrical properties including volumes under clinically applicable conditions.

The conversion of serial slice image data to a 3D form has been accomplished by many methods[5,17,27,32,33,55,61] (Figure 6). 3D display is, in itself, not necessarily useful or desirable for medical imaging unless the additional information provided fulfills some clinical need and influences patient managment. After considering the information requirements for complex craniofacial surgical procedures, a surface display system was designed to meet these requirements.[22,25]

VI. SURFACE IMAGING

In converting the CT scan data (Figure 7) to surface form, the dimensionality of the data is reduced and the spatial information found along the axis orthogonal to the plane of section is restored. Surface reconstruction methods for slice data may be considered techniques for background suppression, redundancy reduction, and contrast enhancement.

A complete CT scan examination of the skull at high resolution (2-mm contiguous nonoverlapping sections, 0.5- to 1.0-mm pixels, 256×256 matrix, with 11 or 12 bits of gray scale) may consist of 100 or more slices if the entire skull is included. We logically divide the skull into three major segments for computer processing: the mandible, the midface and skull base,

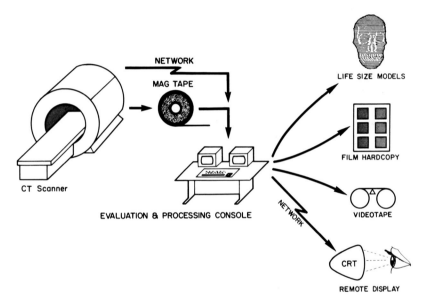

FIGURE 6. Summary of processing steps used to produce 3D images from CT scans. Serial slices are obtained with the CT scanner, transmitted on magnetic tape or through an electronic network to a processing station, and rendered as models, film hardcopy, videotape sequences, or softcopy CRT display.

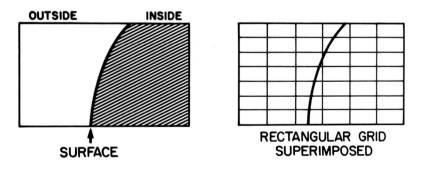

FIGURE 7. A curved surface of an object defines the boundary between elements which lie inside the structure of interest and those which lie outside. For practical reasons, we sample these data using a rectangular grid such as shown here.

and the calvarium. Depending on the clinical circumstances, one or all of these components may be analyzed using surface reconstruction techniques.

The modeling of a solid object for display or for CAD/CAM can be accomplished by many methods:[1,29,49,51] primitive instancing, cell decomposition, computational solid geometry (CSG), boundary representation (B-REP),[6] octree encoding,[16,40] spatial enumeration,[21] and others.

In CT scan analysis, we have found it useful to represent the skull and associated soft tissues by spatial occupancy enumeration for display purposes[17] and by boundary representation when CAD/CAM operations are required.[59]

As explained by Requicha,[42] a list of spatial cells occupied by a solid constitutes a spatial occupancy enumeration scheme. This is precisely the result of a high-resolution CT scan examination of the skull using contiguous nonoverlapping sections. The CT scan data set consists of volume-elements. Imposing the specific spatial scanning sequence, the corresponding ordered sets of three-tuples are called spatial arrays. These spatial arrays are unambiguous, unique, and easy to validate, but potentially quite verbose. The conversion of a spatial array to a 2-dimensional projection is the subject of surface reconstruction.

Spatial enumeration results in a data base that is far too verbose for efficient 3D display in real time when a large number of elements are present, with CT scan data a good example of how complex a data set might be. This verbosity is much less significant, however, when only a small number of predetermined surface views are required, because the raw data need only be accessed once, and this is a minimal requirement for any display method.

We have adopted a convention for the serial CT scan image data geometry. The plane of section (X-Y) and direction of table motion (+Z for head CT scans) are defined. The X axis is oriented in the mediolateral direction. The Y axis is oriented in the anterioposterior direction. The Z axis is oriented in the superior-inferior direction. This convention applies to both surface reconstruction gray scale and calligraphic CAD/CAM image data bases.

The dimensionality of CT scan data can be estimated as follows. A high-resolution skull CT scan examination consists of 100 (or more) slices, each with 256×256 pixels and 11 (or 12) bits of gray scale. This represents nearly 80 megabits of information. If we compute surface images for the soft tissue and bony elements in the frontal, right lateral, and left lateral projections, 6 images are required, each with 256×256 pixels of 8 bits. This represents a reduction in dimensionality to about 3.0 megabits or 4%.

The location of the skull within the transaxial plane of section is subject to variation from patient to patient. The offsets and center information can be used to assist in registering image data within or between serial studies. The software ability to translate and rotate CT image data is useful in characterizing operative intervention on postoperative scans or growth on serial craniofacial CT examinations.

VII. GEOMETRICAL TRANSFORMATION

To obtain surface views efficiently from different directions oriented perpendicular to the plane of section, individual slices are analyzed without influence of their neighbors. The direction cosines needed to produce the geometric projections[12,41] are precomputed. Acceptable reconstructions may be obtained without the use of floating point arithmetic or the use of transcendental function computations for each desired view. This step results in a dramatic reduction in computation time.

VIII. SCALING

The acquisition of CT scan data with slice thickness equal to some integer multiple of the pixel size in the transaxial plane simplifies the scaling of surface views that are oriented

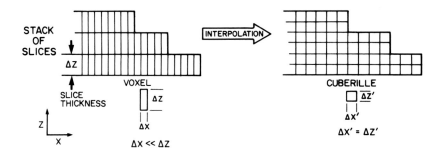

FIGURE 8. A stack of slices comprised of voxels is represented. Z is the direction of table motion, and X lies in the slice plane. Each slice has a thickness ΔZ. Voxels are comprised of 3D elements which have a volume of ΔX by ΔX by ΔZ. Since ΔZ is much larger than ΔX, usually by a factor of 5 or more, data processing is more complex. An alternative scheme is to produce caberilles by interpolation where ΔX′ is equal to ΔZ′.

perpendicular to the plane of section. If the slice thickness to pixel size ratio (e.g., zoom factor) is nonintegral, an additional interpolation step is required to preserve the scale of the output surface images. Typical pixel edge lengths are 0.5, 0.7, 1.0, and 2.0 mm with zoom factors of 4, 3, 2, and 1 and a matrix size of 256×256 and field of view of 128, 180, 256, or 512 mm (approximate) (Figure 8).

Surface images are, in general, accurate within the limits of the CT scanner for geometric measurements. If we use 2-mm sections with a pixel size of 0.5 mm, a frontal view of the facial structures will contain 0.5 mm resolution horizontally and 2 mm resolution vertically.

IX. ANTI-ALIASING

Surface irregularities produced by the large difference in X-Y plane resolution and the slice collimation (e.g., Z-axis resolution) result in an unsatisfactory appearance when contour data are simply stacked for viewing. The aliasing found in the z-direction may be removed by a nonrectangular-median or rank-order filter. The result of this operation is a dramatic cosmetic improvement in the subjective quality of the reconstructed images.

X. OPERATIONS ON SURFACE RECONSTRUCTION IMAGES

The combination of pairs of surface reconstruction images to simulate transparency can be useful in surgical planning. This is often required for simultaneous analysis of the bony and soft tissue surfaces and their relationship, or for right-left comparisons, or for serial studies on the same patient.

A pair of surface images may be converted to a "contour" form by use of a local nondirectional gradient operator. This operation reduces the gray scale dynamic range for the image and makes high-contrast superimpositions feasable (Figure 9).

The organization of computer reconstruction software for the analysis of the skull should reflect a logical scheme. We have developed conventions for the analysis of the skull in patients and have applied them in a large variety of clinical situations (Figure 10).

XI. SYSTEM HARDWARE CONSIDERATIONS

The computer hardware needed to perform the analysis described above exists in modern CT scanners. Using software coded in Fortran and assembler language, acceptable perform-

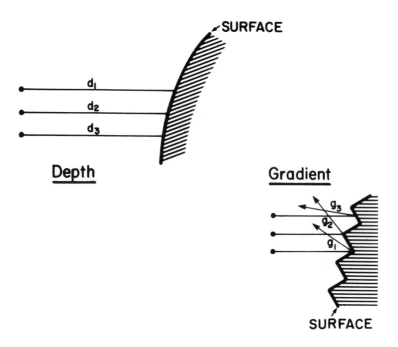

FIGURE 9. *Depth* and *gradient* coding schemes for various surfaces are shown. In the depth coding scheme, distances from a viewer shown as black dots to the left of the surface are measured as distances D1, D2, and D3. With gradient shading, both the distance and the angle at which light returns from the surface are considered in the final image. It is clear that the gradient shading method creates greater local variation in surface appearance than does depth shading.

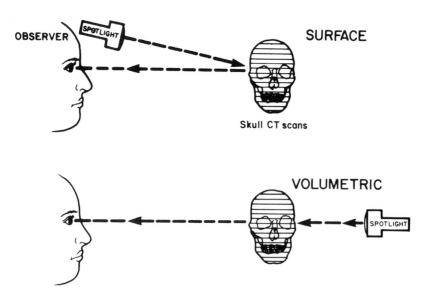

FIGURE 10. Surface imaging is analogous to the viewing of a skull with a light source that returns light to the observer from the external surfaces of the object. Volumetric imaging refers to the translucent or transparent character of scanned objects which transmit light through to the observer. The simulation of these phenomena by computer methods is the basis for visualization techniques which are commonly used in 3D medical imaging.

ance may be obtained on standard CT equipment. The software described above functions without operator intervention and produces timely results with sufficient quality to fulfill most surgical requirements. The modest computational power and display equipment supplied with the CT scaner is sufficient. The final product of the surface reconstruction effort is film copy. We have calibrated the video film recording equipment on our CT scanners to produce life-sized imagery.

XII. VOLUMETRIC VISUALIZATION

An emerging computer graphics technique called "volume visualization" has been applied to the diagnostic evaluation of high-resolution CT scan examinations.[8,31] This technique displays a volume of data, such as a set of CT scans, rather than its surfaces or edges.[43,50] Previously, computer graphics techniques utilized lines or polygonal raster images to approximate the boundaries of an object or other data set. (A polygon is a multisided figure — a triangle has 3 sides, a quadrilateral has 4, a pentagon has 5, etc.) Lines or vectors yield so-called wire-frame representations. Polygons or surface patches are used in more sophisticated applications to give so-called solids models. Volume visualization displays every point in a 3D volume, without surface or edge approximations. Each point in a volume is called a volume element or voxel for short. This is a generalization of the common pixel (picture element).

Traditional computer graphics techniques render 3D surfaces into two dimensions by projecting them into a plane. The final brightness at a point on the display screen is determined by the brightness or color of the surface which is in front of all others at that point. If the surface is partially transparent, then the brightness or color is an average of all surfaces which show through to the point on the display. Most 3D CT reconstructions are made using surface graphics methods with no transparency.

Volume graphics takes the transparency concept one step further. Each voxel in the volume to be displayed is given an intensity value (brightness) or color and a partial transparency (0 to 100%). Then the brightness or color displayed on a display screen is the contribution at that point of all voxels lying along the line of sight to that point. If a ray is sent out from the eye, through the display screen, to the volume behind it, then the brightness or color placed at the intersection point with the screen is an average of the intensities or colors of all voxels intersected by that ray. This averaging is weighted by the transparencies of the intersected voxels. Another way to state this is that all points along the line segment contribute to the final brightness or color of the display pixel intersected by the viewing ray. If this object is tranparent, then points along other such intersecting line segments may also contribute to the display brightness or color.

The volume which is visualized may contain anything: gas, blood, fluid, bone, muscle, brain, etc. The original volume of data we have investigated contains a set of CT scan slices.

Voxels as 3D generalizations of 2D pixels were first mentioned in the literature in the 1970s, particularly in medical imaging based on CT scanning devices. A great deal of effort has been expended on extracting surfaces from voxel collections, because this was the only way of rendering 3D views of them at that time. In the last several years, imaging techniques have emerged which represent the actual contents of volume elements and not just surfaces or edges.

The key technological change inspiring volume graphics is sufficient cost reduction of the memory components to allow devices which store entire volume rasters of memory to be built. It is equally important that processors to compute new images from data stored in these large memories have also become affordable.

Much of the mathematical work already done on 3D computer images is applicable to voxel data sets and can be used as the basis of new volume graphics approaches.

XIII. VOXEL CLASSIFICATION

Separation of the tissue types (bone, fat, muscle, and air) in each CT scan slice and assignment of a brightness or color to each voxel in a visualization is termed "classification". The most typical use of classification in conventional image processing is in environmental remote sensing, such as aerial or satellite photography. For example, Landsat earth resources imagery is classified with the technique called "multispectral classification".[62] These and similar satellites return data of the same land area as imaged through several different band pass filters, which together cover the visible spectrum and more. Each pixel is classified by comparing its relative responses in the different spectral bands. This is equivalent to saying that the satellite approximates the X-ray energy attenuation curve of each pixel imaged. Then classification proceeds by matching each approximate spectrum with known attenuation of tissues. If a pixel's approximate spectrum most nearly matches wheat, then it is classified as wheat and assigned a color. If it matches a spectrum of corn, then it is colored accordingly. The resulting 2D visualization of classified earth resources data is called a thematic map. These techniques have been applied to medical image processing, especially for MRI.[24]

Pseudocolor[10] (a process of assigning arbitrary false colors to an image) has been used to simplify the interpretation of 2D classification, but many find it objectionable. This process effectively transfers the task of segmentation to the eye of the user. Medical imaging applications rarely use pseudocolor, because pseudocolor often contributes artifacts to the data not originally present. In particular, a pseudocolor table without smooth transitions introduces false contours between the areas of sharply distinct colors.

The CT scans we processed were classified into bone, fat, muscle (or soft tissue), and air components on a voxel by voxel basis according to the classification scheme. The gray scale histogram for each CT slice is a summation of the measured attenuations for air, muscle, fat, and bone. These tissue components may be classified for each voxel according to their respective percent contribution to the measured attenuation at each voxel. The presence of each material in a voxel, expressed as a percentage, may be found by a table lookup procedure (Figure 2). As a consequence, a voxel may contain air, fat, muscle, or bone or some combination. In the overlap region noted in the figure, each voxel will contain fractions of fat, muscle, and bone which total 100%.

XIV. COMPUTER AIDED DESIGN

To illustrate the use of computer-aided design (CAD) methods, a midface advancement procedure was chosen. In this procedure, surgical correction of a growth abnormality in the maxilla is performed. The maxillae, paired bones that extend from the upper row of teeth to the orbits, may be hypoplastic (too small) in individuals with certain craniofacial syndromes, including Crouzon's disease. A 7-year-old child with this syndrome was scanned at 2-mm intervals with a 3rd generation CT scanner (Siemens Somatom DR-H). A total of 100 slices of the facial structures were available for analysis.

The correction of maxillary hypoplasia by midface advancement is performed by surgically freeing the maxillae from their attachments, moving them forward as a unit, and immobilizing the result. The surgeon may make use of measurements taken at the level of the teeth (e.g., at the maxillary alveolus) to plan the procedure. The effect of a given advancement at the inferior margin of the maxilla at remote locations cannot be predicted by conventional methods. We performed the analysis of this procedure using a CAD system preoperatively to predict the surgical outcome.

A standard set of gray scale surface reconstruction views was obtained, and with the same raw data, a CAD model was formed with more than 3000 coordinates.

For the production of a CAD model, each CT scan slice was analyzed in turn to extract the surface contour, and these were curve fit using cubic splines.[53] The data were entered into a 3D data base created originally for the design and manufacture of advanced military aircraft (CADD system, for computer-aided design and drafting), operating on a calligraphic real-time vector display terminal (Evans and Sutherland PS-390) with static solid shaded surface display capability.

The application of an interactive CAD system to such a complex anatomic malformation and its surgical correction has been useful in the refinement of the procedure and improvement of the outcome.

XV. MODELS

Computer-aided manufacturing (CAM) techniques can be used to produce life-sized, space-filling facsimiles of the skull from a serial set of CT scans.[46] These models can be useful in fabricating prostheses, performing mock surgery, and improving the surgeon's understanding of particularly complex skull anatomy. The value of a library of anomalous skull facsimilies obtained from CT scans on living patients with rare abnormalities at various stages of growth cannot be overstated. Many of the disorders treated surgically are relatively poorly understood because they are rare, nonlethal, and pathologic specimens generally unavailable. The natural history of many disorders of the skull may be elucidated using these models.

XVI. CONCLUSION

The preliminary clinical experience with 3D image reconstruction can be summarized as follows:

1. 3D-CT reconstructions appear to be less ambiguous than skull X-rays because overlap of structures is not present (Figure 11).
2. Pertinent spatial information concerning aberrant anatomy appears to be easier to understand in a 3D surface format than in the CT slice format.
3. With 3D reconstructions, deep structures which are concealed by superficial ones can be made visible.
4. Measurements of distances, angles, areas, and volumes are simpler to make using 3D surface images than they are using ordinary radiographs or CT/MR slices.
5. The mathematical surface reconstruction techniques for 3D image scans can produce views that are impractical or impossible with ordinary radiography.
6. 3D reconstructions have been used successfully for clinical management and research study of patients with a variety of congenital and acquired craniofacial disorders, musculoskeletal abnormalities (Figures 12 and 13), and other disorders.
7. With little knowledge regarding the programs and methods, the software required for producing 3D-CT reconstructions can be inserted into a CT or MR scanner and successfully operated.
8. The CT scanner has excellent geometrical characteristics that make its image data suitable for input to a CAD and CAM system.
9. CAD/CAM technology can be applied to the registration of 3D-CT images.
10. The major limitations of this work have been the lack of a validated system for providing anatomic fidelity and the inability to register 3D images of organs for the same patient over time or among different patients.

Based on preliminary studies, we conclude that the most significant difference between the medical "display" systems for 3D imaging lies not solely in the quality of pictures, cost, and ease of operation, but in quantitative abilities. The medical systems currently in use have very

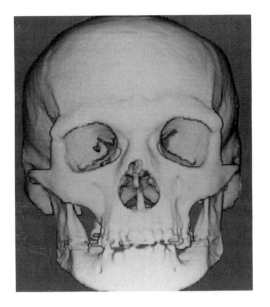

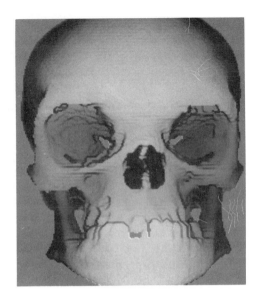

A   B

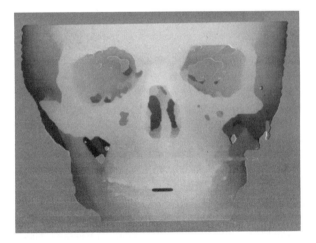

C

FIGURE 11.   3D frontal surface view of the skull. (A) Frontal view of a normal adult skull. Gradient shaded. (B) Frontal view of the skull in a patient with Crouzon syndrome. Contour contrast enhanced. (C) Frontal view of the skull in a patient with Treacher-Collins syndrome. Depth coded.

limited ability to quantify the data that they present — measurements of 3D distances, angles, surface areas, volumes, centroids, moments of inertia are simply not available, or if available have not been validated. The cumulative published application of surface and volumetric reconstruction software for 3D display of CT and MRI scans in thousands of examinations indicates that while simple display of abnormalities is useful and interesting, quantitative study will be increasingly important in the future. Without quantitation, we cannot completely characterize deformities, measure the effects of growth or treatment, or ever achieve a satisfactory means of predicting the outcome of an intervention.

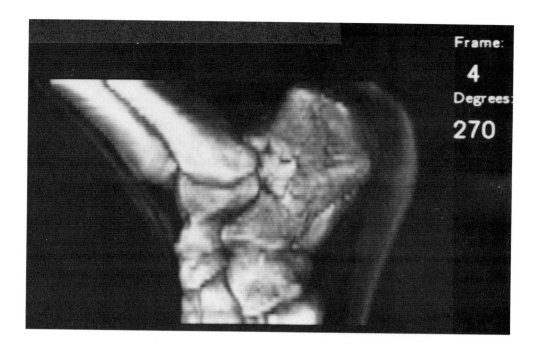

A

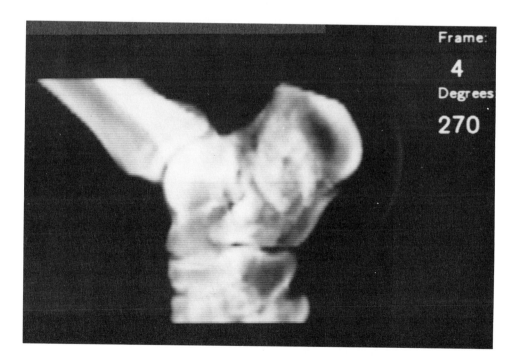

B

FIGURE 12. Lateral view of an adult ankle. (A) Lateromedial view of ankle with comminuted calcaneal fracture. Shaded surface view. (B) Same patient, except volumetric view.

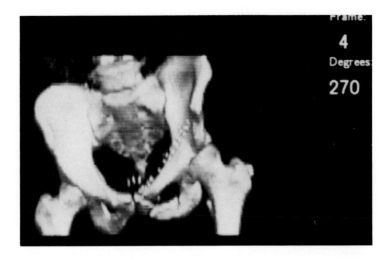

A

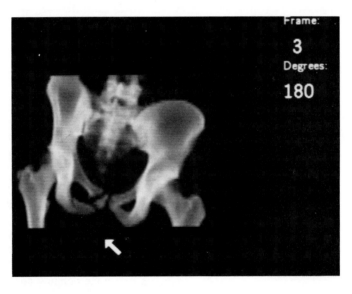

B

FIGURE 13. Postoperative views in a patient with a pelvic fracture. (A) Frontal shaded view of pelvis with pubis fractures. (B) Rear volume view of pelvis in same patient.

REFERENCES

1. Ballard, D.H., and C.M. Brown. Representations of three dimensional structures, in *Computer Vision* (Englewood Cliffs, NJ: Prentice-Hall, 1982), pp. 264–311.
2. Barrett, H.H., and W. Swindell. *Radiological Imaging: The Theory of Image Formation, Detection, and Processing* (New York, NY: Academic Press, 2 vols., 1981).
3. Block, P., and J.K. Udupa. Application of computerized tomography to radiation therapy and surgical planning, *Proc. IEEE* 71:(3)351–355 (1983).

4. Brewster, L.J., S.S. Trivedi, H.K. Tuy, and J.K. Udupa. Interactive surgical planning, *IEEE Comput. Graph. Appl.* 4:31–40 (1984).

5. Lorensen, W.E., and H.E. Cline. Marching cubes: A high resolution 3D surface construction algorithm, *Comput. Graph.* 21:163–169 (1987).

6. Cook, L.T., S.J. Dwyer, and S. Batnitzky. Modeling three-dimensional computer reconstructions from surface contours for diagnostic imaging, *SPIE/Appl. Digital Image Proc.* 359:152–155 (1982).

7. Cook, L.T., S.J. Dwyer, and D.F. Preston. The diagnostic radiological utilization of 3-D display images, *SPIE/ Proc. Display Three-Dimen. Data II* 507:48–51 (1984).

8. Drebin, R.A., L. Carpenter, and P. Hanrahan. Volume rendering, *SIGGRAPH '88 Conference August 1988, Comput. Graph.* 22(3), 1988.

9. Engel, J., B. Salai, and B. Yalle. The role of three dimensional computerized imaging in hand surgery, *J. Hand Surg.* 12B:349–352 (October 1987).

10. Farrell, E.J. Color display and interactive interpretation of three-dimensional data, *IBM J. Res. Dev.* 27:(4)356–366 (July 1983).

11. Fishman, E.K., D. Magid, and D.R. Ney. Three-dimensional imaging in orthopedics, *State of the Art 1988, Orthopedics* 1021–1026 (July 1988).

12. Foley, J.D., and A. Van Dam. Geometrical transformations, in *Fundamentals of Interactive Computer Graphics* (Reading, MA: Addison Wesley, 1982) pp. 245–266.

13. Frieder, G., D. Gordon, and R.A. Reynolds. Back-to-front display of voxel-based objects, *IEEE Comput. Graph. Appl.* January 1985.

14. Fu, K.S., and J.K. Mui. A survey on image segmentation, *Pattern Recog.* 13:65–77 (1981).

15. Fuchs, H., S.M. Pizer, and E.R. Heinz. Design of and image editing with a space-filling three-dimensional display based on a standard rater graphics system, *SPIE/Proc. Display Three-Dimen. Data* 367:117–127 (1982).

16. Heffernan, P.B., and R.A. Robb, A new method for shaded surface display of biological and medical images, *IEEE Trans. Med. Imag.* MI-4:26–38 (March 1985).

17. Hemmy, D.C., D.J. David, and G.T. Herman. Three-dimensional reconstruction of craniofacial deformity utilizing computed tomography, *Neurosurgery* 13:534–541 (1983).

18. Herman, G.T., and H.K. Liu. Three-dimensional display of human organs from computed tomograms, *J. Comput. Assist. Tomogr.* 1:155–160 (1977).

19. Herman, G.T., and H.K. Liu. Three-dimensional display of human organs from computed tomograms, *Comput. Graph. Image Proc.* 7:130–138 (1978).

20. Herman, G.T., R.A. Reynolds, and J.K. Udupa. Computer techniques for the representation of three-dimensional data on a two-dimensional display, *SPIE/Proc. Display Three-Dimen. Data* 367:3–14 (1982).

21. Herman, G.T., and J.K. Udupa. Display of 3-D information in 3-D digital images: Computational foundations and medical applications, *IEEE/Computer Society International Conference on Medical Computer Science/ Computational Medicine* (1982), pp. 308–315.

22. Herman, G.T., and S. Kemp. Clinical three-dimensional display in medicine, *IEEE/Seventh Annual Conference of the Engineering in Medicine and Biology Society* (1985), pp. 604–606.

23. Vannier, M.W., R.L. Butterfield, D. Jordon et al. Multispectral analysis of magnetic resonance images, *Radiology* 154:221–224 (1985).

24. Hoehne, K.H., R.L. Delapaz, and R. Bernstein. Combined surface display and reformatting for the three-dimensional analysis of tomographic data, *Invest. Radiol.* 22:658–664, (August 1987).

25. Huijsmans, D.P., W.H. Lamers, and J.A. Los. Toward computerized morphometric facilities: A review of 58 software packages for computer-aided three-dimensional reconstruction, quantification, and picture generation from parallel serial sections, *Anat. Rec.* 216:449–470 (1986).

26. Kulgus, D.J., G.A. Finerman, and M. Kabo. Three-dimensional imaging techniques for the reconstruction of a complex hip problem, *Orthopedics* 2:1029–1034 (July 1988).

27. Laub, G.A., G. Lenz, and E.R. Reinhardt. Three-dimensional object representation in imaging systems, *Opt. Eng.* 24:901–905 (Sept–Oct 1985).

28. Levialdi, S. Finding the edge, in *Digital Image Processing,* J.C. Simon, R.M. Haralick (Eds.) (Dordrecht, Holland: D. Reidel, 1981), pp. 105–148.

29. Levoy, M. Display of surfaces from volumetric data, *IEEE Comput. Graph. Appl.* 29–37, (May 1988).

30. Lobregt, S., and H.W.G. Kleineshcaars. Three-dimensional imaging and manipulation of CT data, *Medicamundi* 32 (1987).

31. Lorensen, W.E., and H.E. Cline. Marching cubes: A high resolutin 3D surface construction algorithm, *Comput. Graph.* 21:163–169 (November 4, 1987).

32. Macovski, A. *Medical Imaging Systems* (Englewood Cliffs, NJ: Prentice Hall, 1983).

33. Marsh, J.L., and M.W. Vannier. The third dimension in craniofacial surgery, *Plast. Reconstr. Surg.* 71:759–767 (June 1983).

34. Marsh, J.L., and M.W. Vannier. Surface imaging from CT scans, *Surgery* 94:159–165 (August 1983).

35. Marsh, J.L., and M.W. Vannier. *Comprehensive Care for Craniofacial Anomalies* (St. Louis, MO: C.V. Mosby, 1985).

36. Marsh, J.L. Comprehensive care for craniofacial anomalies, *Curr. Probl. Pediatr.* (July 1980).

37. Mazziotta, J.C., and H.K. Huang. Thread (three-dimensional reconstruction and display) with biomedical applications in neuron ultrastructure and computerized tomography, *AFIPS Proceedings, National Computer Conference* (1976), pp. 241-250.

38. Meagher, D. Geometric modeling using Octree encoding, *Comput. Graph. and Image Proc.* 19:129–147 (1982).

39. Newman, W.M., and R.F. Sproull. *Principles of Interactive Computer Graphics*, 2nd ed. (New York, NY: McGraw Hill, 1979), 239–240.

40. Newton, T.H. and D.G. Potts (Eds.) *Technical Aspects of Computed Tomography* (St. Louis, MO: C.V. Mosby, 1981).

41. Pixar Image Computers, Inc., Chap Volumes Volume Rendering Package, *Technical Summary* (San Rafael, CA, 1987).

42. Requicha, A.A.G. Representations for rigid solids: Theory, methods, and systems, *Comp. Surv.* 12:(4)437–464 (Dec. 1980).

43. Rhodes, M.L., W.V. Glenn, and Y.M. Azzawi. Extracting oblique planes from serial CT sections, *J. Comput. Assist. Tomogr.* 4:649–657 (1980).

44. Rhodes, M.L., Y.M. Assawi, and E.S. Chu. Anatomic model and prostheses manufacturing using CT images, *NCGA Proceedings of the Sixth Annual Conference and Exposition/Technical Sessions* (1985), pp. 110–125.

45. Rosenfeld, A., and A.C. Kak. Gray level thresholding, in *Digital Picture Processing*, 2nd ed., v. 2 (New York, NY: Academic Press, 1982) pp. 61–73.

46. Scott, W.W., E.K. Fishman, and D. Magid. Acetabular fractures: Optimal imaging, *Radiology* 165:537–539 (November 1987).

47. Serra, J. Morphology for grey-tone functions, in *Image Analysis and Mathematical Morphology* (New York, NY: Academic Press, 1982) 424–478.

48. Smith, A.R. Volume Graphics and Volume Visualization: A Tutorial, Tech. Memo 176, Pixar Image Computers, Inc., San Rafael, CA (1987).

49. Srihari, S.N. Representation of three-dimensional digital images, *Comp. Surv.* 13:(4)399–424 (1981).

50. Stoffer, J.C. *Graphical and Binary Image Processing and Applications* (Dedham, MA: Artech House, 1982).

51. Sutcliffe, D.C. Contouring over rectangular and skewed rectangular grids, in *Mathematical Methods in Computer Graphics and Design*, K.W. Brodlie (Ed.) (New York, NY: Academic Press, 1980) 39–62.

52. Tessier, P., and D. Hemmy. Three-dimensional imaging in medicine, *Scand. J. Plast. Reconstr. Surg.* 20:3–11 (1986).

53. Udupa, J.K. Display of 3D information in discrete 3D scenes produced by computerized tomography, *Proc. IEEE* 71:(3)420–431 (1983).

54. Vannier, M.W., J.L. Marsh, and J.O. Warren. Three dimensional computer graphics for craniofacial surgical planning and evaluation, *Comput. Graph.* 17:(3)263–273 (July 1983).

55. Vannier, M.W., J.L. Marsh, and M.H. Gado. Three dimensional display of intracranial soft tissue abnormalities, *Am. J. Neuroradiol.* 4:520–521 (May/June 1983).

56. Vannier, M.W., J.L. Marsh, and J.O. Warren. Three dimensional CT reconstruction images for craniofacial surgical planning and evaluation, *Radiology* vol. 149 (January 1984).

57. Wood, S.L., L.L. Fellingham, and J.B. Massicotte. Comparison of image display techniques for solid models in medical imaging, *SPIE Picture Archiving and Communication Systems (PACSIII) for Medical Applications* 536:102–109 (1985).

58. Woolson, S.T., L.L. Ellingham, and P. Dev. Three-dimensional imaging of bone from analysis of computed tomography data, *Orthopedics* 8:1269–1273 (October 1985).

59. Zonneveld, F.W., J.C. Van Der Meulen, and P.F. Van Akkerveeken. Three-dimensional imaging and manipulation of CT data, *Medicamundi* 32 (1987).

60. Jensen, J.R. Thematic information extraction, in *Introductory Digital Image Processing* (Englewood Cliffs, NJ: Prentice Hall, 1986), pp. 177–233.

Chapter 3

A COMPARISON OF MODALITIES

David C. Hemmy and Patricia M. Brigman

Literally, a modality is the form or appearance or disposition of a thing as distinguished from its basic attributes. Therefore, in the context of this chapter, the basic attribute is the production of a three-dimensional (3D) image. The forms or modalities are defined in terms of (a) products providing 3D software, (b) 3D software or hardware included with scanner systems, and (c) workstations incorporating 3D processing or totally dedicated to 3D processing.

The material presented in this chapter was solicited by contacting the **known** producers of 3D products, asking that the following queries be answered:

System Characteristics
1. Product name
2. Product age
3. Installed base
4. Clinical test sites
5. Product availability
6. Price
7. System input interfaces
8. System output
9. Type of product
10. Options available

Major Software Characteristics
11. Algorithm type
12. Maximum number of axial slices permitted
13. Typical study time
14. Cine capability
15. Mearsurements
16. 3D magnification
17. Cutting and restoring anatomy
18. Correlation of 3D anatomy to original axial slice
19. User I/O
20. Processing capability

Image Monitor
21. CRT size
22. Number of windows
23. Multimodality display
24. Type of multi-image display
25. Comments
26. Distribution channels

Of 24 companies questioned, 13 replied. Two new manufacturers also provided information for a total of fifteen surveyed groups (Table 1). Discounting 2 groups who are supported

TABLE 1

Name	Address	Product	Reply
American Diagnostic Medicine	793 Golf Lane Bensenville, IL	*	
AT&T Pixel Machines	1 Executive Dr. Somerset, NJ	W	+
Cemax, Inc.	1705 Wyatt Dr. Santa Clara, CA	W	−
Columbia Scientific, Inc.	8940 Old Annapolis Rd. Columbia, MD	S	+
Computerized Imaging Reference System	2488 Alameda Avenue Norfolk, VA	S	−
Dimensional Medicine, Inc.	10901 Bren Rd. East Minnetonka, MN	W	+
Dynamic Digital Display	3509 Market Street Philadelphia, PA	W	−
Elscint, Inc.	930 Commonwealth Ave. Boston, MA	SCS	+
GE Medical Systems	P.O.Box 414 Milwaukee, WI	SCS	+
H.I.P. Graphics	P.O. Box 23028 Baltimore, MD	S	+
Imatron, Inc.	389 Oyster Point South San Francisco, CA	**	−
I.S.G. Technologies, Inc.	3030 Orlando Dr. Mississauga, Ont.	W	+
Kontron Electronics, Inc.	630 Clyde Avenue Mountain View, CA	W	+
M.P.D.I.	2730 Pacific Coast Hwy. Torrance, CA	S	−
North American Surgical Imaging	9200 West Wisconsin Ave. Milwaukee, WI	S	+
Philips Medical Systems, Inc.	710 Bridgeport Avenue Shelton, CT	W	+
Picker International, Inc.	595 Miner Road Highland Heights, OH	SCS	+
P.I.X.A.R.	3270 Kerner Blvd. San Rafael, CA	W	+
Reality Technologies	P.O. Box 39097 Solon, OH	W	+
Shimadzu Corporation	101 W. Walnut St. Gardena, CA	SCS	+
Siemens Medical Systems	186 Wood Avenue South Iselin, NJ	SCS	−
Tomo, Inc.	15999 W. 12 Mile Rd. Southfield, MI	W	+
Toshiba Medical Systems	2441 Michelle Drive Tustin, CA	SCS	−
Virtual Imaging/PURA	725 Kifer Road Sunnyvale, CA	W	−

Note: * = supported by Dimensional Medicine, Inc.; ** = uses Virtual Imaging Workstation and M.P.D.I. software; S = software only; SCS = software operating on CT scanner system; W = dedicated Workstation; + = reply received; − = no reply.

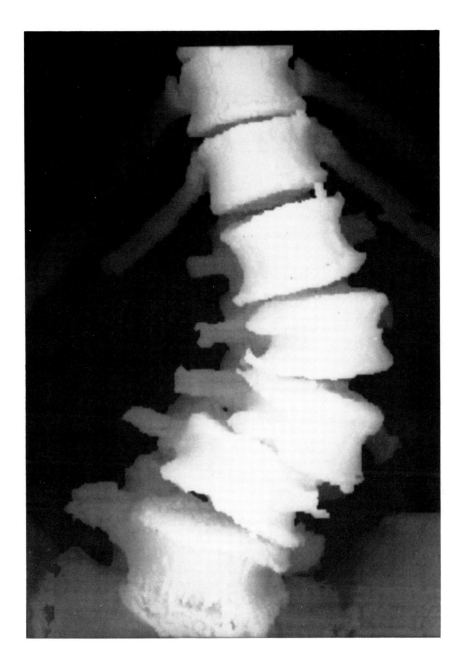

FIGURE 1. Image prepared by North American Surgical Imaging, Inc. Scoliotic thoracolumbar spine.

by other groups on the list, there were 22 companies equally divided among stand alone workstations and software systems.

Two of the companies listed in the software group provide image processing using other systems. Computerized Imaging Reference System (Philips PICS 2000 System) and North American Surgical Imaging (DMI 3200 Workstation) process CT data upon the receipt of a magnetic tape from a remote scanner, providing as a final product a video tape or X-ray-type films (Figure 1).

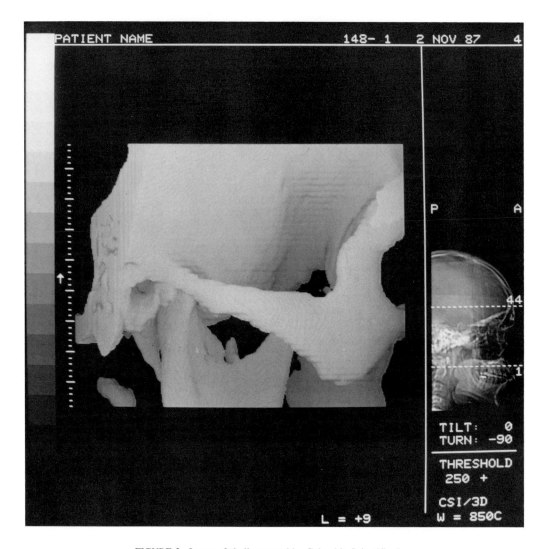

FIGURE 2. Image of skull prepared by Columbia Scientific, Inc.

Columbia Scientific, Inc. markets user-controllable software which resides in General Electric CT host computers using the volume-rendering algorithm of Dynamic Digital Display with gradient shading (Figures 2 and 3). The software operates on both the 8800 and 9800 scanners, filling a void for the operators of 8800 scanners as General Electric Medical Systems has never marketed a software package for the 8800 scanner. Similarly, M.P.D.I. provides a network for GE CT systems or resident software for both the GE 8800 and 9800 scanners.

Elscint, General Electric, Picker International, Shimadzu, Siemens, and Toshiba market software accessible through the CT control console and may provide hardware enhancements to independent display consoles. Prices range from less than $10,000 to nearly $75,000 (Plates 1 and 2* and Figures 4 to 6).

At the upper end of the spectrum are the dedicated workstations. Here prices vary from $33,000 to close to $150,000. Products are offered by AT&T Pixel Machines, Dimensional Medicine, Dynamic Digital Displays, I.S.G. Technologies, Kontron Electronics, Philips Medical

* Plates 1 and 2 appear following page 166.

Systems, P.I.X.A.R., Reality Technologies, Tomo, and Virtual Imaging. The major advantages of the dedicated workstation are primarily the dedication and, generally, increased processing speed, flexibility, higher resolution displays, and more analysis tools although, in some cases, reconstruction times exceed those of CT software programs.

Most of the dedicated workstations permit input from a variety of CT scanners and magnetic resonance scanners via magnetic tape or in some cases by direct link to the scanner. Therefore the choice is no longer limited by the type of scanner in the hospital complex (this includes the Philips PICS 1000 and 2000 series workstations). In most cases, at least 100 axial slices can be accepted.

Output is made in most cases, depending upon the options selected, to film, color monitor, magnetic tape, optical disc, videotape, color hard paper and color film (Figures 7 to 10 and Plates 3 and 4*).

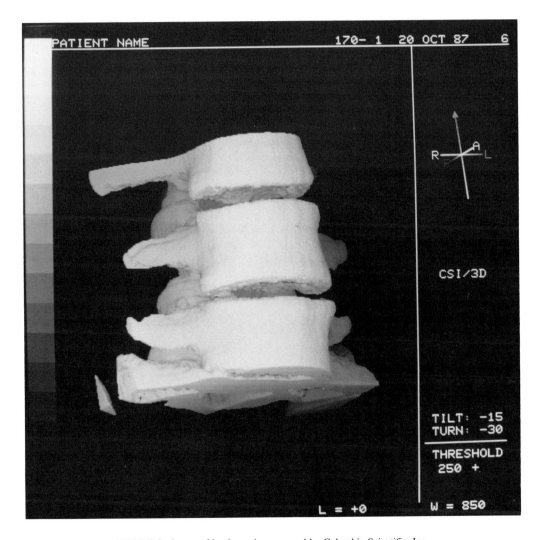

FIGURE 3. Image of lumbar spine prepared by Columbia Scientific, Inc.

* Plates 3 and 4 appear following page 166.

There have been three recent entries into the market: AT&T Pixel Machines, HIP Graphics, and Tomo. AT&T is presently developing medical imaging software for their PXM 900 series and expects it to be available in late 1989. HIP Graphics is marketing a software program providing a 2- and 3D interface to the Sun workstation and the P.I.X.A.R. image computer. It is also sold as a turnkey hardware system. The Tomo workstation is a multipurpose workstation with emphasis placed upon planning for stereotactic procedures.

Most dedicated workstations permit multiplanar reformatting of acquired CT data as well as the ability to "cut the data" along predefined planes, some allowing this to be done interactively. Furthermore, measurements can be made, usually linear, but in some cases, volumetric.

It is not the purpose of this chapter to provide detailed descriptions of every product. These can be obtained directly from the manufacturer (Table 1) or from market surveys which have been prepared.[1]

REFERENCE

1. Wisconsin Medmark, Inc. *Manufacturer's Guide to 3D Systems* (1988).

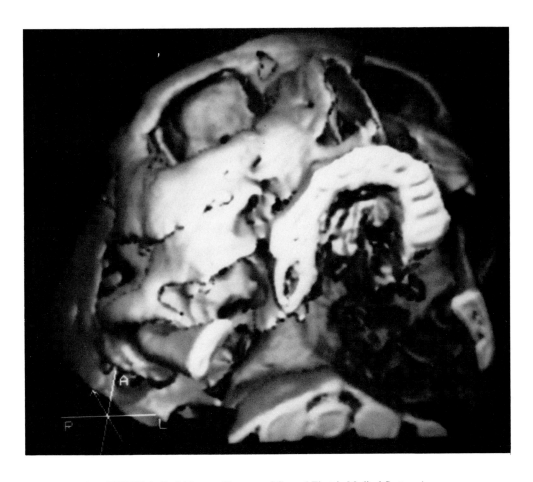

FIGURE 4. Facial bones. (Courtesy of General Electric Medical Systems.)

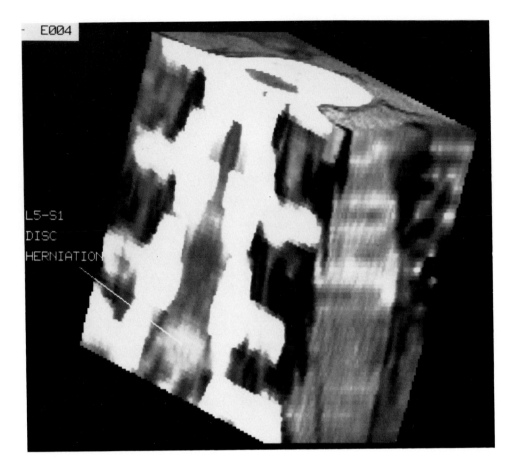

FIGURE 5. Multiplanar "cube" reconstruction of lumbar spine with herniated nucleus pulposus. (Courtesy of Picker International.)

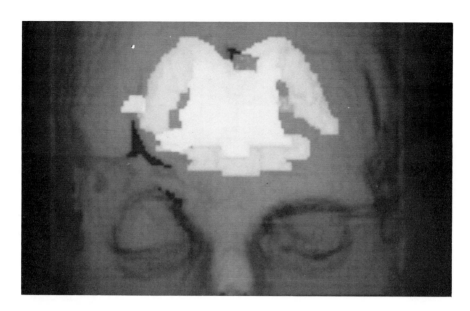

FIGURE 6. Display of ventricular system within head. (Courtesy of Picker International.)

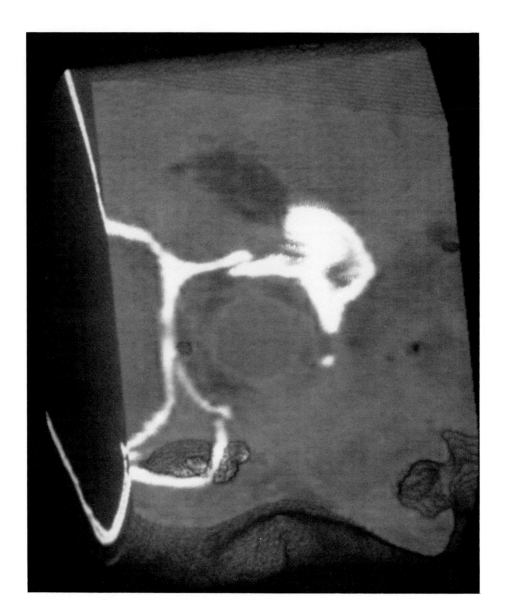

FIGURE 7. Oblique cutplane view through zygomatic fracture. (Courtesy of Reality Imaging.)

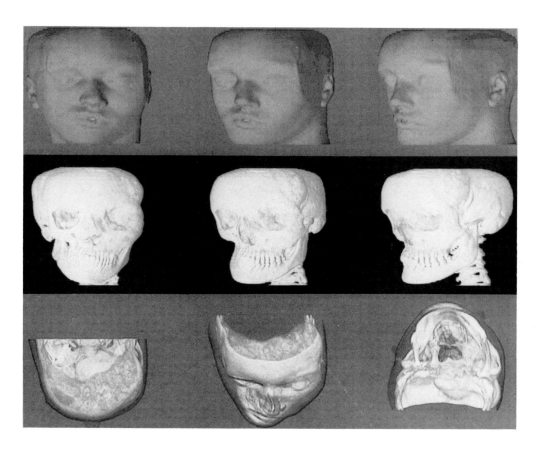

FIGURE 8.   Skull and face and composite views. (Courtesy of P.I.X.A.R. and Pittsburgh Eye and Ear Hospital.)

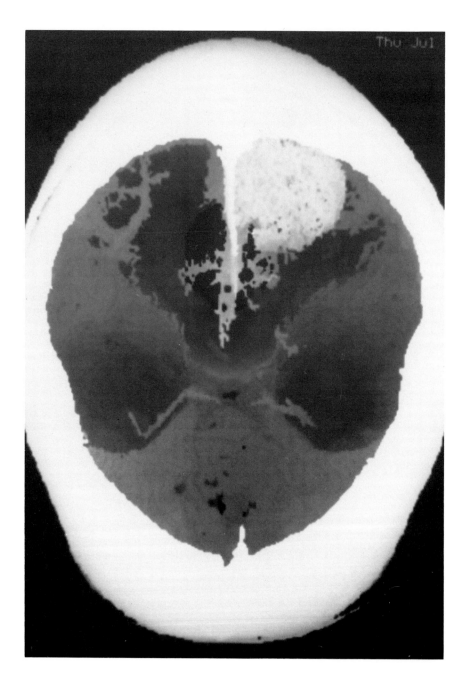

FIGURE 9.   Cutaway view of interhemispheric tumor. (Courtesy of Dimensional Medicine.)

FIGURE 10. Philips PICS 2000 HR workstation.

Chapter 4

THREE-DIMENSIONAL COMPUTED TOMOGRAPHIC REFORMATIONS: ASSESSMENT OF CLINICAL EFFICACY

James E. Gillespie, Anil Gholkar, and Ian Isherwood

TABLE OF CONTENTS

I. INTRODUCTION

In 1984 the Department of Diagnostic Radiology, University of Manchester received the experimental three-dimensional (3D) reformatting software package entitled "3D83"[1] (supplied by GE Medical Systems) for the purpose of assessing its clinical applications. At this time only a small number of references in the literature were available demonstrating the type of image produced by existing 3D software programs.[2,3] Although the clinical potential of these visually stunning images appeared obvious, the need to assess the role of 3D imaging and its relationship to conventional computed tomography (CT) and two-dimensional (2D) multiplanar reformations in a less anecdotal manner was apparent.

Following an initial trial period to gain familiarity with the use and capabilities of the "3D83" package,[4] a number of clinical protocols were established to make this assessment of clinical efficacy. The areas selected for study were congenital, traumatic, and neoplastic abnormalities of the craniofacial region and pelvic trauma. The relative merits of 3D images were to be compared with both conventional CT (including 2D multiplanar reformations) and conventional radiographs. While these comparisons were to be made primarily from the point of view of the diagnostic radiologist, we also intended to assess the referring clinicians' viewpoint as far as was practicable. We hoped to be able to provide the answers to three main questions with regard to 3D reformations, namely: in which clinical situation might they be beneficial, what type of information did they convey, and was this information of value to the radiologist, clinician or both.

II. GENERAL TECHNIQUE

All examinations were performed in the Department of Diagnostic Radiology, University of Manchester in the period 1984–1988. Scanning was undertaken on either a GE 8800 CT/T (until 1986) or a 9800 CT with 3D reformatting performed by the programs "3D83" or "3D98"[5] (developed by the Medical Image Processing Group, Department of Radiology, Hospital of the University of Pennsylvania, under contract from the General Electric Company), respectively. A slice thickness of 1.5 mm was employed routinely on the 8800 CT/T, while on the 9800 CT 1.5- and 3-mm sections were used either exclusively or in combination depending upon the clinical circumstances (see individual protocols).

3D reformatting of CT data was carried out on an independent viewing console (IPDC or IC). Using the "measure" mode pixel display facility, a threshold level of 150 to 200 Hounsfield units (H.U.) was selected for the display of bony surfaces; a level of 60 to 90 H.U. was used to depict enhancing intracranial structures in addition to bony landmarks. Those portions of the input CT sections to be included in the 3D reformation were designated using the region of interest (ROI) facility. 3D images were obtained at 30-degree increments around selected axes of rotation most commonly the X and Z axes. Computer processing time was approximately 1 to 3 hours per patient depending upon the volume to be reformatted and the number of axes required with no operator intervention needed once the program was initiated. Normally the 3D program was allowed to run overnight to maximize use of what was otherwise computer "down-time".

A. DYNAMIC "LOW-DOSE" 3D CT

For craniofacial studies on the 8800 CT/T contiguous 1.5-mm sections were obtained in conventional scan mode using the following technique factors: 120 kVp, 400 mA, PWC2, 576 views. This approach entails relatively long examination times due to the large number of narrow slices (up to 60) required for optimum spatial resolution. Enforced delays to allow X-ray tube cooling further increase the likelihood of movement artefact appearing in any reformatted image whether 2D or 3D.

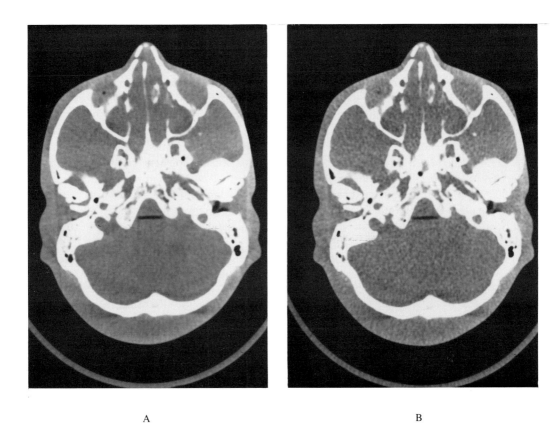

A   B

FIGURE 1. Transaxial CT sections of the skull phantom (same level) obtained at (A) 300 mAs and (B) 80 mAs. Increased soft tissue noise is visible in (B) but bone detail appears identical.

Clinically acceptable CT images can be produced using rapid sequence (dynamic) scanning at reduced exposure settings (160 mAs) in cases of maxillofacial[6] and spinal[7] trauma. Applying this approach in a study using phantom and cadaveric specimens, Andre et al.[8] obtained satisfactory 3D images using very low exposure levels (6 mGy at 50 mAs). After acquiring a GE 9800 CT scanner we conducted a number of studies on a commercially manufactured skull phantom to assess further the low-exposure fracture/rapid sequence scan technique.[9]

Figure 1 demonstrates the effect of reducing mAs from 300 to 80 at 120 kV. Increased image noise was only perceptible in the soft tissue portion of the phantom while the bone detail appeared identical. 3D reformations generated from both sets of scans were indistinguishable (Figure 2). In dynamic scan mode, up to 50 consecutive slices could be performed in less than 5 minutes at 140 and 80 mAs. At 300 or 200 mAs tube heating restrictions limited the number of consecutive scans to 10 or 15. Based on this preliminary data we adopted the dynamic scanning technique for all clinical 3D CT studies using the reduced exposure factors detailed in Table 1. The mAs level chosen was determined by the quality of soft tissue detail required on the conventional CT sections for clinical diagnosis (Figure 3). In patients with facial tumors, for example, it is our experience that the increased image noise in soft tissue is compensated for by the facility to scan rapidly through large tumor volumes when soft tissue enhancement is maximal (Figure 4). A further major advantage of this scanning approach is the considerable reduction in radiation dose to the patient (Table 2).

III. 3D IMAGE ASSESSMENT

In each of the clinical protocols our aim was to compare the information provided by the 3D reformations with that available from both the conventional CT examination, including 2D multiplanar reformations, and standard radiography. To permit as objective an appraisal as possible from the viewpoint of a radiologist, a simple scoring system was devised and implemented in all four craniofacial protocols. Initially we proposed a three-point comparative scale based upon whether the information provided by 3D was inferior, similar, or superior to that obtained from other imaging formats. It soon became apparent that two grades of superiority were needed. In some cases additional conceptual information was being provided; in others the 3D format gave essentially similar information but permitted this to be assimilated much more rapidly than was possible by viewing, for example, a large number of standard CT sections, 2D reformations, or conventional tomograms. A four-point system was, therefore, employed (Table 3).

Clinicians referring cases of traumatic and congenital craniofacial deformities for preoperative assessment were encouraged to complete a clinical proforma containing the following four questions:

1.   Did the 3D images provide additional information?
2.   Did the additional information alter patient management?
3.   Would the conventional CT sections (and multiplanar reconstructions) have been as useful on their own without the 3D images?
4.   In this patient was 3D imaging not useful, useful but not essential, or essential?

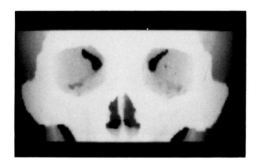   

A   B

FIGURE 2. 3D images of the skull phantom obtained from transaxial sections performed at (A) 300 mAs and (B) 80 mAs. The images are indistinguishable. (From Gholkar, A., J.E. Gillespie, C.W. Hart, D. Mott, and I. Isherwood. *Br. J. Radiol.* 61:1096 (1988). With permission.)

TABLE 1
Radiation Factors for Clinical Applications in 3D CT

80 mAs/120 kVp	140 mAs/120 kVp	200 mAs/120 kVp
Congenital craniofacial anomalies Craniofacial trauma Lesions affecting the axial and appendicular skeleton	Head and neck tumors	Intracranial vascular lesions

From Gholkar, A., J.E. Gillespie, C.W. Hart, D. Mott, and I. Isherwood. *Br. J. Radiol.* 61:1095 (1988). With permission.

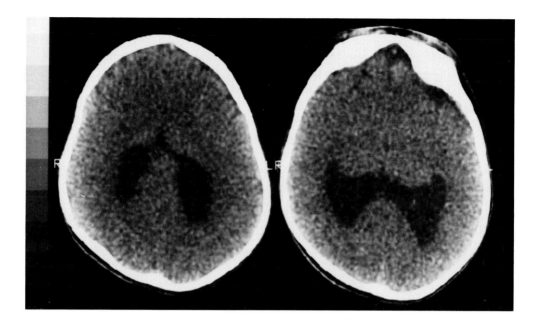

A

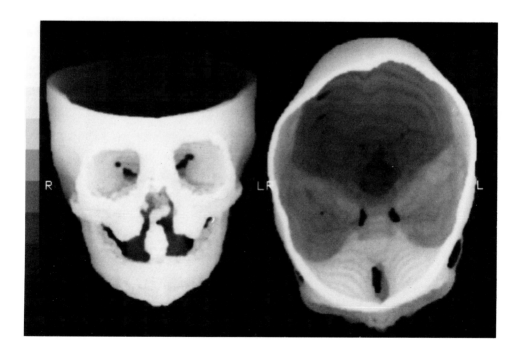

B

FIGURE 3. Congenital abnormalities in a 4-month-old child (A) transaxial sections (80 mAs) through the brain at lateral ventricular level showing absence of the frontal horns and thalamic fusion in semilobar holoprosencephaly. (B) Anterior and superior 3D images demonstrating midline cleft, hypotelorism, and anterior cranial fossa defect due to meningocele. (From Gholkar, A., J.E. Gillespie, C.W. Hart, D. Mott, and I. Isherwood. *Br. J. Radiol.* 61:1096–1097 (1988). With permission.)

These questionnaires were completed by referring clinicians after any surgery was completed. It was anticipated that the perceived value of 3D as determined by the clinical proforma might vary not only with the complexity of the clinical abnormality, but also with the familiarity of the surgeon with CT imaging.

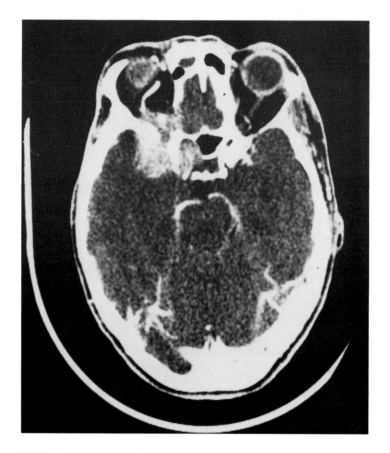

FIGURE 4. Representative CT section from a dynamic craniofacial study performed at 140 mAs. Recurrent adenocystic carcinoma of the paranasal sinuses is seen involving the orbit and intracranial cavity. Note encasement of the right internal carotid artery.

TABLE 2
Radiation Dose to Skin Surface in 3D CT

Examination	mAs/kVp	Radiation dose (mGy) ±6%
Craniofacial	300	40
	140	19
	80	11
Pelvis	300	19
	80	5

From Gholkar, A., J.E. Gillespie, C.W. Hart, D. Mott, and I. Isherwood. *Br. J. Radiol.* 61:1098 (1988). With permission.

TABLE 3
Information Provided by 3D Compared with
Radiography or CT

Score	3D assessment
1	Inferior
2	Similar
3	Superior — similar information more rapidly assimilated
4	Superior — additional conceptual information provided

From Gillespie, J.E., I. Isherwood, G.R. Barker, and A.A. Quayle. *Clin. Radiol.* 38:524 (1987). With permission.

TABLE 4
Clinical Material

Abnormality	No. of patients
Unilateral hypo/hyperplasia	6
Encephalocele	2
Craniofacial synostosis	4
Fibrous dysplasia	1
Frontal pneumosinus dilitans	1
Maxillary recession	6

IV. CONGENITAL CRANIOFACIAL DEFORMITIES

Traditional evaluation of craniofacial deformities has involved the use of routine radiographs, tomograms, and cephalometric studies. CT added a new dimension to this assessment by demonstrating both soft tissue and osseous anatomy using direct multiplanar scanning or 2D multiplanar reformations.[10] Subsequently, the potential value of 3D reformatting has been demonstrated by several groups.[2,3] The following protocol was instituted to assess further the application of 3D CT.

A. PROTOCOL DETAILS

Twenty patients presenting with a variety of abnormalities were included in the protocol (Table 4). Three of those with maxillary recession were also scanned postoperatively, making 23 CT examinations in all. Three children, one with an orbital encephalocoele and two with craniofacial synostoses (Apert and Crouzon syndrome), had previously undergone partial surgical correction of their deformities. Conventional radiographs were available for comparison in 17 of these. Patients were scanned using either contiguous 1.5-mm sections from the supraorbital region down to the maxillary alveolus (GE 8800) or with contiguous 3-mm sections from the forehead to the lower mandible (GE 9800). Scans in young children, or mentally subnormal adults unable to cooperate, were performed under general anesthesia to eliminate movement artefact.

Image assessments were made by one radiologist. The radiographs, including any cephalostats or analytic morphograms,[11] CT scans, and 3D reconstructions, were initially viewed separately and the extent and severity of the deformity noted. Afterwards all films were reviewed together and the 3D images scored using the four-point scale described previously.

In the six patients with maxillary recession the main deformities were malocclusion and mid-facial flattening. The amount of maxillary advancement required to correct this had already been satisfactorily calculated by standard cephalometric techniques. We considered it impractical to scan the entire face using the narrow section width (1.5 mm) necessary to maximize spatial accuracy only to duplicate measurements available from a single lateral cephalostat. 3D evaluation in this group was limited to abnormalities additional to the main facial disharmony which might affect surgical management. We therefore assessed the six patients separately from the other fourteen with more severe and asymmetric deformities.

B. RESULTS
1. Maxillary Recession (Table 5)

3D and radiography — Two patients had facial clefts (one bilateral, one unilateral), one of whom also had a complex developmental anomaly of the craniocervical junction, while the other had unilateral antral hypoplasia. These abnormalities were difficult to appreciate on the conventional radiographs particularly when relatively minor degrees of maxillary asymmetry occurred. The precise nature of the craniocervical junction deformity could only be clearly discerned on the 3D image (Figure 5). In another case the presence of relatively thin facial bones was only apparent on the 3D images which contained unexpectedly large artefactual bony defects or "pseudoforamina" despite using our standard bone threshold level.

3D and CT — Compared with CT and 2D reconstructions, 3D images were less advantageous in depicting the extent of bony anomalies, these generally being well shown on CT. However, the overall effect of these additional deformities on bony architecture was either better or more rapidly appreciated on 3D in the majority of patients. In the three postoperative examinations the path of the osteotomy incision line and localized areas of bone disruption could be appreciated on the 3D image in a manner not possible by conventional CT, thus providing a unique and instructive method of reviewing the effects of the surgical technique employed.

2. Asymmetric/Severe Deformities (Table 6)

3D and radiography — In the majority of patients, radiography was less informative than 3D because of the more generalized and severe nature of the craniofacial abnormality. Superimposition of asymmetric bony structures often made the identification of standard landmarks difficult on cephalometric radiographs. In young children technically satisfactory films were not always obtainable; in others deformities of the external ears precluded accurate head positioning. Thus, 3D was deemed superior to radiography in depicting the extent of the abnormality in the majority of patients and the severity in 10 out of 12 patients (Figure 6).

3D and CT — As with the maxillary recession group, the extent of the craniofacial skeleton involved by the deformity could be seen on CT extremely well and so 3D added little further information. Overall appreciation of the severity of facial disharmony was superior by 3D in 12 out of 14 cases, with 8 rating a maximum score (Figures 7 and 8).

TABLE 5
Assessment of 3D in Maxillary Recession

Comparison 3D score	3D and radiography				3D and CT			
	1	2	3	4	1	2	3	4
No. of examinations								
Deformity								
Extent	0	2	0	3	0	5	4	0
Severity	0	1	0	4	0	3	4	2

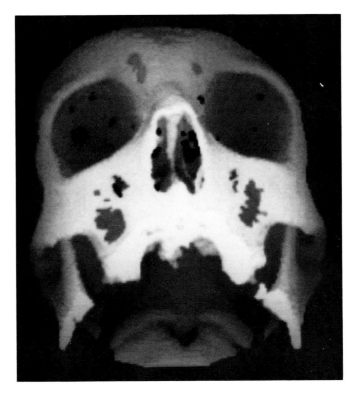

A

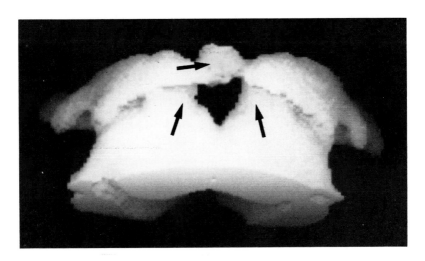

B

FIGURE 5. Congenital abnormalities of the face and upper cervical spine. (A) Antero-inferior view showing bilateral cleft palate and anomalous C1 and C2. Note "pseudoforamina" (artefactual defects) of the anterior antral walls. (B) "Isolated" view of C1 and C2. The three unfused segments of the odontoid process are evident (arrows) while the anterior arch of C1 is incomplete. (From Gillespie, J.E., and I. Isherwood. *Br. J. Radiol.* 59:291 (1986). With permission.)

Surgical view — Clinical proformas were completed for 14 examinations. In all instances, the referring clinician felt that the 3D images provided information additional to that which could be obtained from the standard radiological assessment and that CT and 2D reconstructions would not have been as useful on their own without 3D. This information altered patient management in five cases. Overall, 3D was considered useful in 12 cases and essential in 2. Even when clinical management was not altered, surgeons still valued the 3D format preoperatively as they felt their understanding of the deformity was improved.[12]

C. COMMENT

3D was the best image format for appreciating the severity of facial deformity. From the radiological viewpoint, the usefulness of 3D was most apparent in the more severe, asymmetric anomalies. It is also clear that the clinicians found the spatial information easier to understand when presented in 3D rather than conventional CT format and, hence, of greater benefit preoperatively.

V. CRANIOFACIAL TRAUMA

Patients sustaining facial injuries normally undergo standard radiography as the initial radiological investigation. When further anatomical detail is required because of superimposition of bony surfaces or impaired visualization by overlying soft tissue injury, high-resolution CT is regarded as the procedure of choice over conventional tomography.[13] The application of 3D reformatting techniques in these circumstances was investigated in the following protocol.

A. PROTOCOL DETAILS

Fifteen patients who had sustained a variety of craniofacial injuries were examined (Table 7). Nineteen CT examinations were performed as four of those with orbital "blow-out" fractures were re-evaluated postoperatively.[14] Plain radiographs (plus coronal tomograms in three cases) were available for comparison in 15 of these. The area of injury was scanned using contiguous 1.5-mm sections, usually from the supraorbital margins down to the maxillary alveolus. Occasionally, on the 9800 CT, additional 3-mm sections of the adjacent calvarium were obtained.

Image assessments were made by one radiologist. The conventional radiographs, CT scans, and 3D reconstructions were initially viewed separately and assessed under the headings of fracture detection, extent, and displacement. All images were then reviewed together and the 3D images given a score. Conventional tomograms were not routinely obtained as the additional radiation dose was not considered justifiable in those patients who would shortly undergo high-resolution CT.

TABLE 6
Assessment of 3D in Asymmetric/Severe Deformities

Comparison 3D score	3D and radiography				3D and CT			
	1	2	3	4	1	2	3	4
No. of examinations								
Deformity								
Extent	0	5	1	6	1	10	3	0
Severity	0	2	2	8	0	2	4	8

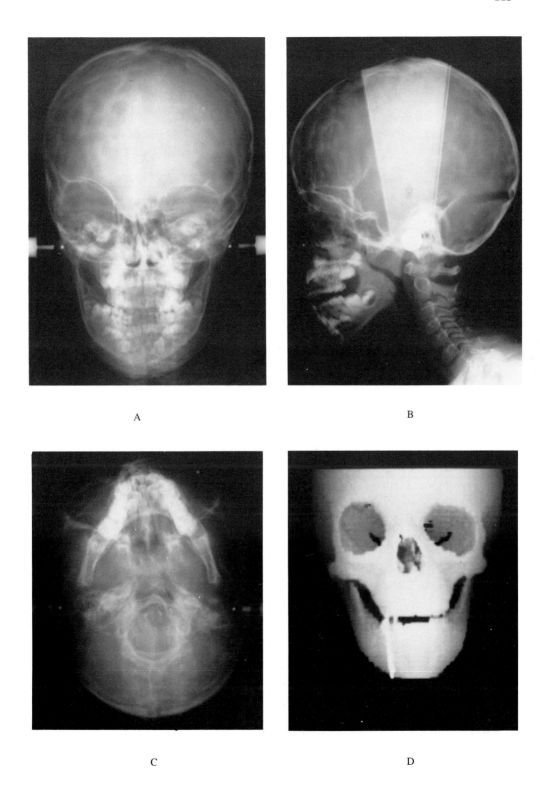

FIGURE 6. Complex deformity with relative hypoplasia of the right craniofacial skeleton in a 5-year-old girl. (A to C) Anterior, lateral, and basal analytic morphograms. (D to G) Anterior, lateral, semibasal, and superior 3D images. The deformity, particularly as it affects the skull base and upper face, is more easily appreciated on 3D.

B. RESULTS
1. Oribital Injuries (Table 8)

3D and radiography — Detection of fractures involving the medial and inferior walls of the orbit was hindered by the presence of "pseudoforamina". These artefactual areas of bone absence are the result of partial volume averaging of thin bone and are particularly prominent around the air-containing paranasal sinuses. Fractures of the orbital floor were only visualized by 3D in three cases where the antero-inferior orbital rim and adjacent antrum were also fractured, examples of the so-called "impure" blow-out fracture. The extent and displacement

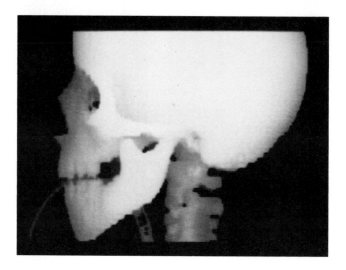

FIGURE 6E

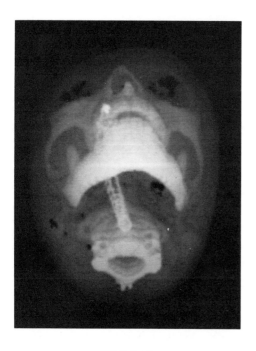

FIGURE 6F

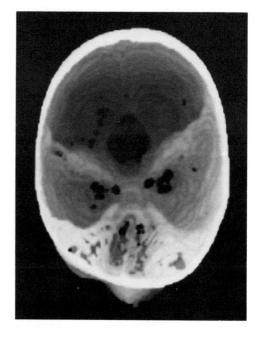

FIGURE 6G

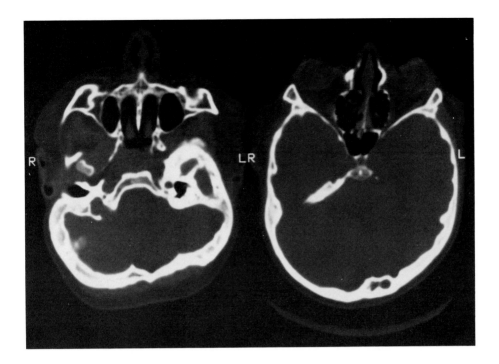

A

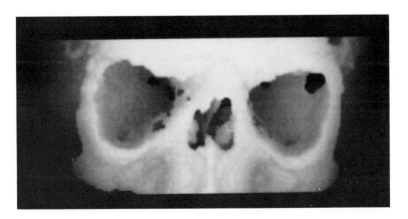

B

FIGURE 7. Severe facial flattening and hypertelorism in a 13-year-old boy with Apert's syndrome. Note asymmetric foramen magnum. (A) Representative CT sections. (B) Anterior and (C) basal 3D views. (B and C from Rabey, G.P. *Br. J. Oral Maxillofac. Surg.* 25:176 (1987). With permission.)

of the fractured floor were better demonstrated by 3D than standard radiographs in these three patients (Figure 9).

3D and CT — Fracture detection by 3D was inferior to CT in all cases. 3D also scored poorly with respect to extent and displacement in most examinations. A maximum score was given to 3D in only one instance where the medial course of a complex anterior antral fracture line could only be clearly appreciated on the 3D image.

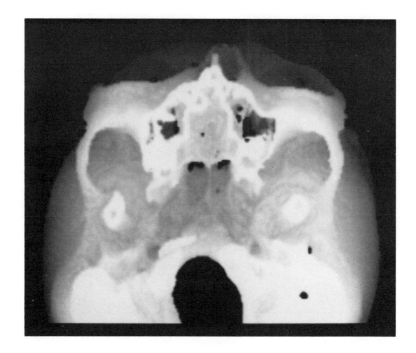

FIGURE 7C

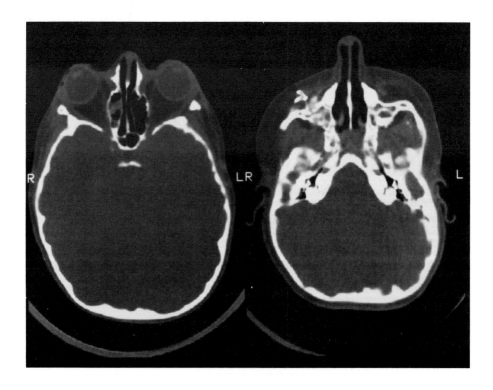

A

FIGURE 8. 6-year-old girl with Crouzon's syndrome displaying hypertelorism and marked irregularity of the right orbital margin. A forehead transposition had previously been performed. (A) Representative CT sections. (B) Anterior 3D image.

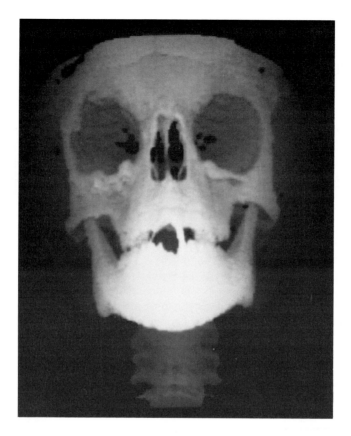

FIGURE 8B

TABLE 7
Clinical Material

Type of injury	No. of patients
Orbital "blow-out" fracture	7
Malar complex fracture	4
Extensive craniofacial trauma	4

From Gillespie, J.E., I. Isherwood, G.R. Barker, and A.A. Quayle. *Clin. Radiol.* 38:524 (1987). With permission.

TABLE 8
Assessment of 3D in Orbital "Blow-Out" Fractures

Comparison 3D score	3D and radiography				3D and CT			
	1	2	3	4	1	2	3	4
No. of examinations								
Fracture								
Detection	6	2	0	0	11	0	0	0
Extent	5	0	0	3	7	1	2	1
Displacement	4	1	0	3	9	1	1	0

From Gillespie, J.E., I. Isherwood, G.R. Barker, and A.A. Quayle. *Clin. Radiol.* 38:524 (1987). With permission.

2. Malar Fractures (Table 9)

3D and radiography — In two patients 3D was superior in detecting anterior antral wall fractures which were not well shown on plain radiography. Assessment of the degree of bone displacement was also considered superior by 3D in two instances.

3D and CT — 3D was considered to be the inferior image format in almost all patients in all three areas of image assessment.

3. Extensive Craniofacial Trauma (Table 10)

3D and radiography — 3D demonstrated either a similar number or fewer fractures than conventional radiographs in all three patients in whom the comparison could be made. However, the extent of any fracture visualized on 3D and the accompanying displacement was always superior on the 3D images (Figure 10).

3D and CT — CT demonstrated more fractures than 3D in all cases. Assessment of the extent of visualized fracture lines by 3D was always at least equal to that shown on CT. Maximum scores were given in all cases with regard to fragment displacement and overall perception of the disruption of facial symmetry (Figures 11 and 12).

Surgical view — Clinical proformas were only completed for seven examinations. Although the results followed the general trend of those in the congenital deformity protocol, we felt the low return was insufficient to draw definite conclusions. At the end of the protocol,

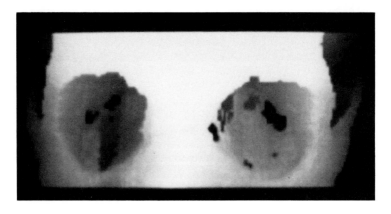

FIGURE 9. Anterior 3D image demonstrating extensive lenticular-shaped disruption of the right orbital floor. The inferior orbital rim was also fractured.

TABLE 9
Assessment of 3D in Malar Complex Fractures

Comparison 3D score	3D and radiography				3D and CT			
	1	2	3	4	1	2	3	4
No. of examinations								
Fracture								
Detection	0	2	0	2	2	2	0	0
Extent	1	2	0	1	3	0	1	0
Displacement	1	1	0	2	3	0	1	0

From Gillespie, J.E., I. Isherwood, G.R. Barker, and A.A. Quayle. *Clin. Radiol.* 38:524 (1987). With permission.

TABLE 10
Assessment of 3D in Extensive Craniofacial Fractures

Comparison 3D score	3D and radiography				3D and CT			
	1	2	3	4	1	2	3	4
No. of examinations								
Fracture								
Detection	1	2	0	0	3	1	0	0
Extent	0	0	0	3	0	1	2	1
Displacement	0	0	0	3	0	0	0	4

From Gillespie, J.E., I. Isherwood, G.R. Barker, and A.A. Quayle. *Clin. Radiol.* 38:524 (1987). With permission.

however, the clinicians were asked for their overall impression of 3D reformations. Their responses indicated a high degree of clinical acceptability as the 3D images conveyed spatial information in the most readily understandable format. As direct visualization of bone fragments during surgery was not always possible, the conceptual information provided by 3D proved helpful in the operative situation.

C. COMMENT

3D was of most value in those suffering extensive craniofacial injuries involving significant bone displacement resulting in severe facial distortion. 3D was the least reliable of the three types of image for identifying fractures around the orbits and antra, due to "pseudoforamina" and difficulties in demonstrating minor fracture lines with little or no displacement. Demonstration of the latter on 3D would require the use of a very high threshold level to highlight the small gap between minimally displaced bone edges on the "measure" mode display, a maneuver which would only exaggerate "pseudoforamina" in much of the facial skeleton.

VI. INTRACRANIAL TUMORS

3D reformatting of specific soft tissues other than external skin has proved difficult due to the relatively similar attenuation values of many adjacent soft tissue structures. In the intracranial cavity, this problem can sometimes be overcome by the use of intravenous contrast injection[4,15] provided enhancement sufficiently greater than that of normal brain occurs in the structure of interest. The feasibility of this approach in patients with mass lesions around the sella region was investigated in the following protocol.

A. PROTOCOL DETAILS

Ten patients with enhancing sellar or para-sellar mass lesions or extensive bone erosion were selected for study (Table 11). Nine of these were referred for conventional high-resolution pituitary examinations and were scanned on the 8800 CT/T using contiguous 1.5-mm slices;[16] three of these were also examined postoperatively. The tenth case, a large trigeminal neurofibroma, was examined on the 9800 CT using contiguous 3-mm sections.[17] Patients received an intravenous injection of either 100 ml of Conray 420 (sodium iothalamate) or 100 to 120 ml of Omnipaque 300 (iohexol) and scanning commenced after 20 to 30 ml had been injected. 3D reformations were obtained at a threshold of 150 to 175 H.U. to display only osseous surfaces and at 60 to 65 H.U. to depict certain enhancing soft tissues and bony surfaces on the same image without normal surrounding brain tissue.

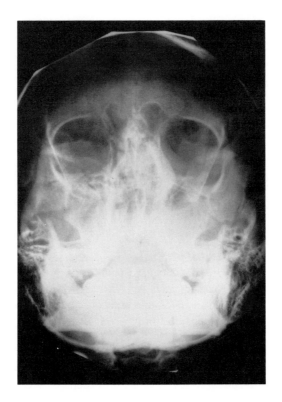

A

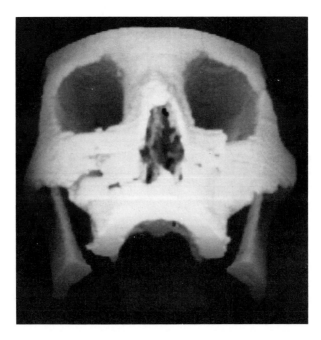

B

FIGURE 10.   Combined Le Fort I and II and left malar complex (tripod) fractures. (A) Plain radiograph. (B) 3D image in similar orientation. Note extensive fractures of both antra, nasal septum, and the inferior and lateral aspects of the left orbit. (From Gillespie, J.E., I. Isherwood, G.R. Barker, and A.A. Quayle. *Clin. Radiol.* 38:524 (1987). With permission.)

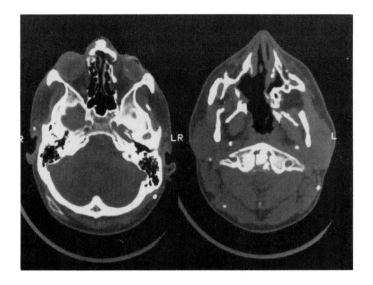

A

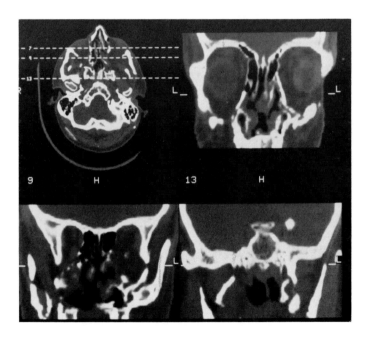

B

FIGURE 11.   Combined Le Fort II and III level fractures. (A) Representative tran-saxial CT sections and (B) coronal reformations showing widespread facial bone disruption. (C) Antero-inferior and basal 3D views. Severe facial distortion is evident. Note the large traumatic cleft palate and displaced left pterygoid plates. (From Gillespie, J.E., I. Isherwood, G.R. Barker, and A.A. Quayle. *Clin. Radiol.* 38:525 (1987). With permission.)

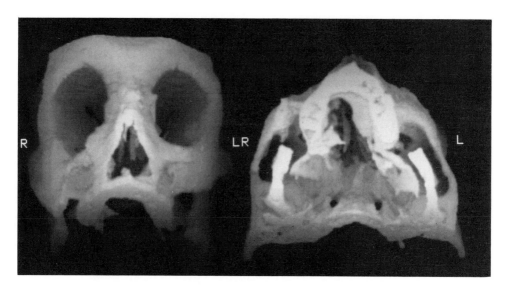

FIGURE 11C

For the first nine cases the 3D images were compared with the CT sections and 2D multiplanar reformations by two independent radiologists and then reviewed jointly; the tenth case was evaluated in a similar manner by one radiologist. Assessment was made under the headings of tumor extent, relation to blood vessels, and bone destruction using the scoring system described earlier. Note was also made of any movement artefact occurring in the 3D reformations.

B. RESULTS (TABLE 12)

Tumor extent — 3D demonstrated any supra-sellar tumor extension well but not in a superior way to CT. Little useful information could be gleaned from 3D regarding the intra-sellar tumor component, while infra-sellar extension could not be assessed. 3D therefore scored poorly under this heading overall.

Tumor relation to surrounding vessels — Displacement of vessels of the Circle of Willis by large supra-sellar tumor masses was readily visible on 3D and was considered superior to CT in six examinations. This advantage was mainly one of time saving (Figures 13 to 15).

Bone destruction — Extensive areas of skull base erosion were easily appreciated on 3D and considered superior to CT in six cases (Figures 13 and 14). Defects of the pituitary fossa floor were not always pathological, however, as a "pseudoforamen" was occasionally seen at this site.

Movement artefact — This was noted in 38% of the examinations, manifested by slice misregistration or "uneven stacking" of the CT slices. The lower incidence in those who received the nonionic contrast medium iohexol may be related to the known reduction in patient discomfort associated with use of the newer low osmolar agents.[18]

C. COMMENT

3D reformatting of intracranial tumors and vascular structures is feasible provided sufficient enhancement occurs. In this study the technique was most useful in permitting a more rapid assessment of the vascular relations of tumors with large supra-sellar components and for the assessment of extensive skull base destruction. Following this study, dynamic scanning at reduced mAs levels using Omnipaque 350 was employed for all intracranial 3D studies requiring contrast enhancement.

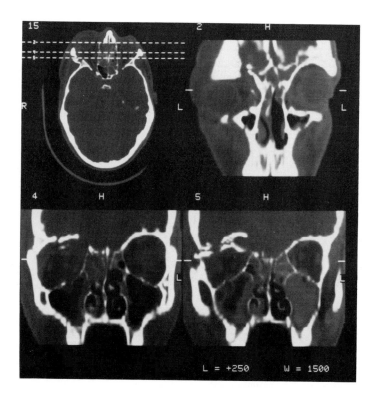

A

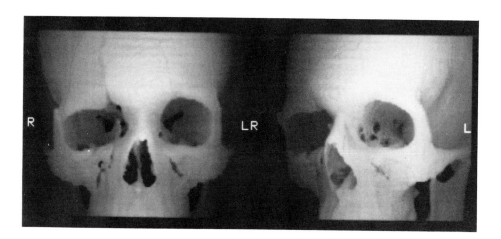

B

FIGURE 12. Extensive craniofacial trauma involving the frontal bone, both antra, right malar complex, and all four right orbital walls. (A) Representative CT coronal reformations. (B) Anterior and left lateral oblique 3D images. (C) antero-inferior and superior 3D images. Downward displacement of the frontal bone fragment into the right orbit has occurred. Note disruption of the right anterior cranial fossa in (C). (From Gillespie, J.E., I. Isherwood, G.R. Barker, and A.A. Quayle. *Clin. Radiol.* 38:525 (1987). With permission.)

3D reformatting in this protocol was performed retrospectively and its influence on clinical management could not be assessed. A further study is currently underway to determine the role of 3D CT in the detection of cerebral artery aneurysms and the planning of neurosurgical procedures.

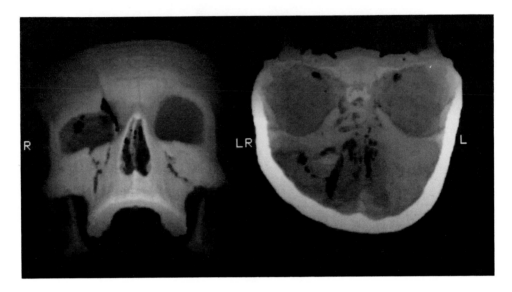

FIGURE 12C

TABLE 11
Clinical Material[16]

Pathology	No. of patients
Pituitary adenoma	5
Chordoma	2
Internal carotid aneurysm	1
Meningioma	1
Trigeminal neurofibroma	1

TABLE 12
Assessment of 3D in Intracranial Tumors[16]

Comparison 3D score	3D and CT			
	1	2	3	4
No. of examinations				
Tumor extent	12	1	0	0
Tumor relation to vessels	2	5	5	1
Bone destruction	2	5	3	3

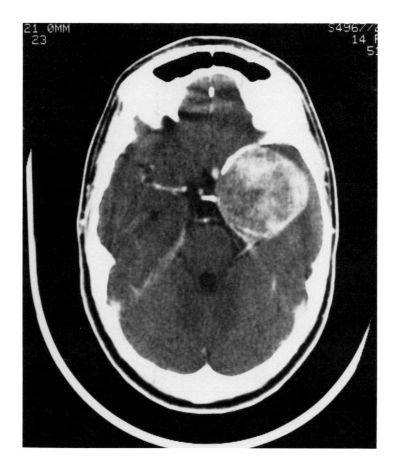

A

FIGURE 13. Large trigeminal neurofibroma in the left middle cranial fossa. (A) contrast-enhanced CT section. (B, C) Superior 3D views obtained on soft tissue and bone thresholds, respectively. Note the stretching and displacement of adjacent arteries by the tumor mass in (B) and the middle fossa floor defect and sellar erosion in (C). (From Lye, R.H., R.T. Ramsden, J.P. Stack, and J.E. Gillespie. *J. Neurosurg.* 67:125 (1987). With permission.)

VII. EXTRACRANIAL NEOPLASMS

Neoplastic lesions of the facial region have traditionally been radiologically investigated using plain radiographs and conventional tomography. CT has been shown to be superior to tomography in assessing disease extension into the orbit, infratemporal fossa, and cranial cavity[19] and is increasingly being used in treatment planning. In the following protocol, the application of 3D reformatting in this clinical area was investigated.

A. PROTOCOL DETAILS

Nine patients presenting with destructive mass lesions involving the extracranial portion of the skull were examined (Table 13). One patient with a chordoma was included in the protocol as the clinical presentation was that of a nasopharyngeal mass. Standard radiographs (plus

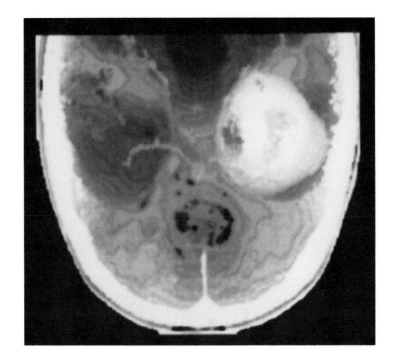

FIGURE 13B

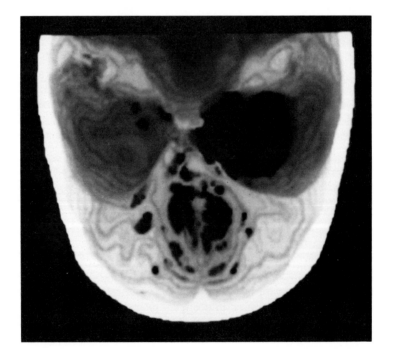

FIGURE 13C

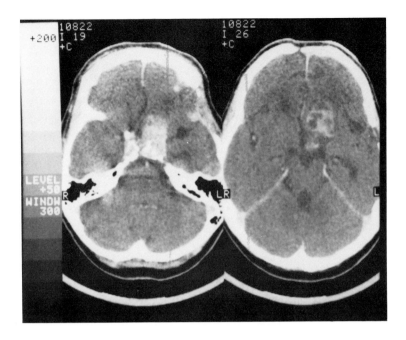

A

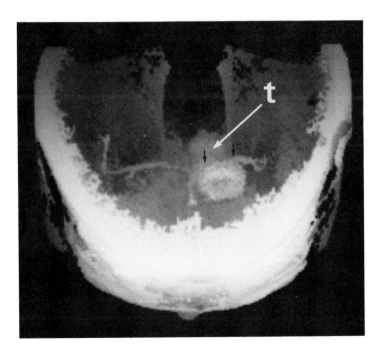

B

FIGURE 14.   Enhancing chordoma with superior cystic component arising from the left side of the clivus and sella. (A) Representative CT sections. (B) Anterosuperior and (C) Posterosuperior 3D views obtained on soft tissue threshold demonstrating the tumor (t) passing below and lateral to the left anterior cerebral artery (short arrows) which is stretched around the mass. (D) Superior 3D view obtained on bone threshold showing extensive destruction of the left sellar region involving the anterior clinoid, fossa floor, petrous apex, and clivus (small arrows). Vault defect from previous craniotomy is indicated (arrowhead). (From Gillespie, J.E., J.E. Adams, and I. Isherwood. *Neuroradiology* 29:33 (1987). With permission.)

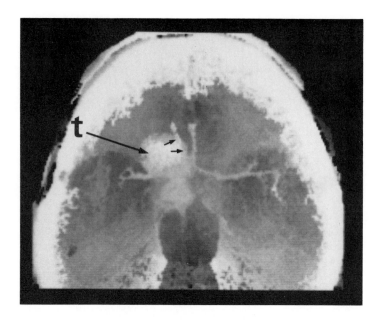

FIGURE 14C

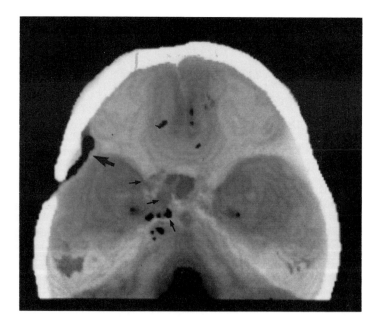

FIGURE 14D

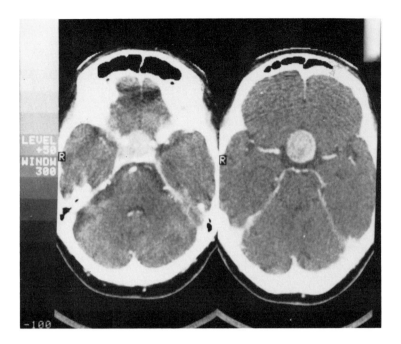

A

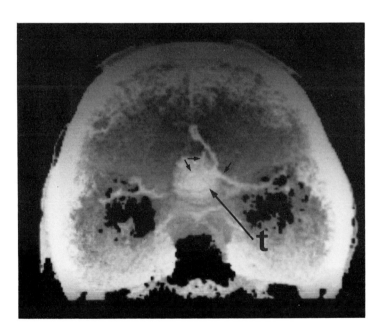

B

FIGURE 15. Pituitary macroadenoma with suprasellar extension. (A) Representative CT sections. (B) Posterosuperior 3D image obtained on soft tissue threshold. Both anterior cerebral arteries (short arrows) are elevated and deviated to the right by the tumour (t). (From Gillespie, J.E., J.E. Adams, and I. Isherwood. *Neuroradiology* 29:34 (1987). With permission.)

tomograms in six cases) were available for comparison in all patients. Scanning was undertaken on the 9800 CT using contiguous 3-mm sections through the relevant anatomical area.

It was our intention initially to obtain 3D reformations on both bone and soft tissue thresholds so that both the soft tissue mass and areas of bone destruction could be evaluated, as in the previous protocol. We were unable, however, to find a soft tissue threshold that would separate satisfactorily tumor from normal surrounding soft tissue as their respective attenuation values were similar, even after intravenous contrast enhancement. 3D evaluation was therefore confined to the assessment of bone destruction.

Image assessments were made by two radiologists. One radiologist evaluated the radiographs (including any tomograms) prior to CT examination. The CT examination and the 3D images were assessed separately by a second radiologist without access to the initial radiographs. All the images were then reviewed jointly and the 3D reformations given a score.

B. RESULTS (TABLE 14)
1. Bone Destruction
3D and radiography — 3D was rated superior in the three cases with intracranial tumor involvement. Two of these patients with nasopharyngeal carcinomas had fairly symmetrical erosion of skull base foramina which was difficult to identify on the tomograms with any confidence. In the third patient with a chordoma, the extent of clival destruction was much better seen on 3D. In the remaining six cases with destruction confined to the facial bones, the 3D reformations offered no overall advantage. Areas of bone destruction were often well shown on 3D (Figure 16) but "pseudoforamina" around the orbits and antra made the detection or exclusion of destruction at these sites difficult (Figure 17).

3D and CT — The scores given to the 3D images were virtually identical to those given above on a case-for-case basis. The presence of intracranial tumor extension from the nasopharynx, however, was always clearly evident on CT even though 3D demonstrated the precise pattern of bone erosion in a superior manner (Figure 18).

C. COMMENT
3D gave the best detailed view of the extent of the skull base erosion. This information did not, however, alter clinical management as CT always provided the essential views of soft

TABLE 13
Clinical Material

Pathology	No. of patients
Carcinoma of paranasal sinuses, nasopharynx, or mouth: 2 with intracranial spread	8
Chrodoma with extracranial spread	1

TABLE 14
Assessment of 3D in Extracranial Tumors

Comparison 3D score	3D and radiography				3D and CT			
	1	2	3	4	1	2	3	4
No. of examinations Bone destruction	3	3	0	3	3	3	1	2

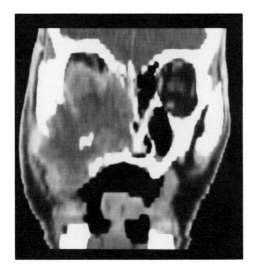

A

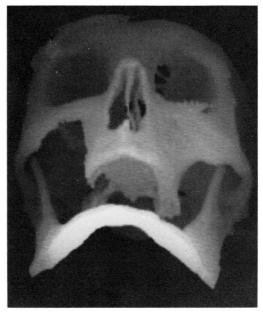

B

FIGURE 16.   Right antral carcinoma with extensive extra-sinus spread. (A) CT coronal reformation showing tumor spread into adjacent orbit, ethmoidal sinus, nasal cavity, and infratemporal fossa. (B) Antero-inferior 3D image demonstrating marked destruction of the right anterolateral antral wall and erosion of the hard palate.

tissue tumor extent. Consequently, 3D was not therefore considered worthwhile in this clinical area.

VIII. PELVIC FRACTURES

Developments in procedures for internal fixation and stabilization of acetabular fractures have resulted in improved long-term function of the hip joint.[20,21] Surgeons undertaking such procedures for complex pelvic fractures require precise characterization of the fractures. Extent and displacement of such fractures need to be assessed fully before contemplating reconstructive surgery. In the past, assessment of pelvic fractures was based on antero-posterior and oblique views of the pelvis. Harley et al.[22] compared conventional radiographs with CT in patients with pelvic fractures and demonstrated that CT was more sensitive in assessing fractures of sacrum, quadrilateral plate, acetabular roof, and posterior acetabular lip. Use of 3D imaging in this clinical situation has been described by Burk et al.[23] and in the present study, we objectively compared the demonstration of complex pelvic fractures using conventional radiographs, CT, and 3D imaging.

A. PROTOCOL DETAILS

Ten patients with pelvic fractures were examined in the supine position on a GE CT 9800. Contiguous 5-mm sections were obtained from the iliac crest to the level of the lesser trochanter, using the low-dose technique.[9] 3D images of the whole pelvis were obtained along the X and Z axes of rotation. In addition, the images of the affected hemipelvis were obtained using a 25 degree "tilt" (T) rotation in which the hemipelvis is tilted with respect to the XY plane and then rotated around the Z axis. The results were assessed under the following headings:

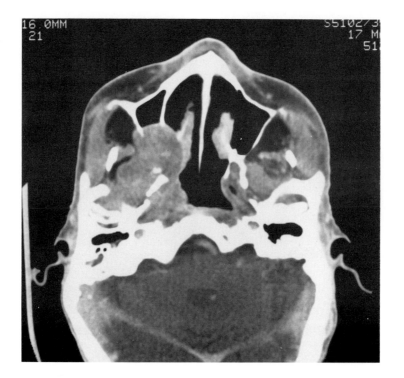

A

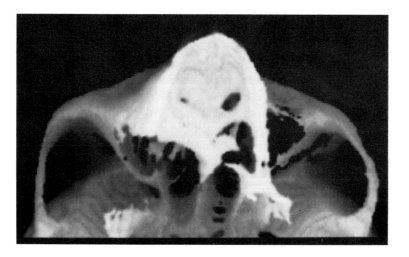

B

FIGURE 17. Unconfined right antral carcinoma. (A) Representative CT section at lower maxillary sinus level. Tumor is only present on the right. (B) Inferior 3D view showing malignant destruction of the lower right antrum and adjacent hard palate. The left antral defect is artefactual but could be mistaken for tumor erosion unless reference to the original CT examination is made.

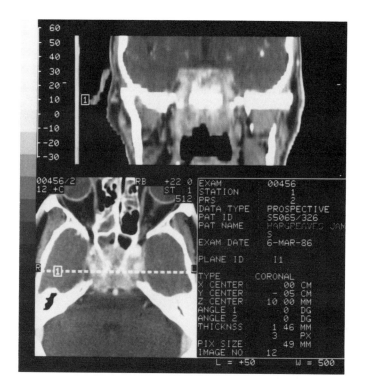

A

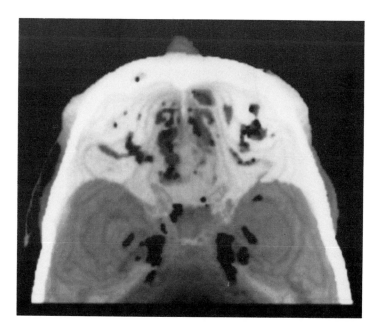

B

FIGURE 18. Nasopharyngeal carcinoma with intracranial spread. (A) Representative CT section and coronal reformation showing tumor extension into the parasellar region via skull base foramina. (B) Superior 3D view demonstrating bilateral erosion of basal foramina and petrous apices.

1. Fracture detection: Each hemipelvis was divided into 13 anatomical regions (anterior column of the acetabulum, posterior column of the acetabulum and acetabular roof, superior pubic ramus, SI joint, etc.) and presence or absence of fractures in each of these regions was assessed separately for all the imaging modalities.

2. Extent and displacement: This was assessed by reviewing all three imaging modalities.

B. RESULTS

Nine patients had fractures involving the acetabulum, while one patient had extensive iliac wing injury but the fracture did not extend to the acetabulum. In five patients the fractures involved both (anterior and posterior) columns of the acetabulum, while in four patients only one column injury was seen. There were also fractures involving the sacrum (two), iliac wing (six), and femur (three). Diastasis of the SI joint was present in seven patients, and similar changes were seen in the pubic symphysis in two patients (Table 15).

Assessment of fractures (Table 16) — The plain radiographs were inferior to the other two imaging techniques with 5 false-positive and 18 false-negative observations. CT was the best technique with no false-positive and only three false-negative observations. Although 3D was inferior to CT in identification of fractures, there were only five false-negatives and no false-positive observations. 3D images were poor in demonstrating intra-articular fragments and undisplaced fractures (Figure 19).

Extent and displacement of fractures — In only one patient who had a single column fracture was this equally shown on all imaging techniques. In most other patients, 3D was the best technique for demonstrating displacement of fractures and relationship of various fragments (Figure 20).

C. COMMENT

Although CT is the best technique for detecting pelvic fractures, it is necessary to evaluate and mentally integrate multiple sections to achieve an overall perspective of complex fracture patterns. In this study and that of Burk et al.,[23] 3D was found to be particularly useful in demonstrating extent, displacement, and degree of rotation of various fracture fragments. Fractures affecting the acetabular columns and quadrilateral plate were optimally demonstrated by 3D (Figure 21). We now routinely use 3D in preoperative assessment of complex pelvic fractures.

TABLE 15
Pelvic Fractures: Clinical Material

Type of injury	No. of patients
Anterior column of acetabulum	2
Posterior column of acetabulum	2
Both columns of acetabulum	5
Diastasis of SI joint	7
Diastasis of pubic symphysis	2

TABLE 16
Pelvic Fractures: Fracture Detection

Imaging modality	False positive	False negative
Plain radiography	5	18
CT	0	3
3D	0	5

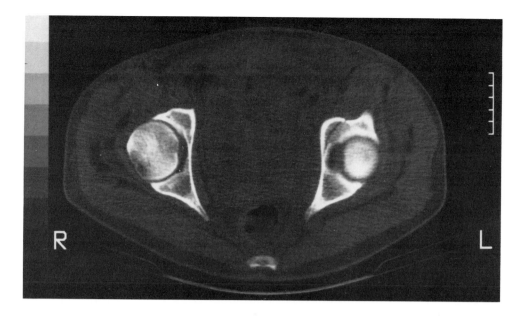

A

FIGURE 19. CT section (A) at the level of the hip joints shows a linear fracture of the anterior column of the acetabulum. This was not visible on either the plain radiograph (B) or 3D images (C and D). Anterior (C) and inferior (D) 3D views otherwise demonstrate the full extent of injury with diastasis of the left sacroiliac joint and pubic symphysis, subluxation of the left hip joint, and fractures of the left femoral head and posterior acetabular lip.

IX. GENERAL DISCUSSION

By definition, 3D CT reformations can only contain information already present in the CT sections from which they are derived. Nevertheless, data which exist but which are neither recognized nor properly understood are of limited value. In certain situations, 2D multiplanar reformations are known to aid interpretation of a CT examination by presenting axially acquired data in alternate planes. 3D reformations further the interpretative process by conveying selected volumetric data in an easily assimilated and familiar anatomical format, facilitating the accurate and rapid communication of complex spatial information to other viewers. While this information is not "new", its appreciation by an observer might be. 3D reformations, therefore, complement rather than replace the conventional CT presentation of anatomy and pathology. 3D cannot display internal architecture or most soft tissue structures satisfactorily at present. Conventional radiography is in many cases still the best starting point in the radiological investigation of disorders involving the skeletal system because of its simplicity, availability, and unsurpassed spatial resolution.

As anticipated, the perceived value of 3D does appear related, at least in part, to the familiarity of the observer with CT and his/her ability to build an accurate concept of 3D relationships by viewing multiple 2D images of whatever type. Our clinicians often found 3D more useful, in certain instances, than we as radiologists did, particularly for purely diagnostic purposes. However, prior to reconstructive facial surgery for congenital or severe traumatic facial deformities, where a surgeon requires a 3D "map" for planning purposes, a 3D image can be uniquely valuable. Furthermore, viewing of a 3D reconstruction can serve an initial orientation role which subsequently enhances understanding of an otherwise difficult CT examination.

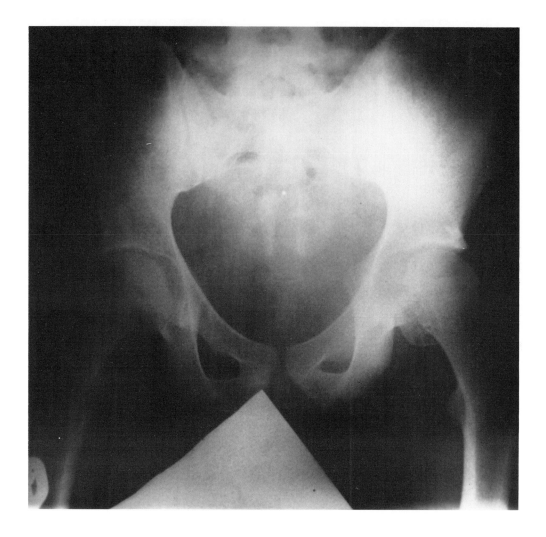

FIGURE 19B

The application of the 3D reformatting process to soft tissue densities presents difficulties. Intravenous contrast injection can overcome the problem intracranially and permit enhancing tumors and adjacent vasculature to be successfully displayed. Arterio-venous malformations may also be depicted in this manner (Figure 22).[24]

Movement artefact was a not infrequent cause of image degradation in the earlier part of this study, particularly when intravenous contrast enhancement was required. Use of dynamic "low-dose" scan techniques to shorten examination times greatly reduces the opportunity for movement to occur during scanning, while at the same time substantially lowering the radiation dose to the patient. The significant advantages of this method suggest that it should be used routinely in most circumstances when 3D reformatting is required.

In conclusion, our experience indicates that 3D imaging does have a useful role to play in a number of specific clinical situations when used in conjunction with CT and other radiological imaging methods.

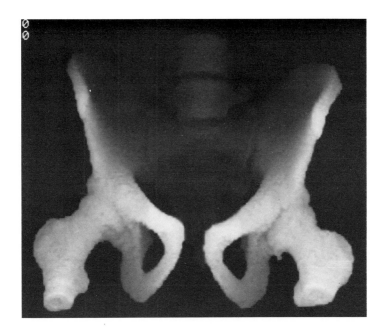

FIGURE 19C

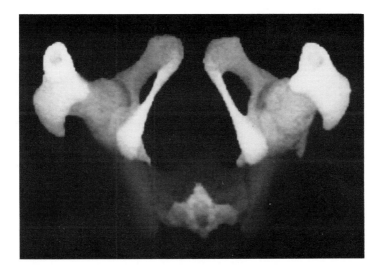

FIGURE 19D

ACKNOWLEDGMENTS

We would like to thank Dr. J. E. Adams, Dr. R. J. Johnson, and Miss C. W. Hart, Department of Diagnostic Radiology, University of Manchester, and Mr. D. Mott, Regional Department of Medical Physics, Christie Hospital, Manchester, for their help and support during this study. We also wish to thank Miss C. Bannister, Mr. J. Lendrum, and Mr. J. Curphey, Departments of Neurosurgery and Plastic Surgery, Booth Hall Childrens' Hospital, Manchester, and Mr. G. Barker and Mr. A. Quayle, Department of Oral and Maxillofacial Surgery, University of Manchester, for their clinical support, Mrs. B. Bates for typing the manuscript, and Mr. D. Ellard for preparing the illustrations.

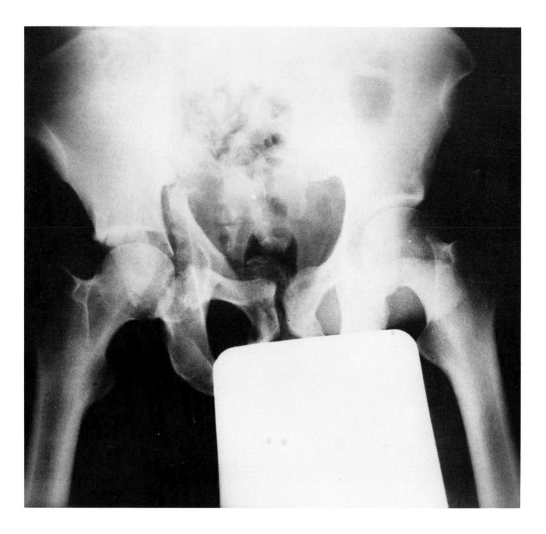

A

FIGURE 20.   Plain radiograph (A) demonstrates a complex right acetabular fracture. CT (B and C) demonstrates fractures of the acetabulum and sacrum (arrow) as well as diastasis of the left sacroiliac joint. Anterolateral (D), posterolateral (E), and medial (F) 3D views showed all fractures and was the best technique for demonstrating the degree of rotation and displacement of the fracture fragments.

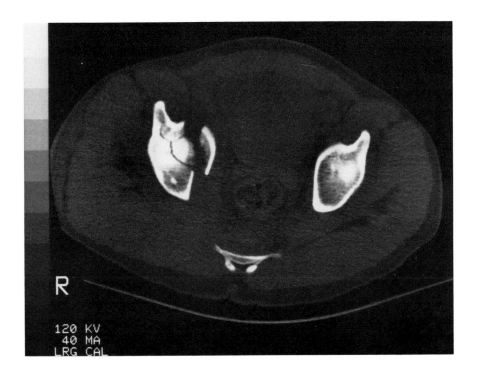

FIGURE 20B

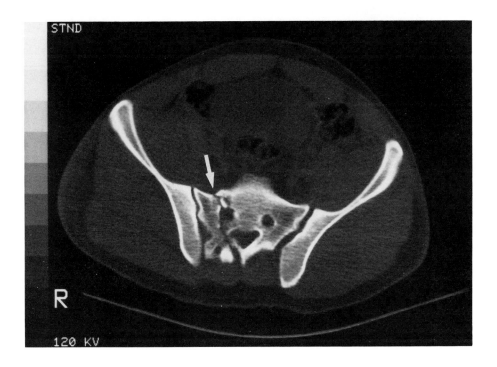

FIGURE 20C

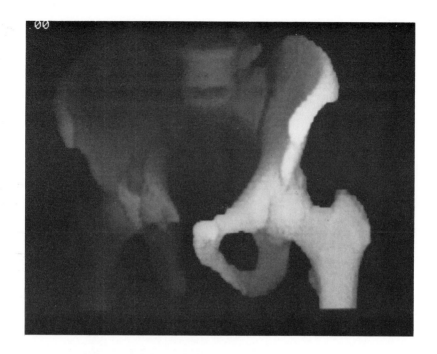

FIGURE 20D

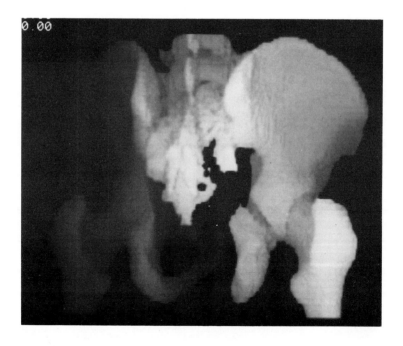

FIGURE 20E

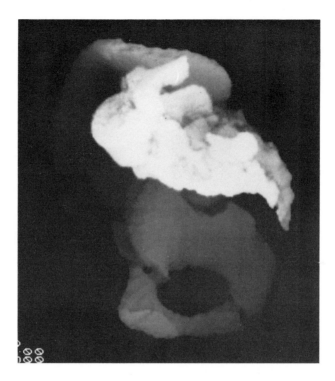

FIGURE 20F

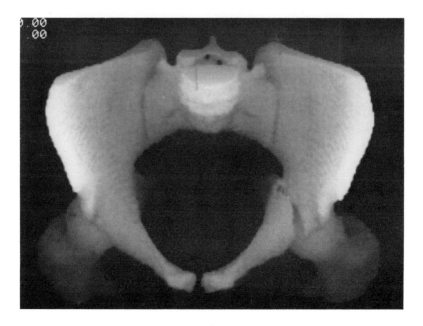

A

FIGURE 21. Anterosuperior (A), superior (B), and medial (C) 3D views showing a "T"-shaped fracture of the left acetabulum.

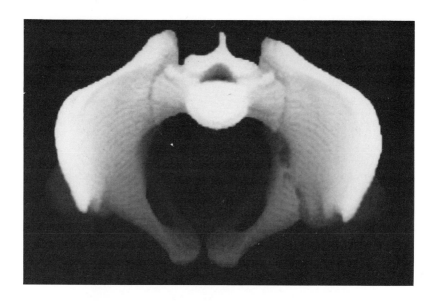

FIGURE 21B

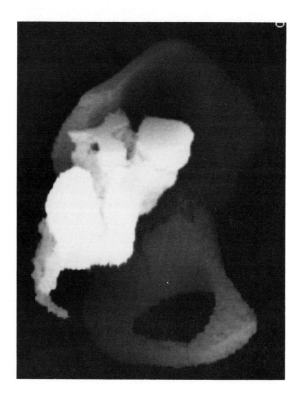

FIGURE 21C

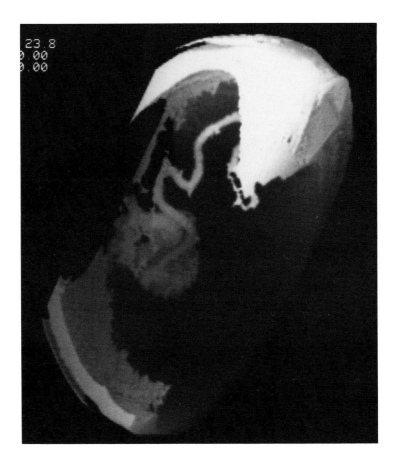

FIGURE 22. "Half-head tilt" 3D view of a left parietal arteriovenous malformation.

REFERENCES

1. Chen, L.S., G.T. Herman, C.R. Meyer, R.A. Reynolds, and J.K. Udupa. 3D83 — An easy-to-use software package for three-dimensional display from computed tomograms. *Proceedings of the IEEE Computer Society International Symposium on Medical Images and Icons* (Arlington, VA: 1984), pp. 309–316.

2. Hemmy, D.C., D.J. David, and G.T. Herman. Three-dimensional reconstruction of craniofacial deformity using computed tomography, *Neurosurgery* 13:534–541 (1983).

3. Vannier, M.W., J.L. Marsh, and J.O. Warren. Three-dimensional CT reconstruction images for craniofacial surgical planning and evaluation, *Radiology* 150:179–184 (1984).

4. Gillespie, J.E., and I. Isherwood. Three-dimensional anatomical images from computed tomographic scans, *Br. J. Radiol.* 59:289–292 (1986).

5. Udupa, J.K., G.T. Herman, P.S. Margasahayam, L.S. Chen, and C.R. Meyer. 3D98: A turnkey system for the display and analysis of 3D medical objects, *SPIE Proc.* 671:154–168 (1986).

6. De La Paz, R., M. Brant-Zawadzki, and L.D. Rowe. CT of maxillofacial injury, in *Computed Tomography in the Evaluation of Trauma*, M.P. Federle and M. Brant-Zawadzki (Eds.) (Baltimore, MD: Williams & Wilkins, 1986), pp. 64–107.

7. Brown, B.M., M. Brant-Zawadzki, and C.E. Cann. Dynamic CT scanning of spinal column trauma, *AJR* 139:1177–1181 (1982).

8. Andre, M.P., R.A. Horn, D. Bielecki, P. Dev, and D. Resnick. Patient dose considerations for three-dimensional CT displays, *Radiology* 157:177 (1985).

9. Gholkar, A., J.E. Gillespie, C.W. Hart, D. Mott, and I. Isherwood. Dynamic low-dose three-dimensional computed tomography: A preliminary study, *Br. J. Radiol.* 61:1095–1099 (1988).

10. Marsh, J.L., and M. Gado. Surgical anatomy of the craniofacial dysostoses: Insight from CT scans, *Cleft Palate J.* 19:212–221 (1982).

11. Rabey, G.P. Current principles of morphanalysis and their implications in oral surgery, *Br. J. Oral Maxillofac. Surg.* 15:97–109 (1977).

12. Bannister, C., J. Lendrum, J. Gillespie, and I. Isherwood. Three-dimensional computed tomographic scans in the planning of procedures for reconstructive craniofacial surgery, *Neurol. Res.* 9:236–240 (1987).

13. Zilkha, A. Computed tomography in facial trauma, *Radiology* 144:545–548 (1982).

14. Gillespie, J.E., I. Isherwood, G.R. Barker, and A.A. Quayle. Three-dimensional reformation of computed tomography in the assessment of facial trauma, *Clin. Radiol.* 38:523–526 (1987).

15. Vannier, M.W., M.H. Gado, and J.L. Marsh. Three-dimensional display of intracranial soft tissue structures, *AJNR* 4:520–521 (1983).

16. Gillespie, J.E., J.E. Adams, and I. Isherwood. Three-dimensional computed tomographic reformations of sellar and parasellar lesions, *Neuroradiology* 29:30–35 (1987).

17. Lye, R.H., R.T. Ramsden, J.P. Stack, and J.E. Gillespie. Trigeminal nerve tumour: Comparison of CT and MRI, *J. Neurosurg.* 67:124–127 (1987).

18. Dawson, P., C. Heron, and J. Marshall. Intravenous urography with low-osmolarity contrast agents: Theoretical considerations and clinical findings, *Clin. Radiol.* 35:173–175 (1984).

19. Eddleston, B., and R.J. Johnson. A comparison of conventional radiographic imaging and computed tomography in malignant disease of the paranasal sinuses and the post-nasal space, *Clin. Radiol.* 34:161–172 (1983).

20. Pennal, G.F., J. Davidson, H. Garside, and J. Plewes. Results of treatment of acetabular fractures, *Clin. Orthop.* 151:115–123 (1980).

21. Tile, M. Fractures of the acetabulum, *Orthop. Clin. North Am.* 11:481–506 (1980).

22. Harley, J.D., L.A. Mack, and R.A. Winquist. CT of acetabular fractures: Comparison with conventional radiography, *AJR* 138:413–417 (1982).

23. Burk, D.L., D.C. Mears, W.H. Kennedy, L.A. Cooperstein, and D.L. Herbert. Three-dimensional computed tomography of acetabular fractures, *Radiology* 155:183–186 (1985).

24. Gholkar, A., and I. Isherwood. Three-dimensional computed tomographic reformations of intracranial vascular lesions, *Br. J. Radiol.* 61:258–261 (1988).

Chapter 5

QUANTITATION USING 3D IMAGES

Gabor T. Herman

TABLE OF CONTENTS

I. INTRODUCTION

Medical three-dimensional (3D) imaging produces "pretty pictures" which, as discussed in the other chapters, can be clinically useful. However, in addition to improved qualitative assessment, 3D imaging provides us with the potential of more accurate quantitation than can be done using the original slice images. In this chapter we discuss available methodologies for quantitation using the 3D surface imaging approach. Specifically, we describe (1) how the location in the original slices of a point on the displayed 3D surface is identified, (2) how the (3D) distance between two such points is measured, and (3) how a volume enclosed by such a surface is calculated. We follow this by a report on some results that have been obtained regarding the reproducibility and accuracy of these methods, especially in the areas of CT-based osseous landmark identification and MRI-based volumetric analysis of the brain. We complete the chapter with a brief discussion of our conclusions.

This is not a survey article; there is a heavy concentration on the approaches taken by the author and his closest colleagues. We do not mean to imply that these are superior to other approaches; they are simply the ones with which we are the most familiar.

II. METHODS

A. 3D SURFACE IMAGE GENERATION

Although this topic is treated in other chapters, to make the rest of this chapter easily understandable, we give here a brief description of the "MIPG perspective" to 3D surface image generation.[1]

The input to a 3D surface image-generating program is a set of slice images. These provide us with one or more values associated with a large number of volume elements (voxels) of known size and location. (For example, it is quite common in MRI to assign more than one value to each voxel by the use of multiple imaging sequences.) The collection of voxels together with their values is referred as a *scene*.

These scenes are often "preprocessed" before going on to the next stage of the program. Typical preprocessing steps are the following. (1) Interpolation: One often uses the original data to estimate what the data would look like if voxels were smaller and/or more densely spaced. Such a preprocessing step is often performed to estimate data for cube-shaped voxels from the typically not cube-shaped voxels of a CT or MRI scanner. (2) Subregioning: This is used to reduce the size of the data set by indicating which region of the original data space contains the object of interest. It can also be used to cut things open, by indicating a subregion which cuts right through the object (3) Derived-scene formation: Here one assigns to each voxel one or more new values based on the previous values. In MRI, the new values may be pure T1, T2 and proton density values calculated from the values obtained by the imaging technique. A generally useful preprocessing step of this kind assigns to each voxel a "gradient" (a measure of the rate of change of the value in the neighborhood of the voxel) based on the prior values of that voxel and its neighbors. Often, the derived scene is a binary one (0–1 valued), obtained, for example, by thresholding the prior values of the voxels.

We illustrate these ideas by an example which is quite relevant to the accuracy of volume quantitation. Suppose we have CT slices in which voxels are 1 mm × 1 mm × 2 mm. Suppose that a CT value of 200 and above is considered indicative of bone. So if a voxel has CT value of 200 or more assigned to it, one can say that it is at least partially occupied by bone. Now suppose that we wish to estimate the total bone volume in the CT slices. One method is to threshold the values at 200, producing a binary scene in which we count the number n of 1-valued voxels and to estimate the bone volume to be 2n mm^3 (since each voxel is 2 mm^3). Another method is to estimate what a CT scanner would do when the voxel size is 1 mm × 1 mm × 1 mm. We can use linear interpolation to estimate CT values for new slices (lying half-way between the old slices) by assigning to each new voxel the average of the two voxel values above and below it. We can then threshold this interpolated scene to obtain a binary scene (of cube-shaped voxels) and count the number of 1-valued voxels as the estimate of bone volume in cubic millimeters. That this sort of approach may indeed improve accuracy has been dramatically visually illustrated by the side-by-side display of binary scenes, one derived from the uninterpolated scene and the other derived from the interpolated scene.[2] A third approach is the so-called "shape-based interpolation".[3] Here we threshold the voxels just as done in the first approach. Then we perform an interesting derived-scene formation. We assign a new value 0.5 to every voxel which is 1-valued and shares a face in the slice with a voxel that is 0-valued and –0.5 to every voxel which is 0-valued and shares a face in the slice with a voxel which is 1-valued. Then we assign (iteratively) values to voxels which have not been previously assigned a value: if they share a face in the slice with a voxel of positive value, they are given a value 1 greater; if they share a face in the slice with a voxel of negative value, they are given a value 1 less. We repeat this process until every voxel has been assigned a new value. The new values indicate distances from the nearest edge (positive for "inside", negative for "outside"). Now we linearly interpolate this derived scene, just as in the second method, to produce a scene with cubed-shaped voxels. Finally, we threshold this interpolated scene at value 0 and count the number of 1-valued voxels to estimate bone volume in cubic millimeters. It has been demonstrated that for complex biological objects this approach is even more accurate than the second approach.[3]

In the MIPG perspective, it is assumed that at the end of the preprocessing stage, the region of the scene is "box-shaped" (i.e., it is a rectangular parallelepiped) and it is subdivided into equal-sized abutting box-shaped voxels, each with the same number of values assigned to it. The next stage in this approach is "surface formation".

We think of a "surface" as a collection of voxel faces which unambiguously divides the region of the scene into two parts (an "inside" and an "outside") such that one cannot get from one part to the other without crossing the surface. In the terminology of geometry, a surface has to be closed and non-selfintersecting (i.e., for each edge of each voxel face in the surface there is a unique "adjacent" face in the surface sharing that edge).

Surface formation is particularly simple if the scene is binary. In such a case, a "surface element" would be defined as a face which separates a 1-valued voxel from a 0-valued voxel. It has been mathematically proven that for any binary scene and for any such surface element, there is a unique surface (consisting of surface elements) containing the given surface element.[4] Furthermore, very efficient computer algorithms have been designed to find such surfaces.[5]

However, it is important to realize that computerized surface formation may be performed on scenes which are not binary. For example, Liu[6] devised a method which operates on the original or on the linearly interpolated scene. The final surface produced by Liu's method is determined globally (rather than locally), i.e., for any voxel face it is impossible to tell until the computer program completes its work whether or not that face will be part of the surface. This is because Liu's program can "change its mind": if it finds that its current guess at a partial surface cannot be completed in a geometrically consistent manner (so that it becomes closed and non-selfintersecting), then it backtracks and tries again.

However, whichever approach is used, at the end of the surface-formation phase we have at hand one or more surfaces, each described as a set of voxel faces.

The final stage of 3D surface image generation in the MIPG perspective is the display of the surface(s). While our programs are general enough to display surfaces at arbitrary orientations with respect to the display screen, in this discussion we restrict our attention to a special approach.

Consider Figure 1. The display screen is assumed to be parallel to the XY plane. We think of the object as lying "behind" the display screen, i.e., in the positive Z direction. We express

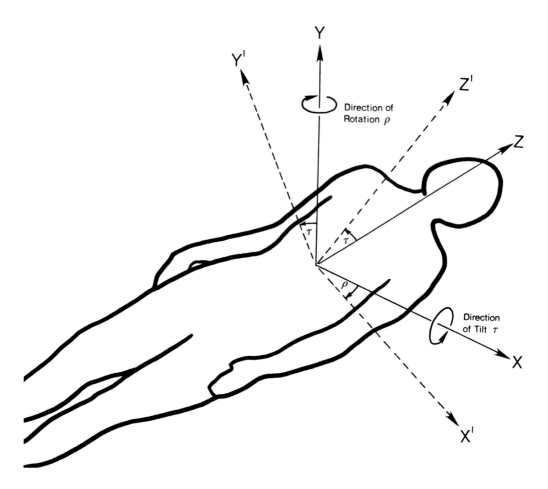

FIGURE 1. The definition of "tilt" and "rotation".

the permissible orientations of the surface (relative to the screen) by a tilt followed by a rotation. As indicated in Figure 1, a tilt is a turning of the scene around the X-axis, and a rotation is a turning of the resulting tilted scene around the Y-axis.

In displaying a surface on the display screen, we use the so-called orthographic projection. This can be thought of as follows: for each point d on the display screen, cast a line in the positive Z direction (i.e., perpendicular to the screen). The first point s on the surface that is met by this line is going to be the one displayed on the screen at point d.

The details of how the display is created are discussed in Chapter 1. Here we restrict ourselves to one observation which will be important later on. For any fixed point s on the surface and for any fixed tilt of the object, the Y-coordinate of the projection d of s on the screen is the same for all rotations. This observation has been successfully used for rapidly obtaining a sequence of displays of a surface as it is rotating (after a single tilt).[7] In what follows we refer to such a sequence of images as a "movie".

B. POINT LOCATION

In the first half of this subsection we discuss an approach taken by a specific software package for 3D display and analysis, namely the package 3D98 which runs on the GE 9800 CT scanners.[7] However, the principles that are described are easily transportable.

When 3D98 generates a movie, it stores the images in that movie in the same way as the original CT slices are stored. Hence these images can be retrieved and displayed by the standard software of the CT scanner. In particular, one can superimpose a cursor on the image of the surface. A natural question is: where in the original slices is the point s on the surface which is displayed at the position d where the cursor is on the screen. For example, if a bone surface is being displayed, we might wish to investigate the appearance of soft tissue in the neighborhood of specific points on the surface. Such information is indeed provided by 3D98; for any image in the movie and for any cursor location d on the screen, the user may request the slice location and the pixel location within the slice of the point s in the surface which projects onto d. Furthermore, the information is delivered practically instantaneously. We now explain how this is done.

The essence of the idea is to make the problem manageable by subdividing the display screen into horizontal strips.[8] Since the tilt in a movie is fixed, we can associate with each strip D a subset S of faces in the surface(s) in such a way that we know that, for any rotation, the surface point s displayed at a point d in the strip D must be on a face which belongs to S. If the strips are narrow, then the subsets associated with the strips will contain a small number of faces (relative to the total number of faces in the surface), due to the property of displayed rotations explained in the last subsection. These subsets are worked out once (at the time of the movie generation) and are stored in the computer of the scanner together with the movie. When the user puts a cursor over an image from the movie and requests the location in the original slices of the corresponding surface point, the computer first identifies the strip D to which the cursor location d belongs and then searches through the associated set of faces S to find the scan coordinates of the face which is visible at d for the given rotation. Since S is relatively small, this is practically instantaneous.

Clearly, such computerized determination of point locations is quite convenient to use. However, even if one has just photographic images from a movie, it is still possible to determine the scan coordinates of surface points. For example, A. H. Abbott[9] created movies with 9° incremental rotations (using the software system 3D83[10]) and then used consecutive images in these movies as stereo pairs to determine the 3D locations of landmarks.

C. DISTANCE CALCULATION

Once the 3D locations of each of a pair of surface points have been identified, the calculation of the (3D) distance between them is quite trivial. Such a capability is provided by many 3D display packages, including 3D83[10] and 3D98.[7]

A more interesting task is the calculation of a curvilinear distance, i.e., the distance along that path on the 3D surface which corresponds to a curve drawn on the screen displaying the surface. Methods for doing this have been reported,[8] but there does not seem to be any reported validation studies, and so we shall say no more about them here.

D. VOLUME CALCULATION

A very desirable byproduct of the surface-display approach to 3D imaging is that the volumes enclosed by the surfaces are obtained essentially free as a result of surface detection.[11] The reason for this is the following.

The total volume in an object is the sum of the volumes which the object occupies in individual vertical columns of voxels. Now consider a single column of voxels. The intersection of the object with that column can be partitioned into separate segments of contiguous voxels occupied by the object. The volume which the object occupies in the column is the sum of the volumes of these segments. Each segment is bordered by two faces in the surface of the object, with one face pointing up and the other pointing down. Assuming a coordinate system in which up is positive, the volume of the segment is the product of the cross-sectional area of the column ("pixel size") with the difference in the coordinate locations of the up-pointing face and of the down-pointing face at the two ends of the segment. Hence, to calculate the complete volume enclosed by a surface while detecting the surface we can do the following. We set an accumulator to be initially zero-valued. During surface detection, every time a new up-pointing face is added to the surface, we add its coordinate location to the accumulator and every time a new down-pointing face is added to the surface we subtract its coordinate location from the accumulator. When surface detection is completed, we multiply the value of the accumulator with the pixel size to get the total volume enclosed by the surface.

An alternative to volume calculation via surface detection is to identify all the voxels which are contained in the object of interest and then multiply the number of such voxels with the voxel volume. Increased accuracy can be achieved by using smaller voxels (recall the discussion on bone volume determination in the previous section) or by estimating partial occupancy of voxels by the object.

III. RESULTS

To give a flavor of applicability, reproducibility, and accuracy of these quantitative approaches, we now discuss two different areas with which we had personal experience: CT-based osseous landmark identification and MRI-based 3D volumetric analysis of the brain.

A. OSSEOUS LANDMARK IDENTIFICATION BASED ON CT

There are a number of reasons why it is desirable to identify the 3D location of osseous landmarks; here we give two of them.

First, the use of standard sets of landmarks opens up the possibility of a mathematical study of craniofacial shape, shape deformity, and shape comparison. As pointed in Chapter 6, having a normative data set of such landmarks allows one to quantitatively analyze deformity. The same approach can be used for growth evaluation[12] and comparison of anthropologic specimens.[9] For these application areas, it is important that researchers use the same set of landmarks.

The second reason is based on the desire to register images of the same patient obtained at different times. In Chapter 10, there is a discussion of the desirability of registering MR, CT, and PET images of the same patient. Our involvement in this area has a different motivation; we wish to carry out a longitudinal study of the postoperative volume changes of craniofacial onlay bone grafts. Due to fusion of the bone graft to the native bone, it is not possible to accurately identify in a follow-up CT study where the native bone ends and the bone graft starts. However, assuming that the native bone is stable, taking a difference of the bony structures based on CT studies made at different times should provide us with an estimate of volume change due to bone grafts. In order to do this we need to register the two CT studies. We do this by the use of landmarks on the 3D displays of the two bone surfaces obtained from them.[13] In this application the landmarks need not be standard. The user displays the movies made from the two CT studies side-by-side and can point at any landmarks which are clearly identifiable in both 3D images.

In the rest of this subsection we discuss experiments that have been carried out to see how well we can identify osseous landmarks based on 3D surface images.

A. H. Abbott[9] carried out a most comprehensive evaluation for her stereo method of landmark identification (discussed in the last section). Five skulls were scanned on the GE 8800 CT scanner, with pixel size 0.8 mm × 0.8 mm, slice thickness 5 mm, and slice spacing (table shift) 3 mm. The package 3D83[10] was used to produce the surface images. In particular, this package performs an interpolation (prior to thresholding for bone) producing estimated values for abutting cubic voxels of size 0.8 mm × 0.8 mm × 0.8 mm. For each of the five skulls, Abbott determined the location of 76 standard landmarks, using her stereo method twice with at least a month in-between. To measure the *reproducibility* of her method she used the figure-of-merit[14]

$$S = \left(\frac{1}{2n} \sum_{i=1}^{n} |x_i|^2 \right)^{1/2}$$

where n is the number of specimens (in this case five) and $|x_i|$ is the 3D distance between the two estimated locations of a particular landmark on the i'th skull. She found that, for the 76 landmarks, S varied between 0.4 mm and 5.2 mm, with a median of 1.7 mm, which is quite impressive considering the 5-mm slice thickness and 3-mm slice spacing. (Another study, involving three patients, yielded similar results.) A study of the accuracy of these landmark identifications was carried out by comparing, for 31 pairs of landmarks, the distances estimated using the 3D approach with those obtained by craniometric measurements using a caliper. The mean difference ranged from 0.2 mm (with a standard deviation of 0.2 mm) to 9.3 mm (with a standard deviation of 1.5 mm). It was observed that while some of the differences were due to inherent limitations of the CT-based approach (e.g., thin bone projections are unlikely to fully image with 5-mm thick slices), all the significant differences were due to changes in the perceived definition of the landmarks during the two measurement processes ("the same landmarks were not being measured although referred to by the same name"). Eliminating such noninherent differences, Abbott found that for the remaining 20 pairs of distances, the mean difference between the two measurement processes was always less than 2.7 mm with a median value of 0.8 mm. She concluded that "three dimensional CT landmark coordinates derived using the multiple stereo imaging technique are consistent with craniometric measurement, provided that the same landmark definitions can be followed for the two measurement techniques."

In order to eliminate the difficulty in landmark definition and to make the use of landmark identification easier for the nonspecialist (such as an X-ray technician) we have attempted to define a number of landmarks (24 to be precise) by unambiguous operational directives.[15] By this we mean the following. First of all, a strict protocol is given which describes precisely the head stabilization and positioning, the scanning technique, and the 3D imaging technique including the subregionings to be used and the number of images in the movies.[16] The following are the directives for the identification of 3 of the 24 landmarks. To identify the left condylon, use the fourth image of the movie of the inferior-anterior subregion of the skull. Find the brightest local point on the lateral part of the condylon head (see Figure 2). To identify the nasale, use the first image of the movie of the inferior-anterior subregion of the skull. Find the brightest point in the middle of the nose at its bottom (see Figure 2). To identify the opisthocranion, use the seventh image in the movie of the superior-posterior subregion of the skull. Narrow the display window to produce an elliptical region on the back of the skull, and find its center (see Figure 4).

To validate the approach, the same skull was scanned twice (with 3 months in-between) on a GE 8800 according to the protocol mentioned in the previous paragraph. (In particular, pixel size was 0.75 mm, slice thickness and slice spacing were both 1.5 mm.) The movies of 3D surface images were generated according to the protocol (using 3D83[10]) and were provided to each of two observers. Each observer had two sessions with each CT study (with at least 2 weeks in-between) in which all of the 24 landmarks had to be identified according to the operational directives. Table 1 shows the intra-observer and inter-observer reproducibility for the three landmarks of Figures 2 to 4.

These results are quite typical. The median of the intra-observer differences in the whole study was 0.75 mm, which is the size of one pixel side and half the size of the slice thickness and slice spacing. This is very good. However, the approach clearly fails to overcome inter-observer variability; apparently the directives are not as unambiguous as we thought.

Our experiment also allows us to comment on the inter-study variability of our approach. This is because we can calculate distances between pairs of landmarks; these distances should be the same for the two CT studies. We have found examples in which, even though inter-observer variability of the estimated distances was negligible, inter-study variability was much more serious. Our worst example is the distance between the right condylon and the right gonion, which based on Study 1 was measured to be 34.1 mm by both observers and based on Study 2 was measured to be 36.9 mm by one and 37.4 mm by the other observer. Such a 3-mm difference between studies may not be acceptable for the applications we have in mind.

In fact, the mean of the differences between the distances we calculated based on the two studies was 1.6 mm. To put this into context, we used the methodology described near the

TABLE 1
Intra-Observer and Inter-Observer Variability of
Landmarks Defined by Operational Directives

	Study i			Study ii		
	O1	O2	IO	O1	O2	IO
Left condylion	0.0	0.0	0.8	0.8	2.4	1.1
Nasale	0.0	0.0	2.4	1.1	0.0	4.1
Opisthocranion	0.0	0.8	4.2	1.7	0.8	2.0

Note: Entries are in mm. Under O1 (resp. O2) we record the distance between the locations identified as the landmarks in two readings by Observer 1 (resp. 2). Under IO we record the distance between the averages of the two locations identified by each of the observers.

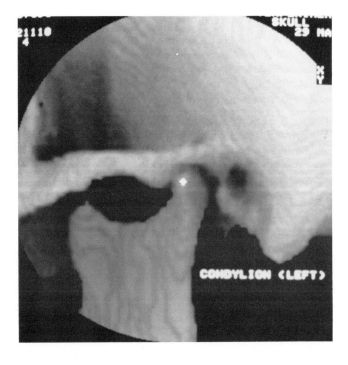

FIGURE 2. The location of the left condylion.

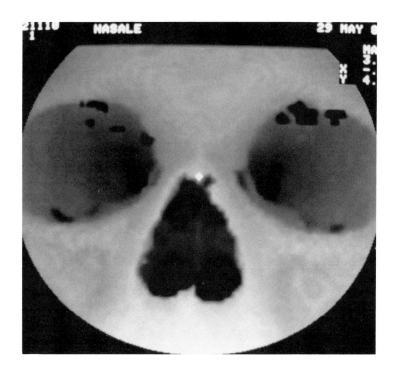

FIGURE 3. The location of the nasale.

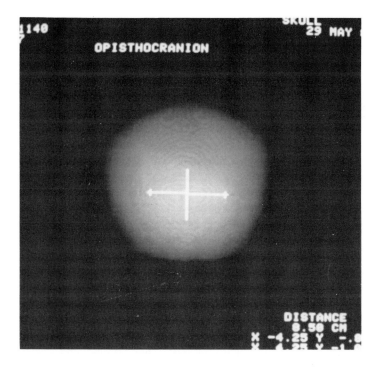

FIGURE 4. The location of the opisthocranion.

beginning of this subsection of displaying the movies made from the two CT studies side-by-side and determining "landmarks" in an ad-hoc way, just because they appear to match. The mean of the differences of distances between these landmarks based on the two studies was less than 0.8 mm. That is about the size of a pixel side and is twice as good as what we obtained based on our operational directives.

It does appear, therefore, that our operational-directive approach, while resulting in excellent intra-observer reproducibility (the size of a pixel side on average), still allows unacceptably large variations between observers (sometimes as much as 4 mm or more). Furthermore, its inter-study reproducibility is not as good as what can be achieved by quite ad-hoc methods. It seems that the very restrictive image-oriented nature of our directives ("find the brightest point" on a certain image) makes them sensitive to minor variations in head positioning and the conciseness of the directives makes them not as unambiguous as is desirable.

We judge the results reported here to be rather encouraging. It appears that even if slice thickness and spacing is twice (or more) the size of a pixel side, variations in estimated landmark locations and inter-landmark distances can be kept down to the size of a pixel side on average and that the same is true regarding accuracy of the distances. However, in order to achieve this level of reproducibility and accuracy for a standard set of landmarks, they and the procedures for identifying them will have to be carefully defined and validated.

B. 3D VOLUME ANALYSIS OF THE BRAIN BASED ON MRI

Our research group got involved in MRI-based regional brain volume analysis as a result of our collaboration with researchers in neuropsychiatry. To indicate our motivation in this area, we quote from Herman et al.[17]

> "More specifically, we conjecture the following. (i) Specific patterns of atrophy, cerebral blood flow and metabolism can help to distinguish the various sub-types of dementia as well as predict the clinical course and prognosis. (ii) Anatomic differences exist between rapid and slow declining Alzheimer's patients. (iii) Three-dimensional analysis can help to identify the earliest changes of Alzheimer's disease in the temporal lobes. (iv) Detection of subtle regional volume alterations of the temporal lobe can help the diagnosis and therapy of unremitting temporal lobe seizures. (v) Sub-classifying clinical variants of schizophrenia requires identification of regional dysfunction. (vi) Clinically valid regional analysis of positron emission tomography images requires correction for spaces occupied by cerebro-spinal fluid, and such correction can be achieved based on appropriately registered and segmented magnetic resonance images."

More recently we have been looking into additional hypotheses regarding various diseases of the brain. We list three examples. (1) In babies with cerebral nervous system injury there will be an increase in ratio of unmyelinated and myelinated white matter for age, with regions of greatest damage showing the greatest increase. (2) The ratio of tumor volume which enhances as a result of intravenous Magnevist to the total tumor volume is directly correlated with survival in gliomas. (3) The ratio of brain volume in multiple sclerosis (MS) to the total brain volume corresponds to clinically defined subgroups of MS.

In order to be able to investigate such hypotheses, we are working on two computer packages for volumetric measurement of brain tissue based on MRI. One package, developed by M. I. Kohn,[17,18] requires a certain amount of human decision making. It is an interactive tool which utilizes the window and graphics techniques available on our Sun workstations. It has been popular with our colleagues in psychiatry, nuclear medicine, and neuroradiology. The other package is intended to be totally automatic. It is currently under development by S. P. Raya.[19] Since only the first of these packages is currently available for routine clinical use, we restrict our attention to roughly describing it, as it can be applied to the last of the hypotheses listed above.

What we need is an estimate of the total brain volume and of the volume involved in MS.

In our approach we attempt to obtain this information from an MRI study which for each voxel provides us with two values, commonly referred to as PD (for proton density) and T2W (for T2-weighted).[18] As a first step we perform a subregioning to restrict our attention to the intracranial content. This we do based on the observation that the gradient (rate-of-change) in T2W is particularly large at the interior surface of the skull and hence we can apply a gradient-based surface detector.[6] From this point on we only deal with voxels inside this surface. The total volume inside the surface is occupied by three types of tissue: normal brain (NB), multiple sclerotic lesion (MS), and cerebrospinal fluid (CSF). We attempt to classify each voxel as being associated with one of these categories. (Partial occupancy is taken care of later on by the use of shape-based interpolation.)

The classification is based on a two-feature scatter plot.[20] This is a function defined over a two-dimensional space whose coordinates correspond to the (PD, T2W) values. The idea is that every voxel has a PD and a T2W value and so corresponds to a unique point in the (PD, T2W) plane, The underlying assumption is that the values associated with each of the classes NB, MS, and CSF cluster together and are separable from each other. To find these clusters, one uses Kohn's software package as follows. One draws over the magnetic resonance slice images little regions which contain voxels all of which are unambiguously in the same one of the three classes, and one indicates which class that is. Based on this information, the package automatically subdivides the (PD, T2W) plane into three regions, associated with NB, MS, and CSF, respectively. We have found that this process classifies some voxels as MS, even though they are clearly not, and a further subregioning (for MS only) is a desirable adjunct. This consists of free-hand tracing boundaries enclosing the lesions in each slice. Voxels within these boundaries will be classified into NB, MS, and CSF; voxels outside them will be classified into NB and CSF only.

We can now think of the result obtained so far as three binary scenes, one associated with each of the classes. For the binary scenes associated with NB and MS, we perform shape-based interpolation[3] (described in the last section) to obtain a binary scene with small cubic voxels. The total volumes occupied by these voxels provide us with out desired estimates. In Figure 5, we show the 3D display of the surface enclosing the brain voxels in one such binary scene.

The total study described above takes less than half an hour for an experienced user. We have carried out a series of experiments to evaluate the reproducibility and accuracy of such approaches. We now describe one of them.

This experiment was designed to check intra-observer, inter-observer, and inter-study reproducibility. Three volunteers with MS lesions were scanned repeatedly (two of them three times and the other one twice). In each study, approximately 30 slices of 5-mm thickness were obtained, both for PD and for T2W. Three readers were used: (1) a radiology resident, (2) a radiology fellow, and (3) a medical student, and they were asked to estimate total brain and MS lesion volumes. Each reader had four reading sessions and these were arranged so that, by the end, each reader read two studies of each patient and each study was read by at least two readers who read both studies twice. In each session three studies were read, one of each patient, in a randomized order. The readers had no way of identifying (other than by the images) the patients from session to session. Table 2 summarizes the experiment and its results.

To analyze these results we define the *variability* between two estimates α and β of a volume by

$$v = \frac{|\alpha - \beta|}{|\alpha + \beta|} \times 100$$

and say that the variability is "$v\%$".

Since every study was read twice by each of at least two readers, we can measure intra-observer variability (see Tables 3 and 4). The entries are the values of v, when α and β are the estimates given by the same reader for the same study at different times. Brain volume variability was below 6.5% in all cases and averaged 1.7% overall; a very respectable result. On the other hand, the results for MS were much less uniform. The variability of the radiology fellow (Reader 2) was always below 10% and averaged below 5%, but the variability of the medical student (Reader 3) averaged over 25% and got up to 81% in one case. Clearly, careful training is necessary if the system is to be used for MS volume determination.

Inter-observer variability is reported in Table 5. Here the entries are the values of v, when α and β are defined as the averages of the estimates by the two observers who are being compared. Pairwise comparisons are given for Patient II, whose scans have been read by all three observers. For brain, the results are again very good, inter-observer variability is always less than 4%, and averages to be 1.6%. In view of the unreliability (as indicated by the large intra-observer variability) of two of the readers for estimating MS volumes, the bad result for inter-observer variability for MS volume estimation is not surprising.

Inter-study variability is reported in Tables 6 and 7. Here the entries are the values of v, when α and β are defined as the averages of the estimates given at the two readings of the two studies that are being compared. Again, the results for brain are excellent, inter-study variability is always less than 3%, with the overall average less than 1%. On the other hand, the inter-study variability of the MS volume estimates is disappointing. Even for the best reader, all we can say is that the average is under 7%. We plan to look into the possibility that this is due to our 5-mm slice thickness (too thick for such small lesions) and investigate whether we get better results with 3-mm thick slices.

Accuracy studies for such applications are much more difficult, since there is no reliable gold standard with which we can compare our brain-volume and MS-volume estimates. Our ongoing accuracy studies[21] therefore involve scanning bottles containing mainly one type of agarous gel into which we inserted a known volume of another type of agarous gel.[22] A sample result from one such study is shown in Figure 6.

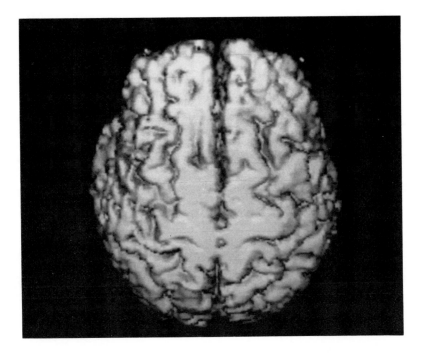

FIGURE 5. 3D display of surface enclosing brain voxels, based on MRI.

TABLE 2
Summary of the Reproducibility Experiment

Reader	Session	Patient I			Patient II		Patient III		
		Study i	Study ii	Study iii	Study i	Study ii	Study i	Study ii	Study iii
1	1		1.1.2			1.1.1	1.1.3		
			1079			1037	1316.5		
			11.5			18.0	28.0		
	2			1.2.3	1.2.1			1.2.2	
				1109	1093			1378	
				17.3	11.6			17.0	
	3	1.3.1				1.3.3	1.3.2		
		1095				1069	1333		
		16.5				15.4	30.2		
	4			1.4.3	1.4.1			1.4.2	
				1067	1126			1326	
				20.9	13.1			34.2	
2	1	2.1.1			2.1.3				2.1.2
		1070			1082				1316
		12.9			10.7				6.2
	2		2.2.1			2.2.3		2.2.2	
			1077			1005		1308	
			11.7			12.3		5.4	
	3	2.3.3			2.3.1				2.3.2
		1051			1011				1335
		15.7			11.0				5.9
	4		2.4.1			2.4.2		2.4.3	
			1051			970		1316	
			14.0			14.0		5.4	
3	1			3.1.2		3.1.1	3.1.3		
				1066		1067	1279		
				16.6		23.3	77.3		
	2	3.2.3			3.2.1				3.2.2
		1086			1083				1399
		12.4			12.2				5.5
	3			3.3.1		3.3.3	3.3.2		
				1091		1068	1456		
				19.6		19.5	8.1		
	4	3.4.1			3.4.3				3.4.2
		1034			1057				1319
		11.2			18.7				10.2

Note: The first line of each entry indicates reader number, session number, and the order in which the study was read. The second line is the estimated total brain volume in cc. The third line is the estimated MS lesion volume in cc.

To summarize, we found the reproducibility of total brain-volume measurements to be excellent. Other experiments, not reported here, showed similar results for regional brain volumes (such as that of the temporal lobes) and CSF volumes. Our experiments with lesion volumes indicate that to obtain good results we have to improve the training of our users and adjust the imaging method used to obtain data. Accuracy studies with phantoms are quite promising. We are looking forward to be able to test with confidence our clinical hypotheses using the method of 3D volume analysis.

TABLE 3
Intra-Observer Variability for Brain

Reader	Patient I			Patient II		Patient III			Average
	Study i	Study ii	Study iii	Study i	Study ii	Study i	Study ii	Study iii	
1		0.7	1.9	1.5	1.5	0.6	1.9		1.4
2	0.9	1.2		3.4	1.8		0.3	0.7	1.4
3	2.4		1.2	1.2	0.1	6.5		2.9	2.4
Average	1.7	1.0	1.5	2.0	1.1	3.5	1.1	1.6	1.7

TABLE 4
Intra-Observer Variability for MS

Reader	Patient I			Patient II		Patient III			Average
	Study i	Study ii	Study iii	Study i	Study ii	Study i	Study ii	Study iii	
1		17.9	9.4	6.1	7.8	3.8	33.6		13.1
2	9.8	8.9		1.4	6.5		0.0	2.5	4.8
3	5.1		8.3	21.0	8.9	81.0		29.9	25.7
Average	7.4	13.4	8.9	9.5	7.7	42.5	16.8	16.2	14.5

TABLE 5
Inter-Observer Variability

	Patient I			Patient II		Patient III			Average
	Study i	Study ii	Study iii	Study i	Study ii	Study i	Study ii	Study iii	
Brain	0.0	1.1	0.4	2.9	3.2	1.6	1.5	1.2	1.6
				1.8	0.7				
				1.1	3.9				
MS	9.6	4.3	2.7	6.5	11.9	18.9	65.2	12.9	16.4
				11.1	12.3				
				17.5	23.9				

TABLE 6
Inter-Study Variability for Brain

Reader	Patient I	Patient II	Patient III	Average
1	0.0	2.6	1.0	1.2
2	0.2	2.9	0.5	1.2
3	0.9	0.1	0.3	0.4
Average	0.3	1.9	0.6	0.9

TABLE 7
Inter-Study Variability for MS

Reader	Patient I	Patient II	Patient III	Average
1	15.4	15.0	6.4	12.3
2	5.3	9.6	5.7	6.9
3	21.1	16.1	68.9	35.4
Average	13.9	13.6	27.0	18.2

IV. DISCUSSION

As stated in the introduction, this is not a survey article. Thus, we have not so far mentioned even such directly relevant works as that of Hildeboldt et al.[23] on osseous landmark identification, Vannier et al.[24] on validation of brain tissue classification based on MRI, McNeal et al.[25] on liver volume measurements and 3D display from MRI, or the papers on estimating cardiac and pulmonary volumes which are discussed in Chapter 11. However, our discussion has given a good impression of the status of quantitation using 3D imaging: it is very promising, but we must proceed with caution.

On the promising side, there have already been a number of good results obtained, and there is every hope that reproducibility and accuracy of quantitation based on 3D imaging will continue to improve. For example, both of the studies on osseous landmark identification that we discussed in the previous section[9,15] used the software package 3D83[10] to produce the images. The quality of these images (Figures 2 to 4) is nowhere near that which can be obtained by the surface-imaging techniques of today (see the illustrations in Chapter 1). There

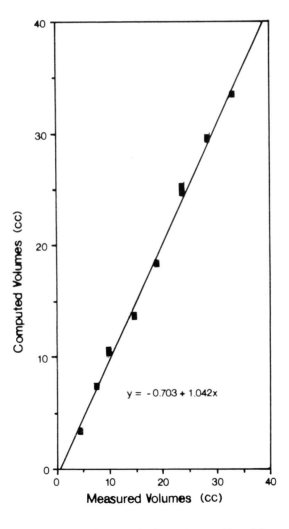

FIGURE 6. Plot of computed volumes (superposition of three readings) versus measured volumes for an agarous gel phantom.[21]

is every reason to believe that improved image quality will result in better performance in landmark identification. Similarly, very little work has been done so far to optimize magnetic resonance image acquisition for the purpose of volumetric analysis of various structures inside the brain. It is hard to imagine that a judicious effort in that direction will not result in significant improvements over the results reported in the previous section. As we acquire experience, we can improve on rule-based approaches,[19] resulting in programs which automatically (and hence reproducibly) deliver clinically relevant numbers based on 3D imaging techniques.

It is worth pointing out that the word "accuracy" was not used in the last sentence. It is indeed important not to have a great hang-up about accuracy; that may unnecessarily impede useful progress. In an area such as MS lesion volume estimation, the accuracy of the methods is just about impossible to test. The important question is not whether the number produced is indeed the "volume", but rather whether it is a reliable indicator of the clinical state of the patient. A number which strongly correlates with the volume would serve just as well as the volume itself.

On the caution side, we must point out that there is a danger in computerized quantitation: programs provide us with numbers without giving us a feel of their reproducibility and accuracy. To use these numbers without having performed careful validation studies, along the lines of those described in the previous section, is foolhardy. Alas, such studies are time-consuming and, frankly, boring. There is a great temptation to simply accept the output of the computer; this temptation must be resisted.

In conclusion, quantitation using 3D imaging can, after proper validation, be a useful tool in various areas of medicine. This has been demonstrated in a few cases already and the potential for future applications is wide-ranging.

ACKNOWLEDGMENT

The research of the author is currently supported by NIH Grant HL28438.

REFERENCES

1. Herman, G.T. Three-dimensional computer graphic display in medicine: The MIPG perspective, in *Pictorial Information Systems in Medicine,* K.H. Höhne (Ed.) (Berlin: Springer-Verlag, 1986), pp. 181–210.
2. Trivedi, S.S. Interactive manipulation of three-dimensional binary scenes, *Visual Comp.* 2:209–218 (1986).
3. Raya, S.P., and J.K. Udupa, Shape-based interpolation of multidimensional objects, *IEEE Trans. Med. Imag.* 9:32–42 (1990).
4. Herman, G.T., and D. Webster: A topological proof of a surface tracking algorithm, *Comput. Vision Graph. Image Proc.* 23:162–177 (1983).
5. Gordon, D., and J.K. Udupa. Fast surface tracking in three-dimensional binary images, *Comput. Vision Graph. Image Proc.* 45:196–214 (1989).
6. Liu, H.K. Two- and three-dimensional boundary detection, *Comput. Graph. Image Proc.* 6:123–134 (1977).
7. Udupa, J.K., G.T. Herman, L.S. Chen, P.S. Margasahayam, and C.R. Meyer. 3D98: A turnkey system for 3-D display and analysis of medical objects in CT data, *Proc. SPIE* 671:154–168 (1986).
8. Trivedi, S.S., G.T. Herman, J.K. Udupa, L.S. Chen, and P. Margashayam. Measurements on 3-D displays in the clinical environment, in *Proceedings of the 7th Annual Conference and Exposition of the National Computer Graphics Association* (Fairfax, VA: National Computer Graphics Association, 1986), v. III, pp. 93–100.
9. Abbott, A.H. The Acquisition and Analysis of Craniofacial Data in Three Dimensions, Ph.D. thesis, University of Adelaide, Adelaide, South Australia (1988).

10. Chen, L.S., G.T. Herman, C.R. Meyer, R.A. Reynolds, and J.K. Udupa. 3D83 — An easy-to-use software package for three-dimensional display from computed tomograms, in *Proceedings of 1984 IEEE Computer Society International Symposium on Medical Images and Icons* (New York, NY: IEEE Publishing Services, 1984), pp. 309–316.

11. Udupa, J.K. Determination of 3D shape parameters from boundary information, *Comput. Graph. Image Proc.* 17:52–59 (1981).

12. Richtsmeier, J.T., C. Hildeboldt, and M.W. Vannier, Data collection from C-T images, in *Proceedings of the 10th Annual Conference and Exposition of the National Computer Graphics Association* (Fairfax, VA: National Computer Graphics Association, 1986), v. I, pp. 199–206.

13. Toennies, K.D., J.K. Udupa, and G.T. Herman. Segmentation of implanted bone grafts using anatomical landmarks, in *Proceedings of the 10th Annual Conference and Exposition of the National Computer Graphics Association* (Fairfax, VA: National Computer Graphics Association, 1989), v. I, pp. 207–214.

14. Dahlberg, G. *Statistical Methods for Medical and Biological Students* (London: George Allen and Unwin Ltd., 1940).

15. Herman, G.T., and A.H. Abbott. Reproducibility of landmark locations on CT-based three-dimensional images, in *Proceedings of the 10th Annual Conference and Exposition of the National Computer Graphics Association* (Fairfax, VA: National Computer Graphics Association, 1989), v. I, pp. 144–148.

16. Herman, G.T. Three-dimensional imaging on a CT or MR scanner, *J. Comput. Assist. Tomogr.* 12:450–458 (1988).

17. Herman, G.T., M.I. Kohn, and R.E. Gur. Computerized three-dimensional volume analysis from magnetic resonance images for characterization of brain disorders, in *Proceedings of a Special Symposium on Maturing Technologies and Emerging Horizons in Biomedical Engineering,* J.B. Mykleburst, and G. F. Harris (Eds.) (Piscataway, NJ: IEEE Service Center, 1988), pp. 65–67.

18. Kohn, M.I. Segmentation of magnetic resonance brain images, in *Proceedings of the 10th Annual Conference and Exposition of the National Computer Graphics Association* (Fairfax, VA: National Computer Graphics Association, 1989), v. I, pp. 168–171.

19. Raya, S.P. Rule-based segmentation of multi-dimensional medical images, in *Proceedings of the 10th Annual Conference and Exposition of the National Computer Graphics Association* (Fairfax, VA: National Computer Graphics Association, 1989), v. I, pp. 193–198.

20. Fukunaga, K. *Introduction to Statistical Pattern Recognition* (New York, NY: Academic Press, 1972).

21. Kohn, M.I., N.K. Tanna, G.T. Herman, S.M. Resnick, P.D. Mozley, R.E. Gur, A. Alavi, R.A. Zimmerman, and R.C. Gur. Analysis of brain and cerebrospinal fluid volumes from magnetic resonance imaging. Part 1: Methodology, reliability and validation, *Radiology* (in press).

22. Mitchel, M.D., H.L. Kundel, L. Axel, and P.M. Joseph. Agarose as a tissue equivalent phantom material for NMR imaging, *Magn. Res. Imag.* 4:263–266 (1986).

23. Hildeboldt, C.F., M.W. Vannier, and R.H. Knapp. Validation of skull 3-D measurements, *Am. J. Phys. Anthrop.* (in press).

24. Vannier, M.W., C.M. Speidel, D.L. Rickman, L.D. Schertz, L.R. Baker, C.F. Hildeboldt, C.J. Offutt, J.A. Balko, R.L. Butterfield, and M.H. Gado. Validation of magnetic resonance imaging (MRI) multispectral tissue classification, in *Proceedings of the 9th International Conference on Pattern Recognition* (Washington, D.C.: IEEE Computer Society Press, 1988), pp. 1182–1186.

25. McNeal, G.R., W.H. Maynard, R.A. Branch, T.A. Powers, P.A. Arns, K. Gunter, J.M. Fitzpatrick, and C.L. Partain. Liver volume measurements and three-dimensional display from MR images, *Radiology* 169:851–854 (1988).

Chapter 6

APPLICATIONS OF COMPUTER GRAPHICS TO THE EVALUATION AND TREATMENT OF MAJOR CRANIOFACIAL MALFORMATIONS

Court B. Cutting

TABLE OF CONTENTS

I. INTRODUCTION

This chapter will address the general applications of computer graphics technology to the study and treatment of craniofacial anomalies. The chapter will not confine itself to three-dimensional (3D) CT, but will cover all of the data sources that are relevant to the study of craniofacial anomalies. It is the belief of the author that simple facial surface sampling will be very important for the longitudinal follow-up of these patients. The psychosocial consequences of having to spend a significant portion of one's childhood with a major uncorrected facial malformation can be devastating. For this reason craniofacial surgeons have moved to correcting many of these problems in infancy and early childhood. The surgical planning problem is thus four dimensional. Abnormal rates of growth of different parts of the facial complex must be measured and fed back into surgical planning. The cost, radiation, sedation, and availability constraints of CT and MR scans as are currently present preclude their use for longitudinal follow-up. Other methods of measuring the facial surface therefore assume greater importance than in other applications areas.

The chapter will begin with the current state of 3D CT scanning as it applies to this problem and will discuss why craniofacial anomalies applications were the principal driving force in the beginnings of 3D CT. We will then appear to digress into cephalometrics which, prior to 3D CT, had been the principal tool used to study these problems. While the reader who is interested in CT might wish to skip this cephalometric digression, I would urge him/her not to. We will use cephalometrics in its current state to point out a number of major deficiencies in functionality of current 3D CT software. A number of major features of cephalometrics still make it more useful than 3D CT for the planning and evaluation related to these clinical problems.

The chapter will then go on to discuss the research which is being done to correct the deficiencies in functionality present in the current generation of 3D CT software. While most of this is unfinished at the time of this writing, it will point out the author's view of where this research should go. While most of this section applies to craniofacial anomalies, it will become clear that further extensions of this work may have significant ramifications in the area of general radiology.

Finally the chapter will discuss the reliable measurement and production of 3D rigid motion of bone fragments. As the surgical planning process becomes more and more precise, an ever-widening gap is being created between a surgical plan and the surgeon's ability to precisely execute that plan. The final section will discuss the possibilities that current methods of rigid motion measurement afford.

II. THREE-DIMENSIONAL CT SCANNING AND CRANIOFACIAL ANOMALIES

Historically clinical 3D CT scanning focused first on craniofacial malformations. This may at first seem peculiar to the reader as other problems in medicine are far more life threatening and vastly more common than the care of craniofacial malformations. 3D processing of what had hitherto been 2D slices provided by CT scanners seems to have begun and was extensively developed by the group at the University of Pennsylvania.[26,27,57] The first application of this idea to the study of craniofacial malformations was performed and extensively developed by Vannier and Marsh at Washington University in St. Louis.[32–36,38,58] Other groups quickly followed. At the present time 3D CT scanning is the standard of practice in every major center that treats craniofacial anomalies. Some units use custom 3D processing hardware and software while others use software resident on the CT scanners themselves to produce the image. 3D imaging services are also available where a surgeon can send the archive tape from a 3D CT study of his patient and receive a full set of 3D images and in some cases a model of the patient's skull produced on a computer-controlled milling machine.[28,52,53]

When considering the various medical problems that 3D CT has been applied to, craniofacial anomalies are a logical first choice. The first reason is that the surgeon is principally interested in bone. This tissue is at the high end of the radiodensity spectrum, making it the easiest to separate from other tissues using simple thresholding. This approach did very well in large areas of solid bone and only produced minor problems where the bone was thin and occupied only a small portion of the volume of a voxel. In these areas "pseudo foramina" were produced where areas of thin bone were seen on 3D CT renderings as large holes in the bone. While these pseudo foramina were a nuisance, the surgeon could still appreciate the overall shape disturbance of his particular patient's skull.

The second major reason that 3D CT was first applied to craniofacial anomalies is that the craniofacial surgeon is often most interested in skull relationships that often "skipped" a large number of slices. To a plastic surgeon the relationship of the lower jaw to the forehead is of paramount importance. The plastic surgeon is interested in the overall shape of the skull, not just a particular feature within it. This is in sharp contrast to the tumor surgeon, for whom "slice-by-slice" analysis has traditionally been adequate. This is not to say that the 3D CT approach is not a major improvement in tumor visualization. This topic will be discussed in greater detail in Chapter 10 by Levin et al. in this volume.

There are a large number of ways to produce a 3D image from a 3D volume sample of the patient's facial anatomy provided by CT or MR scanners. As this will be discussed in great depth in other chapters in this book, I will not delve into these matters but rather focus on the applications of these technologies to the evaluation and treatment of craniofacial anomalies.

Much has been learned about the pathologic anatomy of various craniofacial malformations through the "electronic dissection" that 3D imaging makes possible. Custom editing facilities allow the student of such anomalies to "dissect" an image of the patient's anatomy as though he/she were in a cadaver laboratory. Custom editing requires an interactive computer graphics environment. This will be discussed in the sections that follow. Volume, distance, and angle measurement capabilities have provided useful quantitative information as well.[9,22,33,34,36,50] Plate 5* is a custom edited image of an enlarged ventricular system with part of the calvarium cut away in a patient with Apert's syndrome and concomitant hydrocephalus. The ventricular volume was twenty times the normal value (i.e., approximately 350 cc). Other chapters in this volume will contain other examples of this type of study.

The case seen in Figures 1 and 2 will put the reader into the proper perspective regarding the value of 3D imaging of craniofacial malformations. A preoperative photograph and the corresponding 3D CT scan image are shown on the left of the figure pairs. For the first time a concrete view of the underlying anatomy of a patient with this type of problem was available. Figure 3 shows a view into the anterior and middle cranial fossae of this patient from back to front after cutting away the posterior skull. Note the severely abnormal position of the sphenoid wings. This is useful information to have in planning the various craniotomy incisions which would be required to reshape this patient's skull. On the right sides of Figures 2 and 3 are the postoperative images. While the postoperative appearance is markedly improved, the 3D CT study reveals the large number of pieces which were used to reshape the calvarial vault. The skull did not become mechanically stable for many months after the surgery. We are left to ponder whether a smaller number of larger pieces of cranial bone could have been employed to produce a better result, more quickly, and with better mechanical stability in the immediate postoperative period.

---

* Plate 5 appears following page 166.

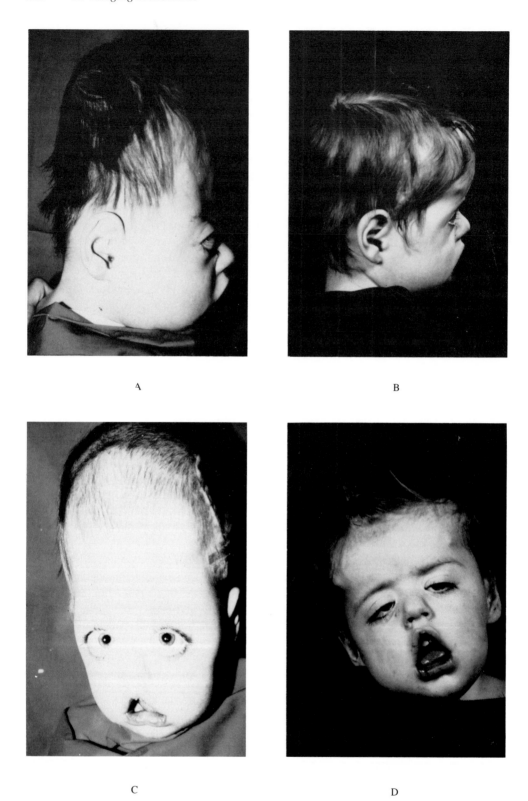

FIGURE 1. (A) Preoperative lateral photograph of patient with severe craniosynostosis due to Pfieffer's syndrome. (B) Postoperative lateral photograph of same patient. (C) Preoperative frontal view. (D) Postoperative frontal view.

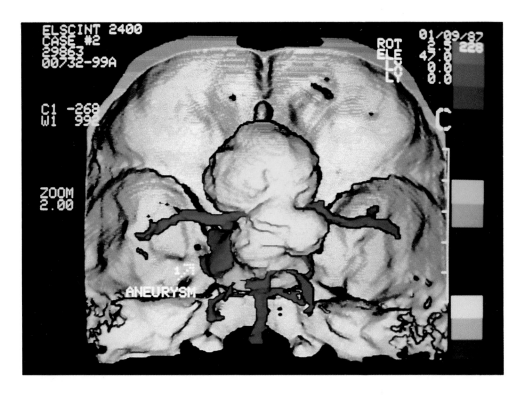

PLATE 1. Composite image of suprasellar pituitary tumor showing relationship to skull base and vessels at Circle of Willis. (Courtesy of Elscint, Inc.)

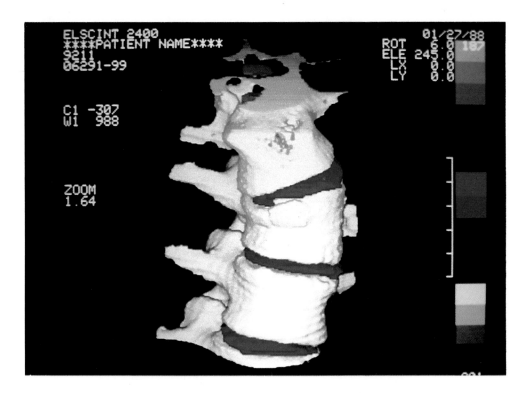

PLATE 2. Scoliotic segments of lumbar spine showing disc, vertebra, and contents of spinal canal. (Courtesy of Elscint, Inc.)

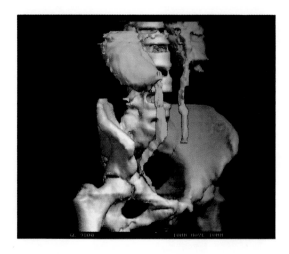

PLATE 3. Kidneys and ureters in pelvis. 10-mm thick slices. (Courtesy of I.S.G. Technologies.)

 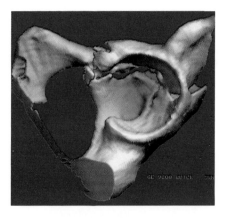

PLATE 4. Removal of femur from acetabulum. (Courtesy of I.S.G. Technologies.)

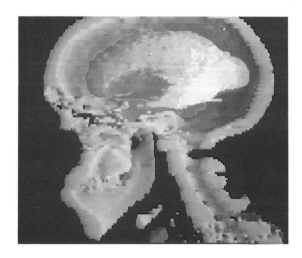

PLATE 5. Enlarged ventricular system inside hemisected skull image of a patient with Apert's syndrome and severe hydrocephalus. Ventricular volume is approximately 350 cc. Image prepared using Cemax 1000 device.

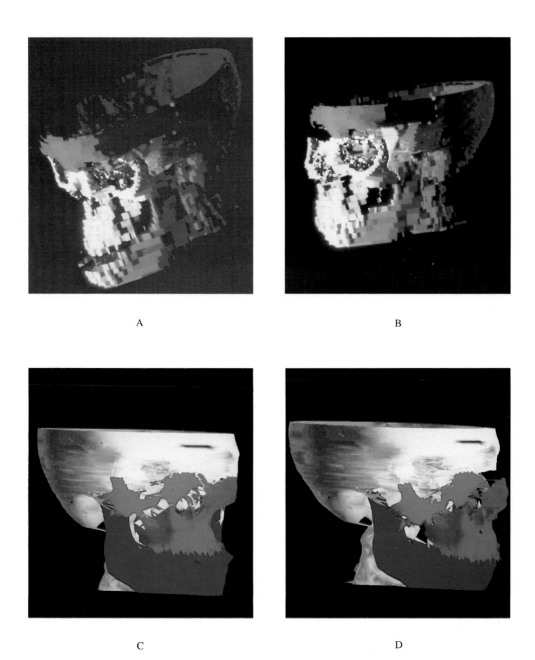

A

B

C

D

PLATE 6. Two 3D CT-based surgical simulation programs written by the author in 1984. In both instances CT data were preprocessed into contour files (bone thresholding) using the Cemax 1000 device manufactured by the Cemax Corporation. The data were then dumped to a Silicon Graphics Iris 2400 workstation where the surgical simulations were performed. In A and B no tesselation was done. The slice contours were simply thickened into "ribbons" and the four-point polygons were flat shaded after the appropriate movements of the bone fragments. In C and D a "triangular tiling" between the slices was performed and Gouraud shading used to improve the image. Note the difficulty in performing a "correct" tiling from stacked 2D contour files. (A) Preoperative left oblique view using the first program. (B) Postsimulation view. (C) Preoperative right oblique view using triangular tiling between slices to represent the sections. A Gouraud shading model is used. (D) Postsimulation view using this image format.

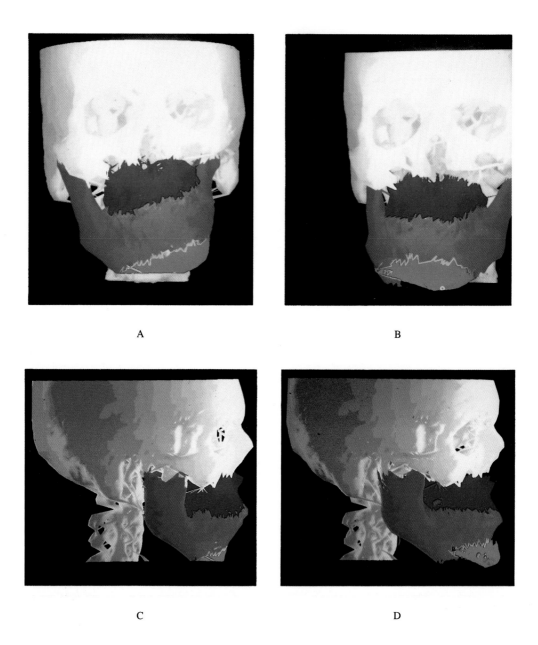

A

B

C

D

PLATE 7. 3D CT surgical simulation of patient shown in Chapter 6, Figure 9. (A) Preoperative frontal image. (B) Postsimulation frontal image. (C) Preoperative lateral image. (D) Postsimulation lateral image. Although the data sample is far richer in the CT-based simulation, the 3D cephalometric simulation in Figure 12 remains more useful. The reason for this is that no comprehensive surface-based cephalometric system has yet been developed which would permit optimization of fragment positioning. CT-based simulation at the time of this writing is still no more than an artist's tool.

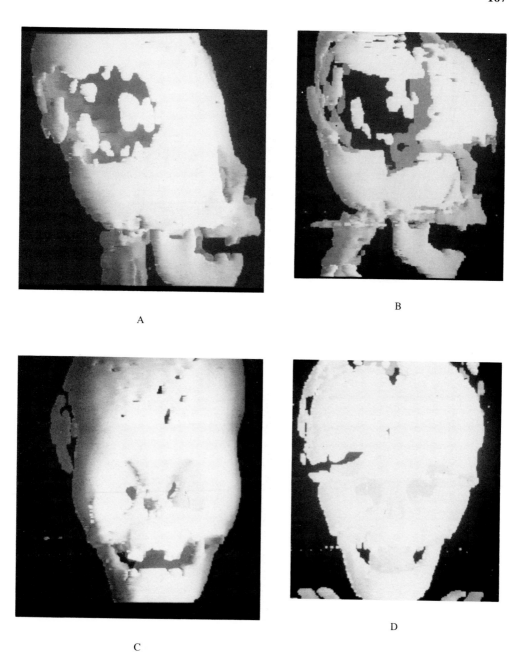

A

B

C

D

FIGURE 2. This figure corresponds to Figure 1 only 3D CT scan images of this same patient with Pfieffer's syndrome are used instead of photographs. (A) Preoperative lateral. (B) Postoperative lateral 3D CT image. (C) Preoperative frontal view. (D) Postoperative frontal view.

III. CT-BASED SURGICAL SIMULATION

The previous example suggests the next logical development in 3D CT software to be surgical simulation. Adapting commercially available computer-aided design (CAD) and manufacturing software seemed like the logical first step in developing surgical simulation. Vannier, Marsh and Warren were the first to pursue this course in 1981.[34] Figure 4 illustrates

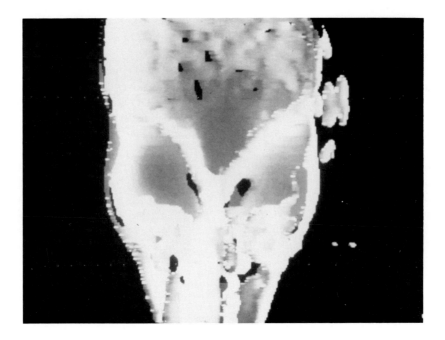

FIGURE 3. Posterior hemisection of the skull of the patient shown in Figures 1 and 2. Note the unusual shape of the anterior middle fossa in association with the extremely highly arched sphenoid wings.

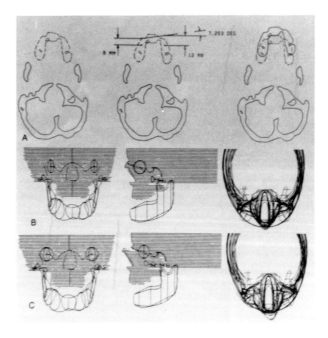

FIGURE 4. First use of commercial computer-aided design software to perform surgical simulation of a LeFort III osteotomy by Drs. Jeffrey Marsh, Michael Vannier, and Mr. James Warren. (A) Movement of dental models to produce a correct occlusion are transferred to the CAD program containing a 3D CT-based model of the patient's skull. (B) Preoperative view of patient showing LeFort III segment not yet moved. (C) Postsimulation view showing the effect on the midface made by the occlusal change specified in (A). (From Marsh, J., M. Vannier, W.G. Stevens, J. Warren, D. Gayou, and D. Dye. Computerized imaging for soft tissue and osseous reconstruction in the head and neck, *Clin. Plast. Surg.* 12:279 (1985). With permission.)

a midface advancement procedure simulated in this way. Following this, other workers began to outline some of the requirements for 3D CT-based surgical simulation and implement some of these features in 3D imaging software.[14,54,55] Cutting et al.[17-19] were the first to write a custom program for the performance of surgical simulation (Figure 5 and Plate 6*) by the operating surgeon. This approach has a number of advantages over the use of CAD software. The problem of taking a geometrically complex structure and cutting and repositioning it seldom occurs in the usual CAD situation. In that environment the designer is able to build his structures from scratch. It must also be remembered that surgical simulation software must be used by a craniofacial surgeon who is seldom sophisticated in the use of a computer. Most human interfaces to CAD software were developed for a design engineer, not a surgeon. Surgical simulation software should have the look and feel of an operating room environment and should be extremely easy to use. Since the time of these papers other surgical simulation programs have been developed for this purpose.[37,41,56] All of these programs have three functional elements: (1) the ability to cut a model of the craniofacial skeleton into various osteotomy fragments, (2) the ability to move the fragments with six degrees of freedom (i.e., translate and rotate about the three Cartesian axes), and (3) measurement capabilities which enable the surgeon to measure distances and angles.

Surgical simulation is an interactive task. Unlike 3D imaging which can be done off line, often overnight, using the CT scan computer, surgical simulation requires the fastest possible computer graphics response and at least a reasonable quality image. This application requires a high-performance graphics workstation. Our unit has been using the devices manufactured by the Silicon Graphics Corporation. The hardware requirements are (a) dedicated graphics hardware for the production of high-speed, high-quality 3D images, (b) sufficient local RAM and CPU power to allow each user to operate sophisticated interactive medical operations independent of a centralized computing resource, and (c) networking connections to the radiology department of the future where a patient's medical image data and its radiological evaluation can be sent to a peripheral workstation in a few moments (currently local mass storage and a tape reader are required). It is certainly possible to provide the hardware requirements for such a workstation in a number of ways. The workstations used in our unit implement most of the graphics operations using dedicated hardware for matrix multiplication, polygon filling, shading, hidden surface removal, etc. This leaves the CPU free for more high-level data-handling tasks. More powerful CPUs and very efficient programming feeding into lower performance frame buffer may also be used to accomplish this task, but it runs contrary to the trend toward parallelism in hardware design. Implementing low-level graphics primitives in hardware is a simple way to maximize throughput through the use of parallel processing. The author expects this trend in hardware design to continue. Every year the computer graphics industry is producing workstations with increasing performance and decreasing price. It is hoped that the development of these workstations will allow 3D medical image processing to be performed by the clinician end user at every nursing station in every hospital at an affordable price in the not-too-distant future. In such a hardware environment, surgical simulation software will provide the clinician with a natural way to study his patient's pathologic anatomy. Currently the only disadvantage to implementing graphics primitives in hardware is the lack of a standard software interface to the graphics subsystem. This often means that graphics function software becomes machine dependent, unlike the rest of the code which is usually implemented in the standard "C" language in the UNIX operating system. This problem will disappear with the emergence of a standard 3D graphics software library. In the meantime the programmer is well advised to keep graphics functions isolated from the rest of his/her code in machine-dependent libraries.

Over the past 6 years our unit has gained considerable experience with CT-based surgical

---

* Plate 6 appears following page 166.

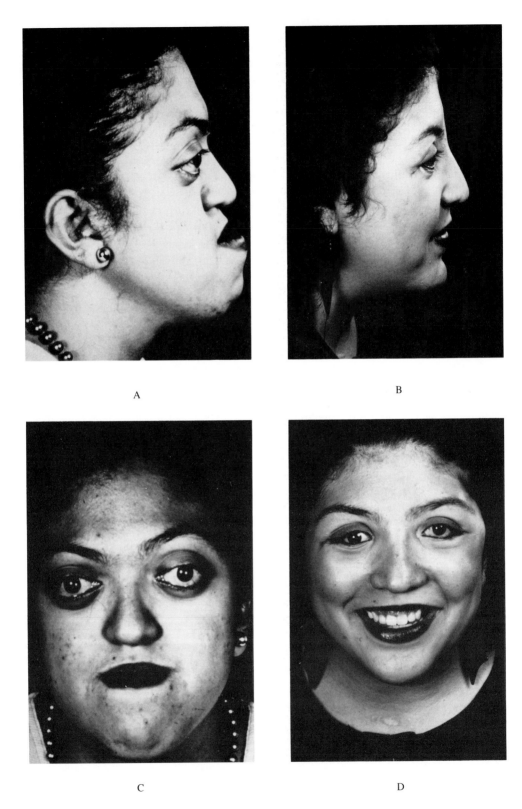

FIGURE 5. Photographs of an adult with Crouzon's disease who underwent multisegment midface osteotomies for correction of the deformity. (A) Preoperative lateral view. (B) Postoperative lateral view. (C) Preoperative frontal view. (D) Postoperative frontal view.

simulation and has concluded that other elements need to be present before such a system will reach its potential. What is missing from the process is a numerical concept of normal. Currently CT-based simulations are an artist's tool. The surgeon cuts and rearranges the facial bones until his/her eye is satisfied. This is not very different from what is currently done in the operating room, with the exception that the surgeon is able to try out several different alternatives at no cost. This approach is most useful in calvarial vault transpositions where surgical options are often radically different in overall design[31] (Figure 6). In this situation the question is more what to do rather than how much to do it. In orthognathic surgery on the other hand it is often easy to decide where to cut the bone, but hard to decide how the bone should be repositioned, particularly in asymmetrical cases where both the upper and lower jaws need to be repositioned. This is where comparing a patient's skeletal structure with normal is particularly useful. This process has been developed to a fine art in traditional cephalometrics, which is the subject of the next section. At the time of this writing such quantitative rigor is absent from 3D CT-based surgical simulations.

IV. THE CEPHALOMETRIC TRADITION

The standard of practice for planning jaw surgery involves the careful evaluation of standard view radiographs called cephalograms. The patient's head is placed in a head holder which assures reproducible postero-anterior and lateral (also sometimes basilar) X-rays with the source and film distances standardized. A large number of standard points can be located on these films which establish a homology between the points of a particular patient and those of any other. The term homology refers to a structural correspondence between individuals. For example, a nasal tip on one individual is homologous to the nasal tip on another and not homologous to the chin. Homology mapping therefore refers to the establishment of a correspondence between like structural features on different individuals. Homology mapping is a key concept in the development of any cephalometric system. Normative databases[15,47] have been produced using this system which allows for a number of statistical operations. In syndromology it allows groups of patients with a particular syndrome to be compared with age- and sex-matched norms. Morphometric systems have evolved wherein distances, angles, and tensors are used to study shape changes between groups.[3,10–12,45,46]

Cephalometrics has also been extensively used in surgical planning for individuals with jaw deformities and abnormal occlusion of the teeth. Usually only the lateral cephalogram is used as these problems are usually bilaterally symmetrical. Skeletal structures are traced and paper cutouts made of the skeletal segments to be moved. After moving the pieces using this pasteboard surgical simulation, the cephalogram is remeasured and compared with normal. This allows a precise "surgical blueprint" to be generated which is used by the orthognathic surgeon in the operating room. This process has been computerized[8,16–18,25,39,59] and is illustrated in Figure 7.

Recently the cephalometric process has been three dimensionalized. At its inception cephalometrics was meant to be considered in three dimensions.[15] Metallic markers have been used to locate 3D points. Baumrind[4,5] used narrow angle X-rays to obtain anatomic point locations without metallic markers. Our group has used the same approach using only postero-anterior cephalograms and standard cephalometric points.[16–18,24] This last approach allows large normative databases derived in two dimensions to be used for 3D analysis. These methods depend on the ray intersection method (Figure 8). If the 3D location of the point source of X-ray is known as well as the location of the shadow of an anatomic point on the X-ray film, these two points determine a line in three dimensions. A second X-ray view of the same structure from a different angle produces a second 3D line. The intersection of these lines in three dimensions is the location of the anatomic point free of the usual magnification artifact. This allows a simple 3D model of a patient's face to be generated which can be used for syndromology or surgical treatment planning.

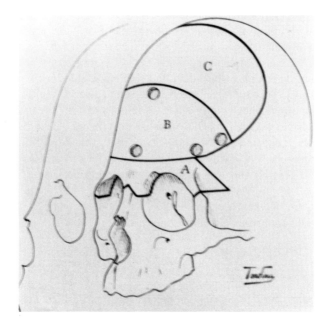

A

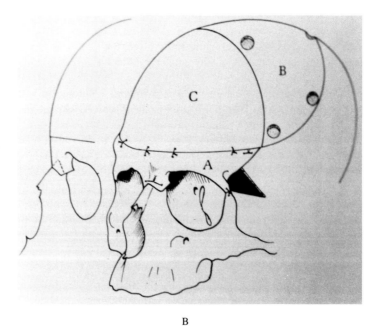

B

FIGURE 6. Calvarial vault remodeling using transposition. (A) Preoperative appearance with osteotomy lines marked. (B) Result following transposition of the two bone fragments. It should be obvious to the reader that a number of different osteotomy and transposition schemes would be possible in this situation. CT-based simulation allows the surgeon to try a number of solutions in order to arrive at the "best." (From Marchac, D., and D. Renier. *Craniofacial Surgery for Craniosynostosis*. (Boston, MA: Little, Brown, 1982), Figure 19–14. With permission.)

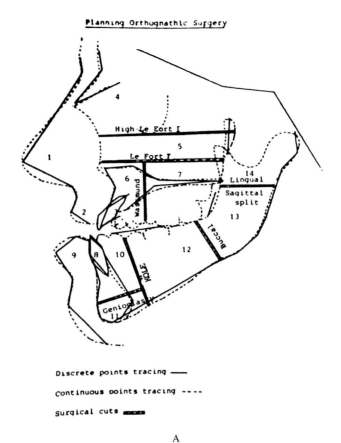

A

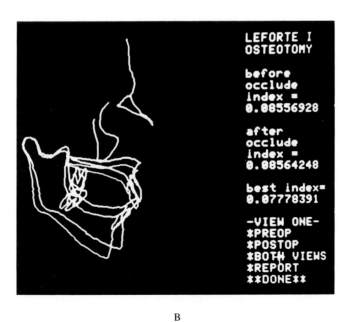

B

FIGURE 7. 2D ceph-based surgical simulations. (A) Surgical simulation program of Bhatia and Sowray's. (From Bhatia, S., and J. Sowray. A computer aided design for orthognathic surgery, *Br. J. Oral Maxillofac. Surg.* 22:237 (1984). With permission.) (B) Author's 2D surgical simulation program including automatic fragment positioning based on optimization to best fit cephalometric norms appropriate for age, race, and sex.

The concept of automatic optimization of bone fragment position has been added to surgical planning based on this cephalometric model.[17] Six numbers are required to specify a rigid motion of a bone fragment, namely a rotation and a translation about each of the cardinal axes (i.e., X, Y, and Z). If there are at least three noncollinear points on each bone fragment, the optimal positions of the bones can be computed automatically if the ideal positions of these landmark points are known. This amounts to a six-variable minimization using the sum of the distances between a patient's landmarks and the norm as the function to be evaluated.[44] Figures 9 and 10 and Plate 7* illustrate surgical simulation for a patient with hemifacial microsomia using this optimization concept. At the present time our group has found cephalometric surgical simulation to be an invaluable aid in the planning of complex orthognathic procedures in asymmetrical cases. This notion of optimization of bone fragment position is dependent on the establishment of a homology mapping between landmark points on the patient and those in a normative data set. It is this requirement that currently prevents the optimization concept to be applied to 3D CT- or MR-based surgical simulation. Much of the current research in our unit is aimed at the development of a 3D surface-based cephalometric system that can be applied to CT and MR data.

V. OTHER DATA SOURCES

Psychosocial considerations have forced the craniofacial community to operate on patients with major craniofacial malformations as infants or young children. In operating on untreated adults with these problems it became evident to workers in the field that production of a normal face was not enough. Suddenly providing aesthetic normality did nothing to reverse the effects of years of social ridicule and deprivation on the psyche. It soon became obvious that the psychosocial consequences of spending childhood with severe facial malformations could be devastating. It also became evident with the passage of time that certain of these

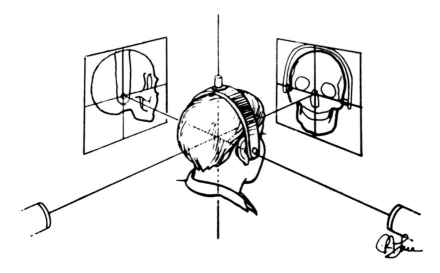

FIGURE 8. Ray intersection method for finding the 3D locus of an anatomic point from two cephalograms taken at an angle to each other. The point source of the X-ray beam and the 3D location of the anatomic point in question determine a line in three space from each view. At the intersection of these two lines is the location of the anatomic point in question free of the usual magnification artifact.

* Plate 7 appears following page 166.

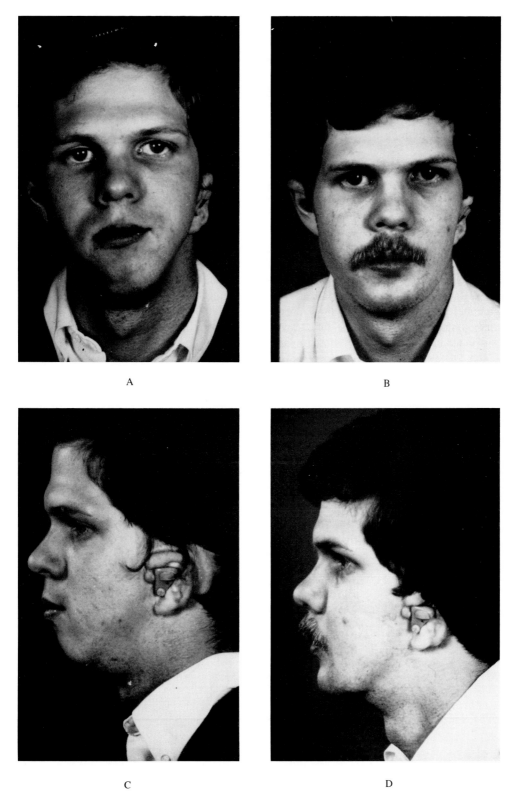

FIGURE 9. Photo series of patient with hemifacial microsomia. Surgical correction consisted of LeFort I osteotomy, sagittal split osteotomy of the mandibular rami, and genioplasty. (A) Preoperative frontal view. (B) Postoperative frontal view. (C) Preoperative lateral view. (D) Postoperative lateral view.

procedures were safer and easier when done on infants. Doing these procedures before the child could generate a significant negative memory of the event also had certain advantages. The only negative consequence of early surgery is that the growth dimension must be taken into account in surgical treatment planning.

Longitudinal follow-up of patients who have already undergone early surgery is the only reliable way to feed back growth information into the surgical planning process for the next generation of patients. 3D facial shape information must be measured postoperatively at regular intervals on all patients who have had early surgery. CT and MR scans are not currently practical for this purpose. High costs, sedation in infants, scheduling difficulties, and radiation hazards limit the use of these modalities to a preoperative study and occasionally a postoperative study if the clinical situation warrants it. Preoperatively these studies are required for neurosurgical evaluation of possible hydrocephalus and intracranial malformation and for the surgical planning process. As the preoperative study has therapeutic impact on the patient, the risks and the cost are justified. It is ethically difficult and financially impossible to justify periodic postoperative studies of this type. With the passage of time it is hoped that nonionizing MR units will replace the cephalostat as instruments of routine postoperative follow-up. At the present time other methods of evaluating facial form must be investigated.

3D facial surface measurements are currently being taken clinically using stereophoto-grammetry[6,49] and automatic light scanners.[1,7,20,40] Stereophotogrammetry involves matching up homologous points on the facial surface using stereophoto pairs. 3D localization is performed using the ray intersection method described in the 3D cephalometric section (see Figure 8). Structured light is sometimes projected on the face to make the point localization process easier. In practice stereophotogrammetry is easy to set up initially and low in cost, but

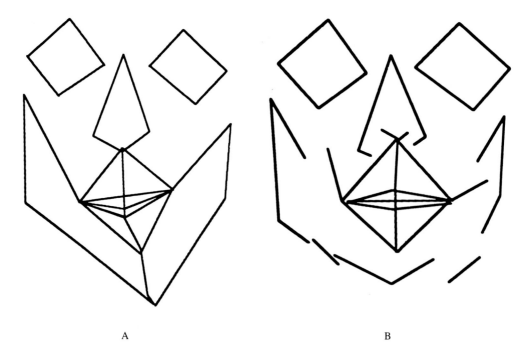

A   B

FIGURE 10. 3D cephalometric simulation series of patient shown in Figure 9. (A) Frontal view of 30 3D points collected as shown in Figure 8 and connected with lines to make the facial image recognizable. Note that the chin deviation to the left and the tilt of the upper and lower occlusal planes. (B) Result following automatic 3D optimization of bone fragment positioning to best fit normative data set.

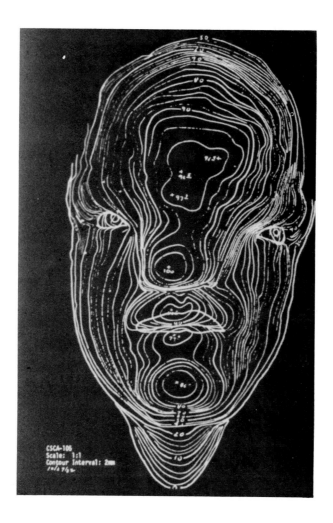

FIGURE 11. Stereophotogram of a patient with hypertelorism. (From Savara, B., S. Miller, R. Demuth, and H. Kawamoto. Biostereometrics and computergraphics for patients with craniofacial malformations: Diagnosis and treatment planning, *Plast. Reconstr. Surg.* 75:495 (1985). With permission.)

the per patient processing time and expense are high (Figure 11). Automatic light scanners offer a viable alternative. Figures 12 and 13 illustrate one such device manufactured by Cyberware Laboratories in Monterey California.[20] A line of laser light is projected on the face and viewed by a video camera at an angle. A contour line of the face can then be captured with every video frame. By slowly rotating the device about the head, very dense 3D surface measurements can be taken without manual intervention. While the initial setup cost is relatively high, the per patient processing cost is very low. The only disadvantage to the automatic scanner is that it takes several seconds to collect the data which are often difficult to come by in an uncooperative child. Stereophotogrammetry has the advantage in this situation as the data can be gathered in a split second. Both methods share the disadvantage that the bone surface can be evaluated only indirectly.

In the near future ultrasound may become increasingly important as a data source in the prenatal diagnosis of craniofacial malformations. At the present time linear array ultrasonic scanners have diagnosed a number of these malformations in utero. By sweeping through the

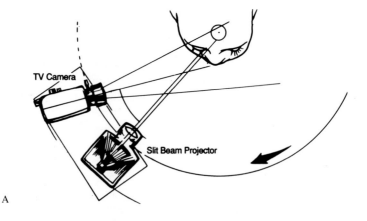

A

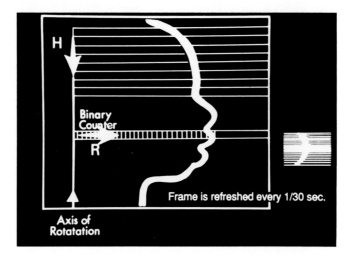

B

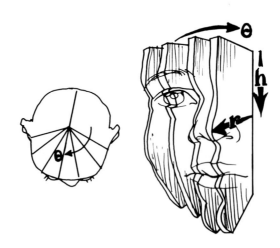

C

FIGURE 12.   Laser scanner explanation series. (A) Top view of slit beam projector projecting a line of light on the face. This line is viewed at an angle by a video camera. (B) Image as seen by the video camera. At the beginning of each video scan line a counter is started. When the light is detected, the counter is stopped and a "radius" measure entered. Each video frame therefore produces several hundred scan radius measures for a given contour. (C) The device is rotated about the head collecting a new contour with each video frame. The facial surface is therefore collected in cylindrical coordinates. (From Cutting, C., J.G. McCarthy, and D. Karron. Three-dimensional input of body surface data using a laser light scanner, *Ann. Plast. Surg.* 21:38 (1988). With permission.)

uterus the ultrasonographer is able to mentally construct a 3D image of the fetus. There is no reason why this process cannot also be carried out using 3D computer graphics in the way the CT and MR "slices" are stacked together to make 3D images. There is of course more noise and motion artifact in ultrasound scans than in CT, making this problem more difficult. We anticipate that in the near future ultrasound will make it possible to view a 3D image of a baby prior to birth. This should be very useful for prenatal screening of craniofacial malformations for mothers at risk.

Ultrasound may also be used to do facial bone imaging for longitudinal follow-up. In 1980 the author placed his head in a water bath while his co-workers Steven Horii, M.D. and James Schimpf, Ph.D. performed a 3D ultrasonic scan of his face. While the problems of noise and standing waves were significant, as well as holding the breath under water, the possible future utility of the method was demonstrated. While the author's interest in CT stopped further development, it is anticipated that advances in 3D ultrasound will be forthcoming.

VI. THE FUTURE OF CT-BASED SURGICAL SIMULATION

In the future surgical simulation programs in infants and children will automatically generate a "surgical blueprint" of what osteotomy fragments are necessary and how they

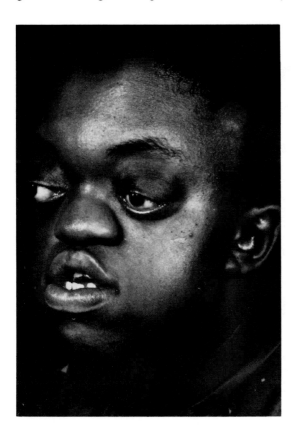
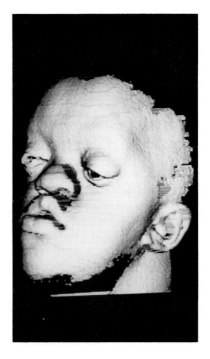

A   B

FIGURE 13. (A) Photograph of patient with Apert's syndrome. (B) 3D image of same patient generated from laser scan data collected as shown in Figure 12 and postprocessed on a Silicon Graphics 4D70 workstation. (From Cutting, C., J.G. McCarthy, and D. Karron. Three-dimensional input of body surface data using a laser light scanner, *Ann. Plast. Surg.* 21:38 (1988). With permission.)

should be positioned to produce a normal face when the child reaches 18 years old. If it is impossible to achieve such a result, the machine will generate a color-coded 3D image of the projected appearance of the patient's face when he/she is full grown if nothing else is done. The color coding will point out what parts of the facial surface are statistically significantly abnormal with respect to a normative data set. While this is science fiction at the present time, it is useful to point out the areas where development will be necessary in order to achieve this objective and then proceed in an orderly fashion toward it.

It is necessary for all 3D imaging modalities to be placed in a common data format to facilitate data transmission, archiving, and the proliferation of alternative software packages to process it. There has been a lot of secrecy on the part of the CT and MR scan manufacturers as regards the publication of their data formats to software developers for proprietary financial motives. By keeping the format secret they hope to keep further lucrative software development within the confines of their corporate structure. This has been nothing but an obstacle to progress. The CT scan computers are in constant clinical use. These machines cannot be expected to do time-consuming postprocessing of CT studies. These data must be post-processed on dedicated workstations. Further, the data should be available over an electronic hospital network which must handle data from scanners of different manufacturers. A joint committee of the American College of Radiology (ACR) and the National Electrical Manufacturer's Association (NEMA) is developing standard formats for storing CT data. It should not be too long before scanner manufacturers universally adopt such standards or risk a negative effect on sales. At the present time any scanner that our software cannot read is useless to us. Patients can only be sent to a facility where the manufacturer publishes his archive tape format. On our part this is simple necessity. Hopefully the other manufacturers will realize the wisdom of adopting a common data format. Currently all of the basic CT and MR data handling in our unit is done using a program called QSH. QSH was written by Gerald Maguire, Ph.D. and Marilyn Noz, Ph.D.[29,30,42] It reads scans from a number of different manufacturers into a common data format from which the three-dimensionalization process proceeds. QSH also provides a number of utilities for general-purpose CT and MR handling and display. It is a general-purpose frame buffer software suite which is in the public domain.

There is also a need for a standard format for the storage of surface data. We have adapted Bruce Baumgart's winged edge library for this purpose.[2] This library has been rewritten in the C language and is in the public domain. It contains all the facilities necessary for the creation, storage, and modification of topologically connected surfaces which surround solid objects. There is also a general-purpose database capability which allows any type of information to be attached to any object, vertex, face, or edge in the structure. Mr. Alan Kalvin under the direction of his faculty advisor Dr. Robert Hummel has added segmentation software to Harlyn Baker's "weaving wall" algorithm for the construction of a surface-extracted solid model from CT scan data. This kind of geometric extraction of surfaces from volume data is the first step in the automatic recognition of an anatomic geometric hierarchy in such data. The importance of this will be discussed in the sections that follow.

The finite element method will probably be the technique used to predict the soft tissue response to changes in the underlying bone position.[48,61] Volumetric "tiling" of the soft tissue over the bone along with shells for the periosteum and skin may provide a reasonable way to model the soft tissue. The parameters of shear and elasticity of the finite elements would be adjusted until an optimal model for each finite element model is achieved. The model may then be made more complex as necessary to provide a good correlation with the actual result measured some months following surgery. The laser light scanners mentioned in the next section should allow much of this validation, although a repeat CT scan would be preferable.

As alluded to previously, growth prediction will be an essential element of surgical planning in the future. Unfortunately this will require years of careful longitudinal follow-up of the current generation of patients who have undergone surgery as infants and children. The

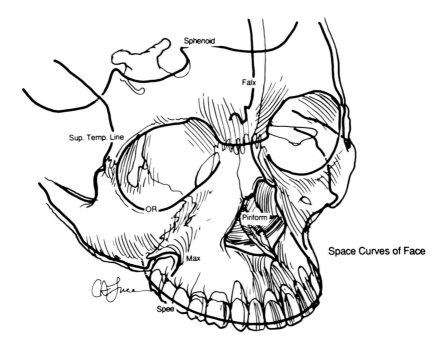

FIGURE 14.   Principal ridge curves of the face. Using a combination of heuristic and differential geometric methods the author expects to quickly find these structures with minimum technician time. (From Bookstein, F.L., and C. Cutting. A proposal for the apprehension of curving craniofacial form in three dimensions, in *Craniofacial Morphogenesis and Dysmorphogenesis,* pp. 127–40 Vig, K. and A. Burdi (Eds.) (Ann Arbor, MI: Center for Human Growth and Development, 1988), pp. 127–140. With permission.)

finite element method may be used to examine the probable effects of growth following surgery. We feel it will take a number of years before surgical planning of craniofacial procedures reaches the high level of precision which is possible.

VII. SURFACE-BASED CEPHALOMETRICS AND STATISTICS

Surfaces extracted from CT and MR scans provide a far richer data set than that of traditional cephalometrics, but they continue to be less useful in the surgical planning process due to the lack of a quantitative cephalometric system for analyzing them. What is needed is a comprehensive system for comparing all aspects of facial surface form to a normative data set.[13] The first necessary ingredient in developing such a morphometric system is the establishment of a homology between every point of a patient's facial surface with a homologous point on a normative data set.[51] To do this requires that we continue the geometric simplification process two steps further. Thus far we have gone from volumetric data to surfaces. On these surfaces, ridge curves must be found. A diagram of the most important ridge curves on the skull is given in Figure 14. Curvature and torsion maxima on these ridge curves can be defined as landmark points. Differential geometric methods can be applied to a topologically continuous surface model to establish this geometric hierarchy.[23] At this point we intend to use heuristic methods to establish homology between a patient's landmark points and those of a normative data set in the cephalometric tradition. It is hoped that this process can be made as automatic as possible. It is hoped that such pattern recognition software will allow anatomic identification of structures to be done automatically.

From the point of heuristic landmark identification we expect that the machine can then

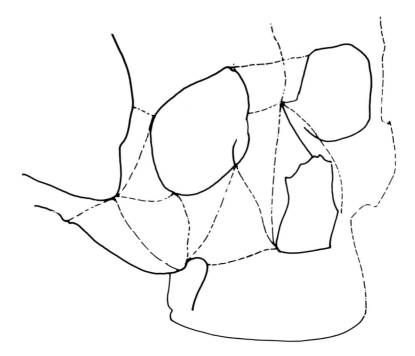

FIGURE 15. From ridge curves and the landmarks which are curvature and torsion maxima on them, it is hoped that a standard decomposition of the facial surface can be developed using geodesics between points to bound out standard "regions." This schema may provide the basis for a new surface-based cephalometric system. (From Bookstein, F.L., and C. Cutting. A proposal for the apprehension of curving craniofacial form in three dimensions, in *Craniofacial Morphogenesis and Dysmorphogenesis*, Vig, K. and A. Burdi (Eds.) (Ann Arbor, MI: Center for Human Growth and Development, 1988), pp. 127–140. With permission.)

again take over and proceed back down the geometric hierarchy. Ridge curve and geodesics between landmarks can be used to bound the surface into standard surface regions as shown in Figure 15. Parametric point assignments on these regions can be used to establish a homology between the points on the skull surface and a normative data set. This is the goal we sought a few paragraphs before.

Once surface homology mapping has been accomplished, certain problems in 3D CT become tractable. Statistical operations such as averaging and the measurement of deviations in shape become possible. Optimization of osteotomy fragment positioning can now be done based on all of the information available on the skull surface. This principle can be applied to laser scan data as well as surface data derived from 3D CT.

This type of semi-automatic homology mapping has implications for the radiologic community as a whole. The plastic surgery community will probably be content with the statistical comparison of surfaces. Being able to show a computer-generated image of an average postoperative result from a given surgical procedure with color coding of the surface to denote statistical deviation in shape from the norm will bring science into what is currently an artistic process. The radiology community will not be content to stop at surfaces but will require that statistical evaluation be done on the entire body volume. This is simply one more step down the geometric hierarchy. For example, assume a parametric patch on the outer surface of the brain and a similar patch at the grey matter-white matter interface across from it sensed by an MR scan. By the addition of another parameter for depth, a homology mapping can be made between the brain volume element of a patient in relation to that of a homologous normative data set. Differences between the acceptable Hounsfield numbers or density gradients for a voxel in the normal and that from a particular patient could be displayed using a color-coded volumetric display.

Fortunately the plastic surgeon has the easier task of only needing to perform statistical evaluations of surfaces. Currently the evaluation of the therapeutic efficacy of a given surgical procedure for a given common plastic surgery problem is not done in a rigorous way. The proponent of the procedure usually presents slides of his best results in support of the method. The average result and the variability of the procedure cannot be reported in this way, much less any statistical comparisons to the currently accepted treatment. Surface homology mapping makes these goals possible. Just being able to perform an average will be a major improvement. Variability will be less easily reported in an easily understood way to the clinician. Simple positional variability requires the use of some standard, global frame of reference which we feel should be avoided. Under the direction of the statistician in our group, Dr. Fred Bookstein, we will avoid the use of such superimposition methods and use instead coordinate system independent shape descriptors for this purpose. Principal curvatures and directions at each surface point would seem to be the most likely quantities used for this evaluation. A color-coded 3D surface graphic would be used to report the results of the evaluation to the clinician.

Growth adds another dimension to the problem of intuitive statistical reportage. For reasons alluded to previously in this chapter it is not acceptable to wait until a child's face is full grown prior to performing corrective surgery. Most craniofacial units agree that the psychosocial consequences of such a delay can be devastating to the patient. Stereophotogrammetry and automatic light scanners (and occasionally CT and MR scans) will provide 3D longitudinal growth data on these patients. Four-dimensional statistical analysis will not be easy to present to the clinician in an intuitive way. Color-coded 3D graphics are a major expense in publishing in current journals. The dimension of time would probably be best presented using animation. Videotape, film, or direct computer graphics output may become required media for the presentation of such data.

Another area of major importance in the study of craniofacial malformations is syndromology. Thus far we have been concerned about the comparison of an individual patient with the norm. Syndromology concerns itself with the grouping of patients with various congenital syndromes. How are Crouzon's, Pfeiffer's, and Apert's syndrome patients different? What numerical shape descriptors characterize the facies of each syndrome? This kind of study will be facilitated greatly by the development of a comprehensive surface-based cephalometric method. The light scanners will allow first-degree relative studies to be performed without risk and at no significant cost.

VIII. SURGICAL ROBOTICS

Thus far we have discussed computer modeling and planning of craniofacial surgical procedures, without any reference to how these procedures are to be carried out. Extremely precise planning will mean nothing if the surgeon is unable to perform the procedure with precision. Fortunately the majority of the acts the surgeon is called upon to perform are done well using current surgical methods with the exception of one. Making the incisions, performing the osteotomies, mobilizing the bone fragments, and fixing them in position with screws and plates is a well-developed craft. The only part of the surgical execution equation which is not well controlled is **bone fragment positioning.** With most of the operations currently in vogue, the surgeon uses large one- and two-piece osteotomy designs. The fragment positioning operation is often aided greatly by the orthodontist whose requirement of a functional dental occlusion determines the rigid motions of the bone fragments. Fortunately in many cases the dental and aesthetic requirements of fragment positioning are in relative harmony. In other cases however these requirements are quite disparate. Further, the intrinsic shape of the large fragment used may be incorrect and result in an undesirable aesthetic compromise.[21]

In these more severely malformed patients, multisegment craniomaxillary osteotomies are called for. Theoretically if the face were broken up into a thousand pieces, it would be possible to reposition them such that a normal composite skull shape would result. In reality this would be a surgical nightmare. Positioning and fixating each fragment would be impossible and the operation intolerably lengthy. The appropriate approach would seem to be to use the fewest number of bone fragments which would bring the patient's postoperative facial appearance within the normal range. In many cases this can be achieved with a single "monobloc".[43] In other cases cutting the midface into five or six separate fragments may be required. It is this circumstance which makes it difficult for the craniofacial surgeon to perform the operation with precision. Fixation has been facilitated recently by the adoption of small screws and plates. Unfortunately no similar technical breakthroughs have been forthcoming in the precise intraoperative positioning of the fragments. The reader may ask why this cannot simply be done by inspection. During the performance of these procedures, biological necessities prevent the surgeon from appreciating the craniomaxillary skeleton as a totality. The cheek-bones and paranasal regions present the most difficulty. In order to avoid large incisions on the face, the midfacial skeleton is approached from inside the mouth and through the scalp behind the hairline. The soft tissue is then peeled off the bone allowing exposure to the midface. Unfortunately the optic nerves and the muscle cone must remain attached to the orbital apex. This prevents a clear view of all of the zygomata and paranasal area from above. The intraoral route affords a view of these structures from below but alas at an odd angle and with the loss of the forehead as a reference frame. At all times the usually edematous soft tissue of the midface overlies the bone to some degree. This is not simply sculpture. The artistic method is being hampered by biological necessity. It is the author's belief that this is why obviously more flexible midface shaping using multiple-segment maxillary osteotomies is not common practice at the present time.

The tools of robotics offer some benefits in the bone fragment positioning problem. Robotics involves the execution and verification of precise, reproducible, 3D rigid motions. Our laboratory has been experimenting recently with the verification of correct 3D rigid motions using a Polhemus Navigation Systems 3 Space Digitizer. This device employs electromagnetic fields to report the 3D position (X, Y, and Z coordinates) and the 3D rotational attitude (i.e., Euler's angles) which completely specify rigid motions of a pencil-like stylus. Thus far we have used this device to link dental model surgical simulations to cephalometric and CT-based simulations described previously by our group. This involves first establishing coordinate system identity between the 3D cephalometric model of the teeth and those same teeth as found on the dental model. Matching of at least three noncollinear points is required for this part of the procedure. Then the stylus is rigidly linked to the upper dental model so that movements of the model are in synchrony with the cephalometric simulation. This makes it possible to appreciate what effect establishing the correct occlusion will have on the shape of the midface in large one-piece craniomaxillary procedures. Alternately the device can be used for the construction of intermediate occlusal splints. It was our original intention to take the device to the operating room to verify correct bone fragment positioning (Figure 16). Unfortunately the device is only accurate when operated in an environment free of ferrous metal and stray electromagnetic fields. For this reason intraoperative rigid motion measurement may be better done with another instrument.

Recently we have been considering the use of an active robot to correctly position multiple osteotomy fragments in collaboration with Dr. Russell H. Taylor of the IBM T. J. Watson Research Center. Prior to completing the osteotomies, percutaneous pins could be placed in

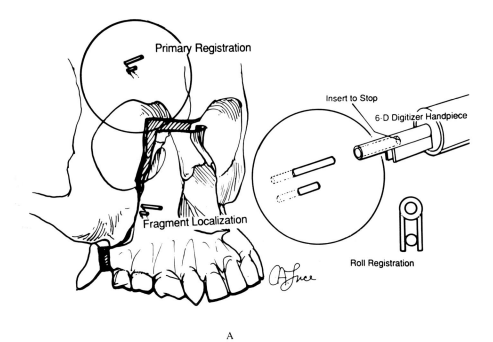

A

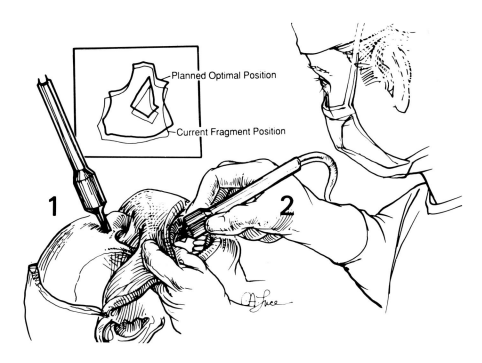

B

FIGURE 16. 6d digitizer as it might be used in an operating room. (A) Two small K-wires would be placed in each bone fragment prior to completing osteotomies. Baseline readings are taken of the relationships between the K-wires, then the osteotomies are completed. (B) After moving the fragments, the position of the midface (2) is compared with that of the skull base for reference (1). This electronic caliper could be used to check the accuracy of the bone positioning operation.

each of the bone fragments and their resting orientations measured. Using the skull base as a frame of reference, a small robotic arm would grasp the fragment and position it correctly following completion of the osteotomy. The surgeon may then concentrate on placing bone grafts and applying screw and plate fixation. In this way multisegment midface osteotomies could be reproducibly performed with great precision.

Robotics has already been used for the milling of custom implants. The Cemax Company currently offers clinicians this capability as a service. This process is most useful in tumors and trauma where a defect has been produced on one side of the face while the other side is normal. Using mirror imaging techniques, a subtraction image of the defect is produced. Using a computer-controlled milling machine the missing part can be manufactured precisely. As surgical planning improves, deficient faces could be built up in this way in congenital anomalies as well. Currently the absence of a completely satisfactory implant material for use in the face has been a stumbling block to this approach. These implants are also being used as models for the shaping of autogenous bone graft implants.

Computer-controlled milling also allows the production of a prototype skull of a patient. Surgical simulation can then be performed on a lifelike model in the most natural way. Relatively high cost, destruction caused by simulated surgery, and the abandonment of a digital environment which allows the measurement of the result and the optimization of fragment positioning are drawbacks. A model of a milled skull using this process is shown in Figure 17.

IX. CLOSING REMARKS

This chapter has tried to cover the history of computer graphics applications to the treatment of craniofacial malformations, discussed current research, and passed into ideas for

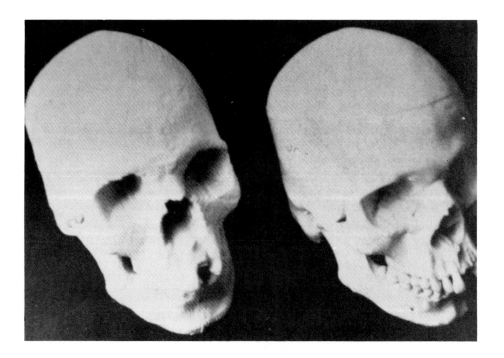

FIGURE 17. Milled skull on right from CT-scanned data of skull on left. (Photo courtesy of the Cemax Corporation.)

the future. While much of the chapter is simply science fiction at the time of this writing, the author is convinced that it will not remain so and that care of patients with severe craniofacial malformations will improve as a result.

ACKNOWLEDGMENTS

The author wishes to thank his collaborators in various parts of the research discussed herein. Many of the ideas expressed in this chapter are theirs or partly theirs. Dr. Joseph G. McCarthy (plastic surgeon), Dr. Barry Grayson (orthodontist), and Mr. Hiechun Kim (cephalometric technician and computer programmer) at the Institute of Reconstructive Surgery at New York University are at the core of the craniofacial research team. Dr. Fred Bookstein (University of Michigan-Center for Human Growth) is responsible for most of the morphometrics and statistical ideas expressed herein. Dr. Marilyn Noz (New York University) and Dr. Gerald Maguire (Columbia University) contributed their QSH software to this project. Dr. Kenneth Perlin and Dr. Robert Hummel of the New York University Department of Computer Science provided the principal computer science faculty support for the project. I would also like to thank Dr. Russell H. Taylor of the IBM T. J. Watson Research Lab in Yorktown Heights, New York for participating with us and generating most of the ideas expressed in the robotic bone fragment positioning research. Last, but not at all least, I would like to thank the graduate students, Mr. Alan Kalvin, Mrs. Betsy Haddad, Mr. Daniel Karron, and Ms. Patricia Bedard, for the coding of their original software.

REFERENCES

1. Arridge, S., J. Moss, A. Linney and D. James. Three dimensional digitization of the face and skull, *J. Maxillofac. Surg.* 13:136 (1985).
2. Baumgart, B.G. Geometric Modelling for Computer Vision, Ph.D. dissertation at Stanford University, University Microfilms, August 1974.
3. Baumrind, S., and D. Miller. Computer-aided head film analysis: The University of California San Francisco Method, *Am. J. Orthod.* 78:41 (1980).
4. Baumrind, S., F. Moffit, and S. Curry. Three-dimensional x-ray stereometry from paired coplanar images: A progress report, *Am. J. Orthod.* 84:292 (1983).
5. Baumrind, S., F. Moffit, and S. Curry. The geometry of three-dimensional measurement from paired coplanar x-ray images, *Am. J. Orthod.* 84:313 (1983).
6. Berkowitz, S., and J. Cuzzi. Biostereometric analysis of surgically corrected abnormal faces, *Am. J. Orthod.* 72:526 (1977).
7. Berkowitz, S. Automatic laser scanning of dental models, in Abstracts of the Meeting of the American Cleft Palate Association (San Antonio, TX, 1987).
8. Bhatia, S., and J. Sowray. A computer aided design for orthognathic surgery, *Br. J. Oral Maxillofac. Surg.* 22:237 (1984).
9. Bite, U., I. Jackson, G. Forbes, and D. Gehring. Orbital volume measurements in enophthalmos using three-dimensional CT imaging, *Plast. Reconstr. Surg.* 75:502 (1985).
10. Bookstein, F.L. The Measurement of Biological Shape and Shape Change, Lecture Notes in Biomathematics, v. 24 (Berlin: Springer-Verlag, 1978).
11. Bookstein, F.L. Tensor biometrics for changes in cranial shape, *Ann. Hum. Biol.* 1413–1437 (1984).
12. Bookstein, F.L. A statistical method for biological shape comparisons, *J. Theor. Biol.* 107:475 (1984).
13. Bookstein, F.L., and C. Cutting. A proposal for the apprehension of curving craniofacial form in three dimensions, in *Craniofacial Morphogenesis and Dysmorphogenesis* Vig, K. and A. Burdi (Eds.) (Ann Arbor, MI: Center for Human Growth and Development, 1988), pp. 127–140.
14. Brewster, L., S. Trevedi, H. Tuy, and J. Udupa. Interactive surgical planning, *IEEE Comput. Graph. Appl.* 4:31 (1984).
15. Broadbent, B.H., B.H. Broadbent, and W. Golden. *Bolton Standards of Developmental Growth* (St. Louis, MO: C.V. Mosby, 1975).

16. Cutting, C., F. Bookstein, B. Grayson, and J.G. McCarthy. Computer aided planning of orthognathic surgery, in Abstracts Plastic Surgery Research Council Meeting, Detroit, April 1–4, 1984.

17. Cutting, C., F. Bookstein, B. Grayson, L. Fellingham, and J. McCarthy. Three-dimensional computer assisted design of craniofacial surgical procedures: Optimization and interaction with cephalometric and CT-based models, *Plast. Reconstr. Surg.* 77:877 (1986).

18. Cutting, C., B. Grayson, F. Bookstein, L. Fellingham, and J.G. McCarthy. Computer-aided planning and evaluation of facial and orthognathic surgery, *Clin. Plast. Surg.* 13:449 (1986).

19. Cutting, C., F. Bookstein, B. Grayson, L. Fellingham, and J.G. McCarthy. Three-dimensional computer-aided design of craniofacial surgical procedures, in *Craniofacial Surgery,* Marchac, D. (Ed.) (Berlin: Springer-Verlag, 1987), p. 17.

20. Cutting, C., J.G. McCarthy, and D. Karron. Three-dimensional input of body surface data using a laser light scanner, *Ann. Plast. Surg.* 21:38 (1988).

21. Cutting, C., J.G. McCarthy, B. Grayson, F. Bookstein, and H. Kim. Case for multisegment osteotomies in Crouzon's disease, manuscript in preparation (1989).

22. Dufresne, C., J. McCarthy, C. Cutting, F. Epstein, and W. Hoffman. Volumetric quantification of intracranial and ventricular volume following cranial vault remodelling: A preliminary report, *Plast. Reconstr. Surg.* 79:24 (1987).

23. Frobin, W., and E. Hierholzer. Analysis of human back shape using surface curvatures, *J. Biomech.* 15:379 (1982).

24. Grayson, B., C. Cutting, F. Bookstein, H. Kim, and J.G. McCarthy. The three-dimensional cephalogram: Theory, technique, and clinical application, *Am. J. Orthod. Dentofac. Orthop.* 94:327 (1988).

25. Harradine, N., and D. Birnie. Computerized prediction of the results of orthognathic surgery, *J. Maxillofac. Surg.* 13:245 (1985).

26. Herman, G., and H. Liu. Display of three-dimensional information in computed tomography, *J. Comput. Assist. Tomogr.* 1:155 (1977).

27. Herman, G., and H. Liu. Three-dimensional display of human organs from computed tomograms, *Comput. Graph. Image Proc.* 9:1 (1979).

28. Kaplan, E. 3-D CT images for facial implant design and manufacture, *Clin. Plast. Surg.* 14:663 (1987).

29. Maguire, G., and M. Noz. Standardizing the raster display for medical images using a fixed set of frame buffer primitives, *J. Med. Sys.* 10:209 (1986).

30. Maguire, G., J. Jaeger, L. Farde, and M. Noz. Use of graphic techniques for error evaluation, *J. Med. Sys.* 11:277 (1987).

31. Marchac, D., and D. Renier. *Craniofacial Surgery for Craniosynostosis,* (Boston, MA: Little, Brown, 1982).

32. Marsh, J., and M. Vannier. The "third" dimension in craniofacial surgery, *Plast. Reconstr. Surg.* 71:759 (1983).

33. Marsh, J., and M. Vannier. *Comprehensive Care for Craniofacial Deformities* (St. Louis, MO: C.V. Mosby, 1985).

34. Marsh, J., M. Vannier, W.G. Stevens, J. Warren, D. Gayou, and D. Dye. Computerized imaging for soft tissue and osseous reconstruction in the head and neck, *Clin. Plast. Surg.* 12:279 (1985).

35. Marsh, J., M. Vannier, and R. Knapp. Computer assisted surface imaging for craniofacial deformities, in *Advances in Plastic and Reconstructive Surgery,* v. 2 Hebal, M. (Ed.) (Chicago: Year Book Medical Publishers, 1985).

36. Marsh, J., S. Celin, M. Vannier, and M. Gado. The skeletal anatomy of mandibulofacial dysostosis (Treacher Collins syndrome), *Plast. Reconstr. Surg.* 78:460 (1986).

37. McEwen, C., I. Jackson, and R. Robb. Craniofacial surgical simulation additions to the Mayo Clinic ANALYZE software, Personal communication (1987).

38. Merz, B. Computer guides facial reconstruction [news], *JAMA* 249:1409 (March 18, 1983).

39. Moshiri, F., S. Jung, A. Sclaroff, J. Marsh, and W. Gay. Orthognathic and craniofacial surgical diagnosis and treatment planning: A visual approach, *J. Clin. Orthod.* 16:37 (1982).

40. Moss, J., A. Linney, S. Grinrod, S. Arridge, and J. Clifton. Three-dimensional visualization of the face and skull using computerized tomography and laser scanning techniques, *Eur. J. Orthod.* 9:247 (1987).

41. Moss, J., S. Grinrod, A. Linney, S. Arridge, and D. James. A computer system for the interactive planning and prediction of maxillofacial surgery, *Am. J. Orthod. Dentofac. Orthop.* 94:469 (1988).

42. Noz, M., and G. Maguire. QSH: A minimal but highly portable image display and handling toolkit, in *Computer Methods and Programs in Biomedicine* (in press, 1988).

43. Ortiz-Monasterio, F., A. Fuente del Campo, and A. Carrillo. Advancement of the orbits and the midface in one piece, combined with frontal repositioning, for the correction of Crouzon's deformities, *Plast. Reconstr. Surg.* 61:507 (1978).

44. Press, W., B. Flannery, S. Teukolsky, and W. Vetterling. Downhil simplex method in multidimensions, in *Numerical Recipes* (Cambridge: Cambridge University Press, 1988), p. 305.

45. Ricketts, R., R. Bench, J. Hilgers, and R. Schulhof. An overview of computerized cephalometrics, *Am. J. Orthod.* 61:1 (1972).

46. Ricketts, R. Divine proportion in facial esthetics, *Clin. Plast. Surg.* 9:401 (1982).

47. Riolo, M.L., R.E. Moyers, J.S. McNamara, and W.S. Hunter. An Atlas of Craniofacial Growth, Monograph No. 2, Craniofacial Growth Series, Center for Human Growth and Development, University of Michigan, 1974.

48. Rockey, K., H. Evans, D. Griffiths, and D. Nethercot. *The Finite Element Method* (New York, NY: John Wiley & Sons, 1983).

49. Savara, B., S. Miller, R. Demuth, and H. Kawamoto. Biostereometrics and computergraphics for patients with craniofacial malformations: Diagnosis and treatment planning, *Plast. Reconstr. Surg.* 75:495 (1985).

50. Stalnecker, M., L. Whitaker, H. Rosen, G. Herman, and J. Upuda. Applications of volume determination and three-dimensional imaging, in *Craniofacial Surgery,* Marchac, D. (Ed.) (Berlin: Springer-Verlag, 1987), p. 25.

51. Thompson, D. On the theory of transformations, or the comparison of related forms, in *On Growth and Form,* Bonner, J.T. (Ed.) (Cambridge: Cambridge University Press, 1961), pp. 268–325.

52. Toth, B., W. Stewart, and L.F. Elliot. Computer designed prostheses for orbitocranial reconstruction, in *Craniofacial Surgery,* Marchac, D. (Ed.) (Berlin: Springer-Verlag, 1987).

53. Toth, B., D. Ellis, and W. Stewart. Computer designed prosthesis for orbitocranial reconstruction, *Plast. Reconstr. Surg.* 81:315 (1988).

54. Tuy, H., and J. Udupa. Representation, display and manipulation of 3D objects using directed contours, Proceedings of 18th Hawaiian International Conf. on System Sciences 2:397 (1983).

55. Udupa, J. Interactive segmentation and boundary surface formation for 3D digital images, *Comput. Graph. Image Proc.* 18:213 (1982).

56. Udupa, J. Computerized surgical planning: Current capabilities and medical needs, *Soc. Photoopt. Instr. Eng. Proc.* 626:474 (1986).

57. Udupa, J., G. Herman, P. Margasahayam, L. Chen, and C. Meyer. A turnkey system for the display and analysis of 3D medical objects, *SPIE Proc.* 671:154 (1986).

58. Vannier, M., J. Marsh, and J. Warren. Three-dimensional CT reconstruction images for craniofacial surgical planning and evaluation, *Radiology* 150:179 (1984).

59. Walters, H. et al. Computerized planning of maxillofacial osteotomies: The program and its clinical applications, *Br. J. Oral Maxillofac. Surg.* 24:178 (1986).

60. Willmott, D.R. Soft tissue profile changes following correction of class III malocclusions by mandibular surgery, *Br. J. Orthod.* 8:175 (1981).

61. Zienkeiwicz, O. *The Finite Element Method* (London: McGraw-Hill, 1977).

Chapter 7

THREE-DIMENSIONAL IMAGING OF THE MUSCULOSKELETAL SYSTEM

W. Gregory Wojcik and John H. Harris, Jr.

TABLE OF CONTENTS

I. INTRODUCTION

The application of three-dimensional (3D) reformation of axial computed tomographic (CT) data in radiologic diagnosis had its genesis in the work of Herman and Liu[1] in 1977. Marsh and Vannier[2] and Herman and Udupa[3] both in 1983 demonstrated its practical clinical application. Subsequently, and as a result of software and hardware refinements, 3D CT has become generally, and world-wide, accepted as an essential component of diagnostic imaging of anatomically complex areas of the skeleton and to a lesser, but increasing, degree in preoperative surgical planning. The reader is referred to the Selected Readings for a list of references citing the applications of 3D CT to various skeletal parts.

3D reformation of axial CT data is a logical and extremely useful extension of the concept of multiplanar reformation. The compelling advantage of 3D CT is that the picture of the *in vivo* skeletal part being examined is displayed as though it were a dried skeleton (Figure 1). This unique display of surface anatomy demonstrates spatial relationships of complex skeletal anatomy and pathology unmatched by any other imaging modality. The information provided by the numeric integration of axial CT data avoids the ambiguity, or potential for ambiguity, inherent in the mental integration required in analysis of two-dimensional (2D) multiplanar images.

We have directly and personally investigated three systems for 3D imaging. The first was that developed by the Medical Image Processing Group at the University of Pennsylvania, 3D83 and 3D98 for use on GE 8800 and/or 9800 scanners.[3] The second was the Cemax 1000[4] imaging system developed by Cemax, Inc. of California. The final system we evaluated, and in which we actively participated in its development, is the "Quick 3-D" algorithm developed and marketed by GE Medical Systems, which has been in use in our department for the past 2 years.[5] For the past 12 months this algorithm has been used on a routine clinical basis.

In comparing the system we use to the others, the most obvious improvement is the remarkable speed of processing the data using only standard CT equipment. Because this algorithm is designed to be nonsequential, the software permits parallel processing. The array processor, inherent in the CT scanner, reduces processing time to the degree that reformations may be done between consecutive examinations. The program is user friendly. Informative menus are displayed at every step of the reformation prescription, which greatly facilitates user training. This program can also be implemented on the GE 9800 independent console (when it is equipped with an array processor) and on the CT and MR independent workstation currently under development using SUN Microsystems equipment.

II. METHOD AND TECHNIQUE

The method described here is that preferred at The University of Texas Medical School at Houston and Hermann Hospital.

The original CT examinations are performed on a GE 9800 "Quick" CT scanner using routine CT techniques appropriate to the part being examined. The 3D reformations are derived from axial CT data obtained from 3-mm thick contiguous slices.

Prescription of the reformation requires approximately 1 minute, surface data generation approximately 4 minutes, and a new image of the surface in different projection is generated every 3 to 4 seconds. While selecting the surface to be displayed in 3D reformation, the radiologist also establishes the combination of angle (ambient light) shading and depth shading that optimally shows the area of interest.

Generation of the 3D surface and of standard 3D images can be performed by a CT technologist following specific diagnostic protocols established for different anatomic parts.

Our routine protocols result in multiple images depicting the surface from different perspectives with the object being rotated, usually in 30 degree increments. Depending on the

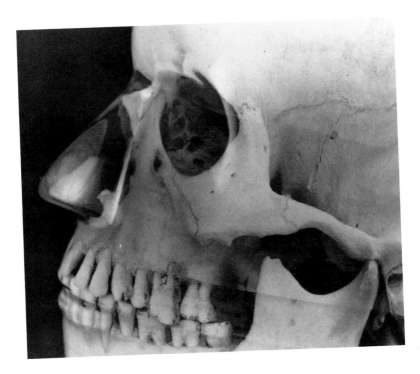

A

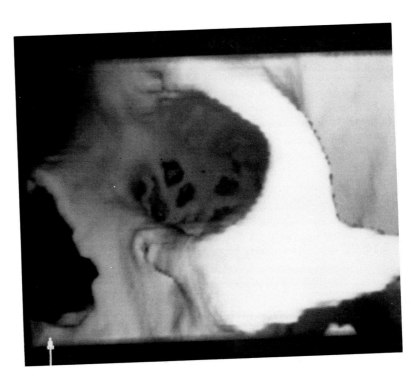

B

FIGURE 1. (A) Skull encased in lucite representing a head phantom. (B) 3D reformation of the area of the left orbit obtained from the CT study of the phantom (A) using 1.5-mm thick contiguous axial cuts.

anatomic area and the clinical question, the average study consists of 18 to 30 images in different projections. The examination includes 3D cross-sectional and/or other images in which portions of the anatomy have been numerically deleted to visualize underlying structures. Such special projections are prescribed by the user as dictated by each clinical problem. The algorithm we employ provides the capability of obtaining cross-sectional images without the need to prescribe a new surface. All images should be reviewed by a radiologist on the monitor so that additional images may be generated, if necessary, to optimally demonstrate the area in question.

Depending upon patient throughput and/or clinical urgency for obtaining additional 3D images, the additional images may be obtained at the operator's console, the physician's console, or the axial data may be transferred to a magnetic tape for manipulation at a remote independent console. Thus, 3D reformations are obtainable on a clinically timely basis. The current software version does not permit simultaneous scanning and manipulation.

3D reformation is essentially a conceptual extension of the process that provides sagittal and coronal reformation from axial CT data. 3D reformation requires no modification of the radiographic technique used to obtain multiplanar CT (MPCT) images. Since there is no clinically relevant improvement in 3D diagnostic accuracy by using 1.5 mm thick "slices" in the vast majority of clinical examinations, 3D reformation from routine 3 mm axial data does not prolong the initial CT examination. The "dividing cube"[6,7] algorithm used in this system essentially interpolates the data contained in consecutive axial images and creates many axial data sets which would correspond to slices intervening between original axial cuts. The localization of the elements of the 3D surface and the angulation of the plane of the surface in every point of the object (defined by "surface normal") is calculated as a product of interpolated densities in a cube surrounding each point. The "dividing cube" algorithm, in our experience, results in improved accuracy of the surface representation in comparison to the previous versions of 3D software.

Many differences exist between various 3D display systems which may make one system appear superior to another. However, user-dependent factors, such as mastering optimal technique of the examination, are essential regardless of the equipment used. Application of the factors listed below have vastly improved image quality and the diagnostic quality of the examination in our clinical experience.

Slice thickness of the axial CT examination from which the 3D reformation is derived is important. Optimum visualization of intricate, complex structures requires thin axial slices. However, the image quality improvement derived from 1.5-mm axial slices (Figure 2) usually does not justify the extended examination time. In our practice, the vast majority of 3D reformation is performed using 3-mm thick contiguous slices. This protocol establishes a clinically appropriate balance between image quality and time of data acquisition. The slice interval (overlapping vs. contiguous slices) also influences quality of 3D images.[8] In our opinion, the overriding reason for use of contiguous (nonoverlapping) slices is the ease of detection of artifacts related to change of patient position between consecutive slices. When such change is slight, it may be misinterpreted and lead to diagnostic errors. Such motion-related artifacts are more apparent when contiguous slices are used; thus making misinterpretation less likely.

Size of the region of reformation—The area to be reformatted in 3D should be tightly cropped to display the anatomy which is clinically relevant in order to obtain optimum high-resolution images. The influence of the size of the area of reformation on the image resolution is demonstrated in Figures 3 to 5.

Surface shading—The illusion of three-dimensionality is created by varying the shade of the image depending on the angulation of the elements of the surface and their distance in relation to the source of light. Quick 3D allows much flexibility in the surface display. By

A

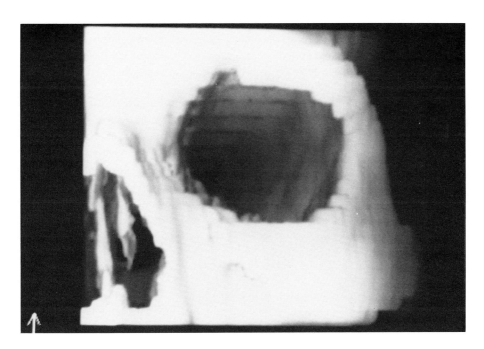

B

FIGURE 2. Effect of slice thickness on 3D image. 3D reformations of the region of the left orbit obtained from the CT examinations of the head phantom with 5-mm thick contiguous axial slices (A), 3 mm (B), and 1.5 mm (C). All other parameters were unchanged. The thicker the axial slices, the more obvious are the stepladder artifacts. These artifacts are much more apparent on structures which are nearly parallel to the plane of the slices (as the roof of the orbit) than on structures which are perpendicular (lateral wall of the orbit). Although image (C) is of clearer resolution than image (B), in our experience the improved resolution afforded by the 1.5-mm thick slices is usually not clinically significant.

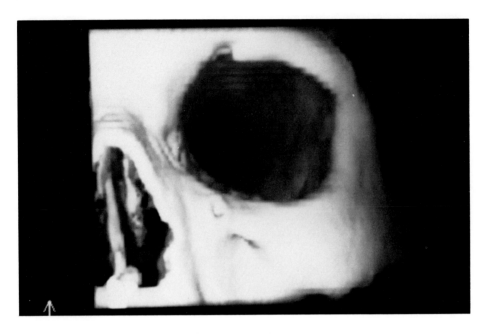

FIGURE 2C

varying the component of depth and angle (ambient light) shading the character of the surface display can be tailored to the clinical question. Figure 6 demonstrates images of the same surface displayed using different combinations of ambient light and depth shading. Figure 7 demonstrates differences of the surface display by different 3D algorithms. Combined effect of different surface shading and different 3D algorithms is demonstrated.

Radiographic exposure techniques—3D CT is utilized to demonstrate almost exclusively bone-soft tissue and air-skin interface. Because of their high contrast these structures can be well demonstrated even in the presence of a high noise (low signal-to-noise ratio) (Figure 7 and 4). This fact permits 3D images to be obtained with as much as 80% reduction of the conventional CT examination mAs. This reduction in radiation dose has practical application in pediatric patients, pregnant patients, and others who require repeated 3D imaging.

Using such low technique is appropriate when the images of soft tissues are clinically irrelevant and obtaining 3D images of bones is the main reason for the examination.

III. RATIONALE FOR 3D IMAGING AND CLINICAL APPLICATIONS

The primary purpose of radiologists is to provide the most accurate diagnostic information possible within the capabilities of available imaging modalities. That tenet has motivated the development of MPCT and, more recently, MRI. However, even with this remarkable level of imaging sophistication, the end-product image was only a 2D presentation of a 3D anatomic part until recent CT software modifications made 3D reformation practical and clinically useful.

Many circumstances occur in skeletal radiology when diagnostic accuracy and conceptual understanding of the skeletal abnormality would be greatly improved if the part in question could be displayed as if the surface of the skeleton were being observed directly and in those projections that demonstrate the abnormality and its spatial relationships to best advantage. The ability to view CT images in any plane about any axis is not unique to 3D CT. However, the ability to observe the skeletal surface and spatial anatomy and pathology are unique features of 3D CT that make it diagnostically superior to other imaging systems.

Conversion of the 2D data of conventional images into useful clinical information requires mental integration of these data into the 3D perspective. Mental integration of all of the images frequently obtained to evaluate a complex skeletal problem (i.e., plain radiographs, polydirectional tomography, and MPCT) is frequently difficult even for experienced radiologists and is even more difficult for less experienced radiologists or nonradiologists.

Mental integration of multiple images is subjective and characterized by individual variation or interpretation related to factors other than direct observation of the image such as

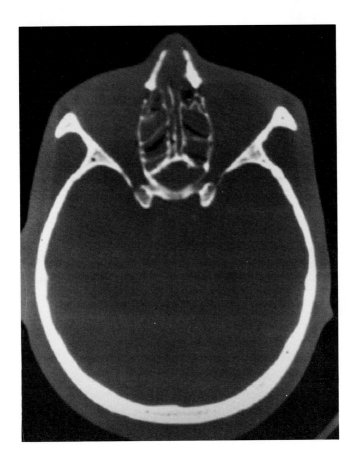

A

FIGURE 3. Effect of size of area of reformation. Images from a study of the head phantom obtained using 3-mm thick contiguous slices. The axial image (A) demonstrates the area of 3D reformation (*rectangular cursor*) containing entire skull. Part (B) is the 3D reformatted image of the skull from the same study using area of reformation shown in (A). Present version of the Quick 3D algorithm in large areas of reformation somewhat compromises resolution for increased speed for processing. Quality of a 3D image of a large object is only slightly degraded (B) because in large objects, the large pixel size of the final 3D display is the limiting factor in image resolution. Part (C) is the numeric magnification of the selected area from image (B) obtained on the operator's console which highlights inaccuracies of surface representation of image (B).

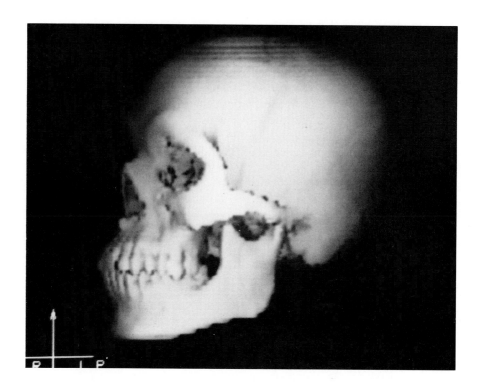

FIGURE 3B

FIGURE 3C

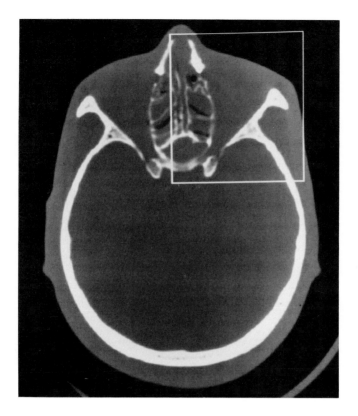

A

B

FIGURE 4. Effect of size of area of reformation. Part (A) is the same axial slice as 3(A). The cursor defines a smaller area of 3D reformation. Part (B) is the 3D reformatted image derived from this small area of interest. Although image 4(B) was derived from the same axial data as the image 3(C), there is remarkable improvement of resolution of image 4(B) compared to Figure 3C. The size of the area of 3D reformation should be as small as clinically feasible.

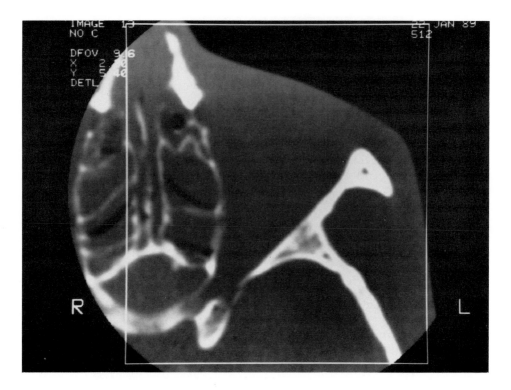

A

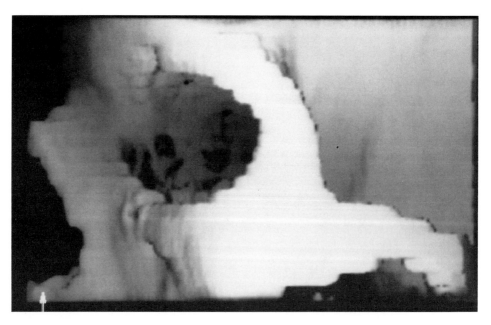

B

FIGURE 5. Effect of size of region of reformation. Part (A) is an image obtained from the same raw data as the images 3(A) and 4(A). Axial slices from the same study of the head phantom were retrospectively reconstructed using a smaller field of view. The rectangular cursor defines the area for 3D reformation. (B) 3D image obtained from the set of axial images as is shown in (A). The improvement of the resolution is clinically insignificant compared to image 4(B).

knowledge of the anatomy of the part, knowledge or understanding of the skeletal pathology, and understanding of displayed pathophysiology. These vagaries may be the genesis for ambiguities in roentgen interpretation that may occur even between experienced radiologists viewing the same set of radiographs simultaneously. This problem is compounded when one experienced in mental integration of multiple images attempts to transmit the clinically relevant information derived from the mentally integrated images to another individual or group.

The cases displayed in Figures 8 to 16 have been selected to emphasize and illustrate the importance of surface display and spatial relationships afforded by the 3D images of 3D CT imaging in the resolution of complex clinical problems involving the skeleton.

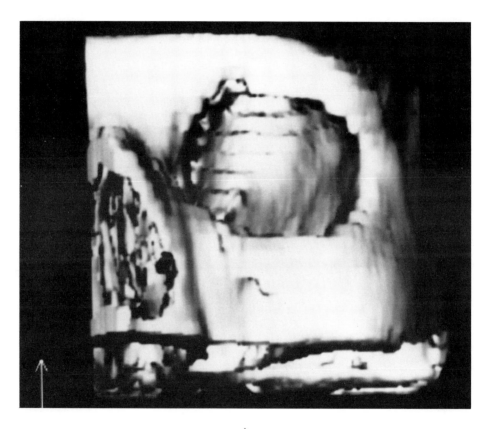

A

FIGURE 6. The character of the surface shading is directly effected by the amount of either ambient light or depth shading selected by the user. The effect of various combinations of each are illustrated in the following figure parts. The ambient light is inversely related to angle shading. The higher value selected for ambient light, the more angle shading is diminished. (A) Minimal allowable ambient light and minimal allowable depth shading. This kind of display accentuates influence of angulation of the surface (angle shading) giving a "metallic" quality to the surface and generating electronic artifacts. (B) Minimal allowable ambient light and maximal allowable depth shading. (C) Maximal allowable ambient light and maximal depth shading. This combination suppresses the influence of the angulation of the surface contour rendering the image flat and diminishing the 3D illusion. (D) Maximal allowable ambient light and minimal depth shading. The influence of surface angulation and depth are both suppressed. Real and pseudo-foramina stand out. (E) Combination of angle (ambient light) and depth shading that has proved to be optimum for most initial skeletal 3D examinations.

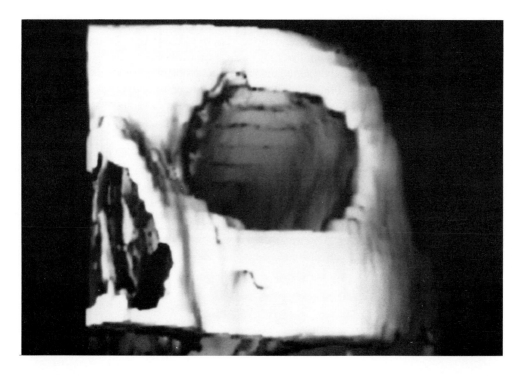

FIGURE 6B

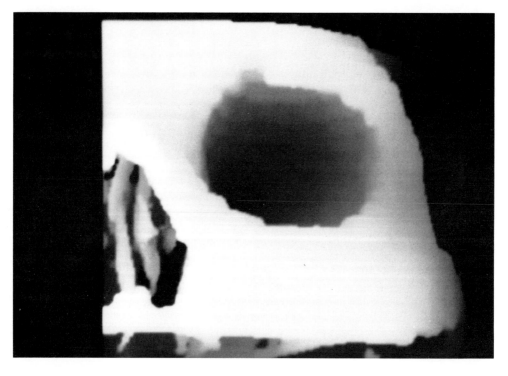

FIGURE 6C

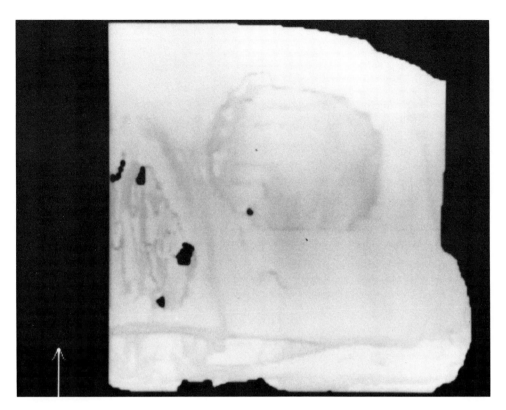

FIGURE 6D

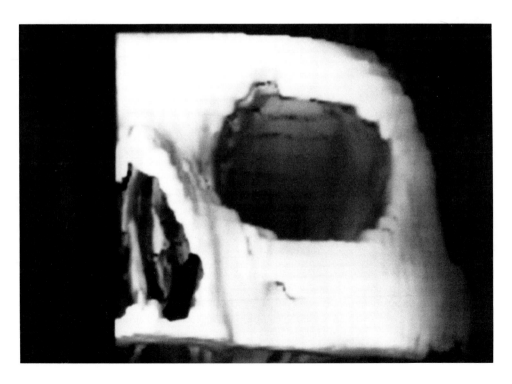

FIGURE 6E

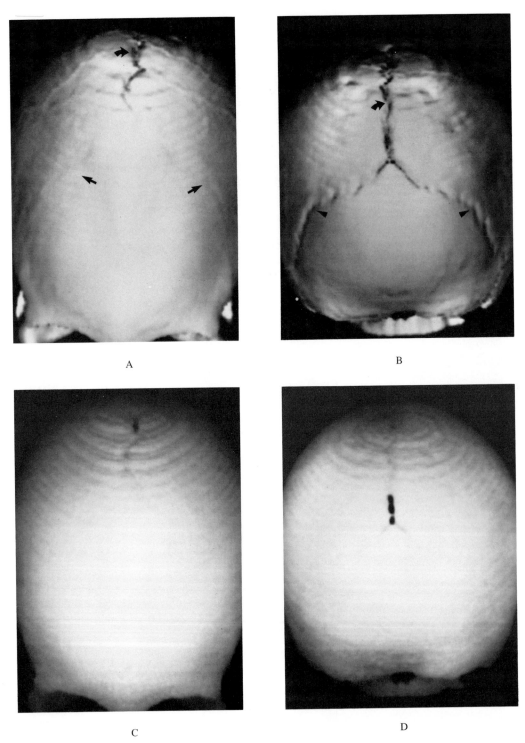

A

B

C

D

FIGURE 7. Improved accuracy of surface display afforded by dividing cube algorithm augmented by the flexibility of shading is demonstrated in figures (A and B). Images A to D were all obtained from the same data using the same threshold level but with different algorithms. The algorithms used to obtain images (A) and (B) permits operator manipulation of surface shading to optimally address the clinical question. The second program in its present version does not. In this example, the normal sagittal (*curved arrows*) and lambdoidal (*arrowheads*) sutures (A, B) and the prematurely fused coronal sutures (*arrows*) (A) are well demonstrated with the user-flexible algorithm whereas the same structures are barely visible in the fixed algorithm (C and D) images. Consequently, the diagnosis of coronal craniostenosis is made with greater confidence in images (A) and (B).

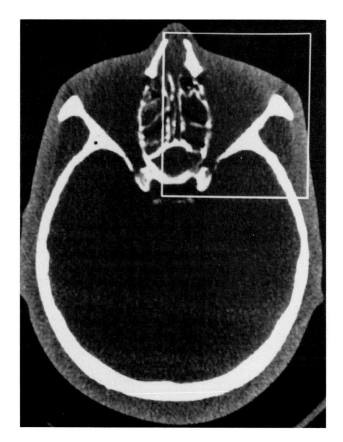

A

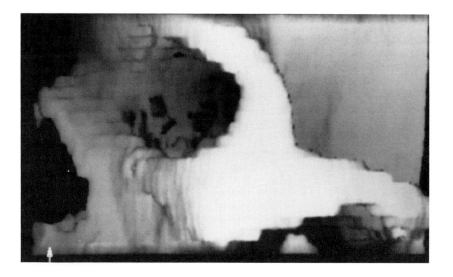

B

FIGURE 8. Effect of radiographic exposure. Images of the head phantom obtained using 40 mA, 120 KVp, and 2 sec scan (A) demonstrate the "grainy" quality of the simulated soft tissues caused by the high noise. The 3D image derived from this examination (B), however, is almost as good as image 4(B) obtained using 170 mA, 120 KVp, and 2 sec scan. In spite of the degraded axial image obtained from the low mAs, the 3D image derived with these technical factors very closely approximates the image quality of that derived from high mAs.

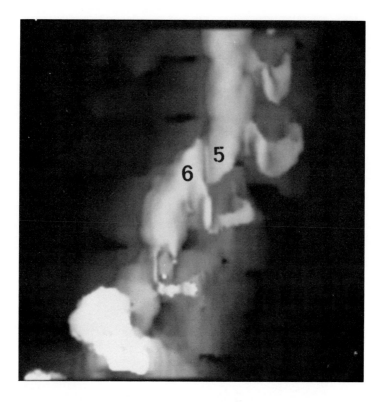

FIGURE 9. C_{5-6} right unilateral interfacetal dislocation. The articular mass of C_5 is unambiguously shown to be anteriorly dislocated with respect to that of C_6 on the 3D image.

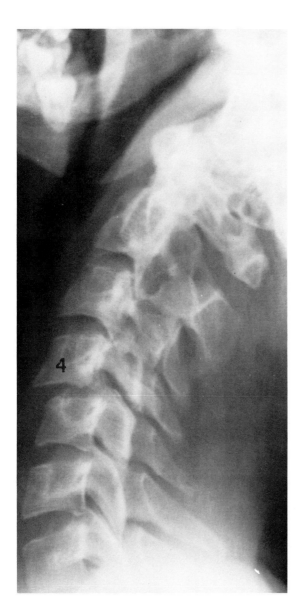

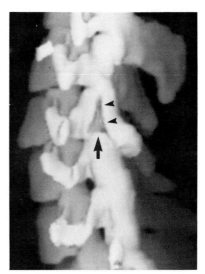

A   B

FIGURE 10.   Unilateral interfacetal fracture dislocation of C_4 on C_5 is clearly evident in the plain lateral radiograph (A). Only 3D CT (B) demonstrated the clinically significant vertical fracture (*arrowheads*) of the incompletely dislocated articular mass of C_4 and the transverse fracture (*arrow*) of the superior articular anterior process of the lateral mass of C_5.

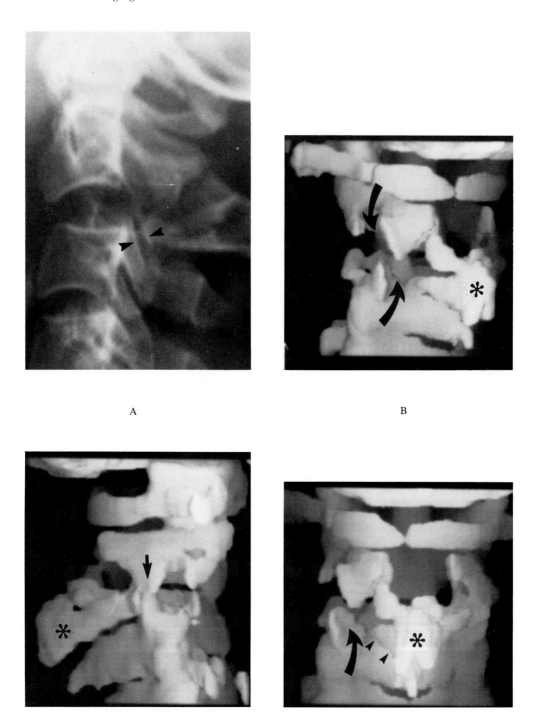

FIGURE 11.   Effendi Type II traumatic spondylolisthesis with caudal displacement of the posterior fragments. On the plain lateral radiograph (A), the C_{2-3} interfacetal joint spaces (*arrowheads*) appear to be normally maintained. The 3D images (B to D) all confirm caudal displacement of the comminuted posterior fragments (*). A major diastasis of the C_{2-3} interfacetal joint (*curved arrows*) is recorded on the right (B) and an incomplete C_{2-3} interfacetal fracture-dislocation (*arrows*) on the left (C). The amount of left interfacetal joint diastasis (*curved arrows*), caudal displacement of the comminuted posterior fragments (*), and the unrecognized C_3 right laminar fracture (*arrowheads*) are all clearly evident on the posterior view (D).

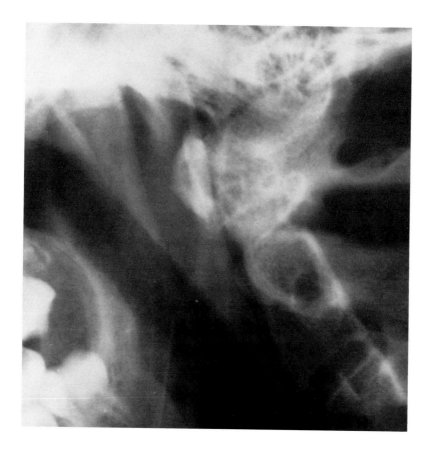

A

FIGURE 12.   Comminuted high (Type II) dens fracture. The fracture is not evident on either the lateral radiograph (A) or the axial CT (B), but was clearly evident (*arrowheads*) on the frontal (C) and lateral (D) polydirectional tomograms obtained after the routine examination. Subsequent 3D reformation of the axial CT data with the electronic deletion of the posterior arch of C_1 (E to G) clearly delineates the dens fracture lines (*arrowheads*). Had the 3D reformation been obtained at the time of the CT examination, the diagnosis would have been established in a timely fashion and the patient spared the inconvenience and expense of an unnecessary study.

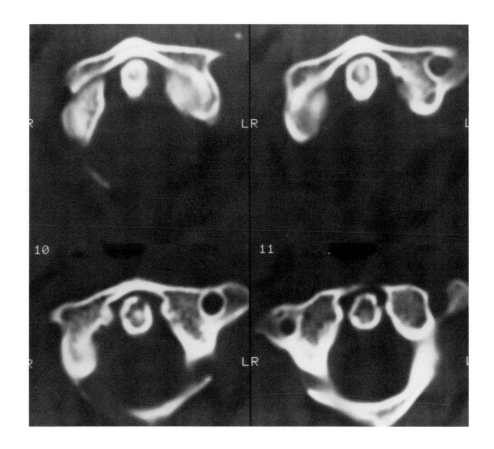

FIGURE 12B

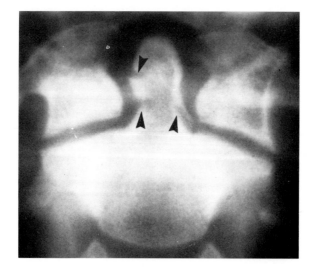

FIGURE 12C

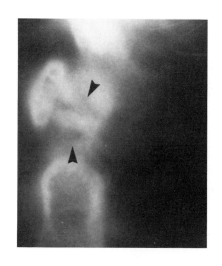

FIGURE 12D

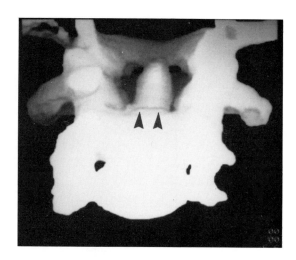

FIGURE 12E

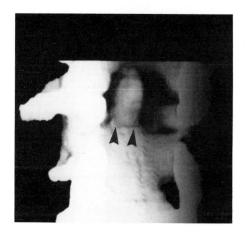

FIGURE 12F

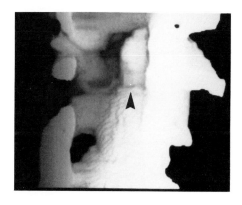

FIGURE 12G

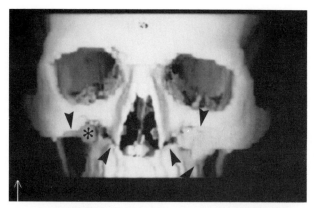

A

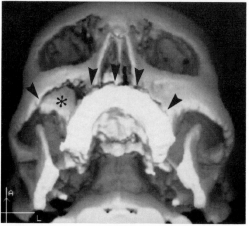

B

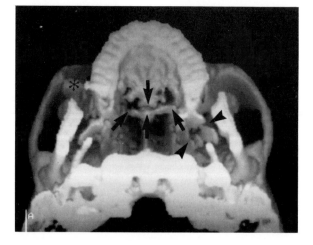

C

FIGURE 13.   Comminuted LeFort I mid-face fracture (*arrowheads*) seen in 3D CT straight frontal (A), Waters (B), and submento-vertical (C) projections. The 3D images not only clearly state the concept of the LeFort I injury, but illustrate the size and degree of depression of the large right anterior anteral wall fragment (*) and extension of the fracture to involve the posterior aspect of the hard palate (*arrows*) and the fracture of the left pterygoid wings (*arrowheads* [B, C]).

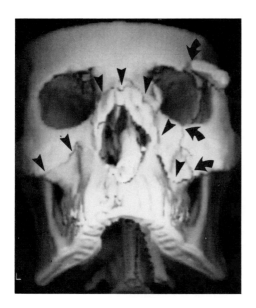

A

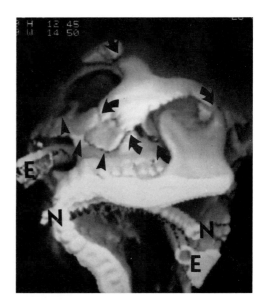

B

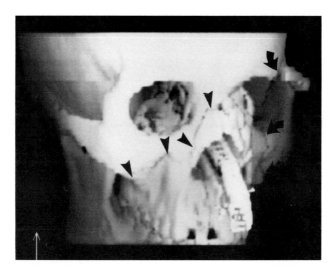

C

FIGURE 14. The concept of concomitant LeFort II (*arrowheads*) and comminuted left zygomatico-maxillary fracture (*curved arrows*) is graphically illustrated by 3D reformation in frontal (A) and oblique (B and C) perspectives. The concept of maxillary anteral involvement is beautifully depicted in (B) and the degree and extent of displacement of the zygomatico-maxillary fragments clearly defined in (A to C). An endotracheal tube (*E*) is in place transnasally and a nasogastric tube (*N*), transorally.

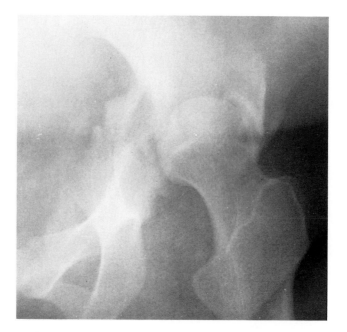

A

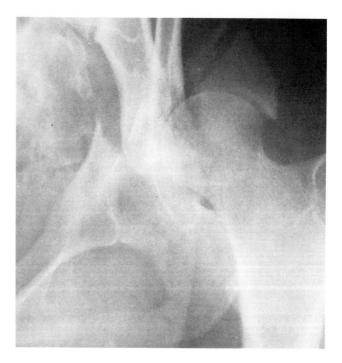

B

FIGURE 15.   The diagnosis of comminuted left acetabular posterior fracture-dislocation is easily made in the antero-posterior (A) and oblique (B) radiographs of the hip. However, conceptualization of the distribution of the "T" fracture (*arrows*) and the size and position of the large posterolateral acetabular fragment (*) and posterior dislocation of the femur is much more clearly and graphically illustrated in the frontal (C), left anterior oblique (D), lateral (E), and posterior (F) 3D reformation, as is the transverse component of the acetabular fracture (*arrowheads*). The vertical component of the "T" fracture was evident only on the 3D image of the inner surface of the left hemipelvis (G).

215

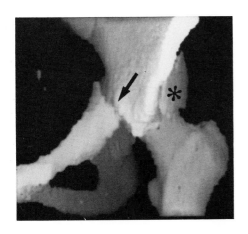
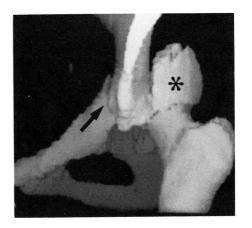

FIGURE 15C   FIGURE 15D

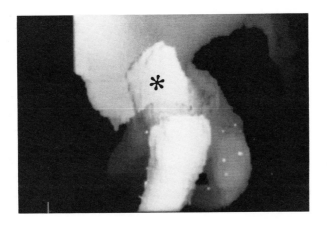

FIGURE 15E

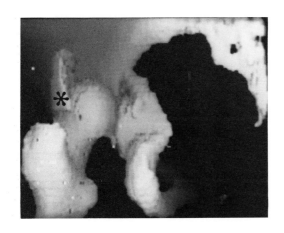
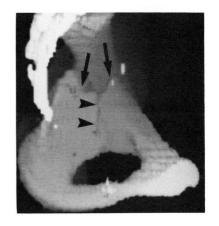

FIGURE 15F   FIGURE 15G

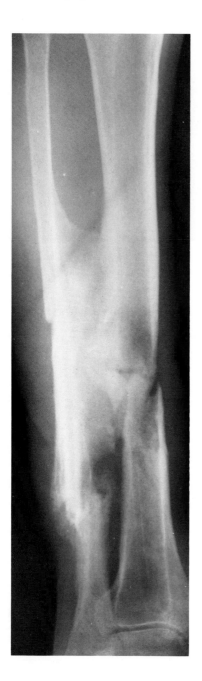
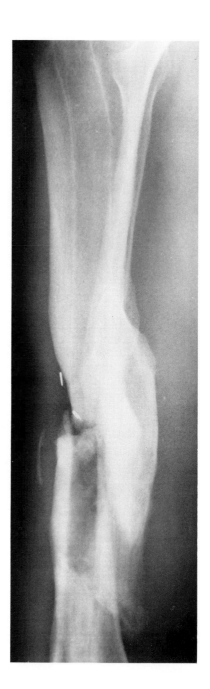

A   B

FIGURE 16. The clinical question to be answered in this patient with comminuted fractures of the mid and distal tibia and fibula is the status of bony union. The frontal (A) and lateral (B) plain radiographs are not definitive. The axial CT scans (C and D) demonstrate many areas of incomplete or nonbony union. Conceptualization of the pathology of the axial images defies imagination. Electronic integration of the axial data visualized in the frontal (E), lateral (F), and each oblique (G and H) 3D images clearly defines the location and extent of fibrous and/or nonbony union.

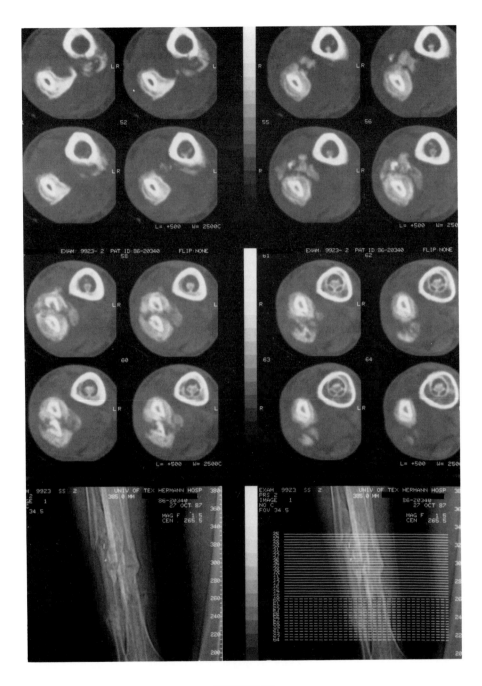

FIGURE 16C

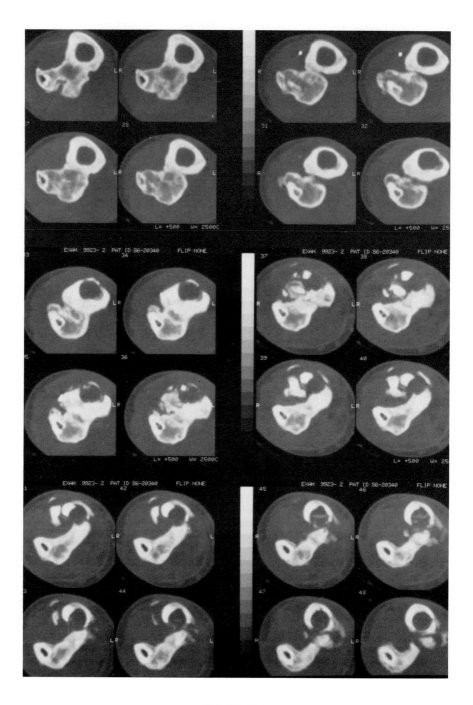

FIGURE 16D

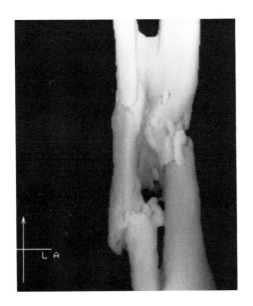

FIGURE 16E

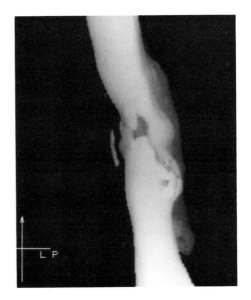

FIGURE 16F

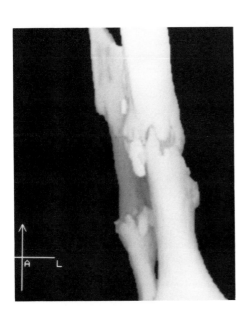

FIGURE 16G

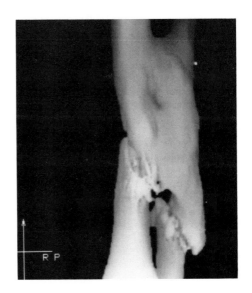

FIGURE 16H

ACKNOWLEDGMENT

This work was supported, in part, by the John S. Dunn Foundation and by GE Medical Systems. The authors gratefully acknowledge the superb photographic reproductions prepared by Jay Johnson, B.S., Radiographic Photographer and the excellent and unstinting secretarial support provided by Kathy Norred.

REFERENCES

1. Herman, G.T., and H.K. Liu. Display of three-dimensional information in computed tomography, *J. Comput. Assist. Tomogr.* 1:155 (1977).
2. Marsh, J.L., and M.W. Vannier. Surface imaging from computerized tomographic scans, *Surgery* 94:149–165 (1983).
3. Herman, G.T., and J.K. Udupa. Display of 3D digital images: Computational foundations and medical applications, *IEEE Comput. Graph. Appl.* 3:39–46 (1983).
4. Woolson, S.T., L.L. Fellingham, P. Dev, and A. Vassiliadis. Three-dimensional imaging of bone from computerized tomography, *Clin. Orthop. Rel. Res.* 202:239–248 (1986).
5. Wojcik, W.G., B.S. Edeiken-Monroe, and J.H. Harris, Jr. Three-dimensional computed tomography in acute cervical spine trauma: A preliminary report, *Skeletal Radiol.* 16:261–269 (1987).
6. Cline, H.E., W.E. Lorensen, S. Ludke, C.R. Crowford, and B. Teeter. Two high resolution surface construction algorithms, *Med. Phys.* 15:320–327 (1988).
7. Lorensen, W.E., and H.E. Cline. Marching cubes: A high resolution 3D surface construction algorithm, *Comput. Graph.* 21:163–169 (1987).
8. Udupa, J.K. Display of 3D information in discrete 3D scenes produced by computerized tomography, *Proc. IEEE,* 71(3):421–431 (1983).

SUGGESTED READINGS

Three-Dimensional CT Diagnostic Imaging
1. Fishman, E.K., D. Magid, D.R. Ney, J.E. Kuhlman, and A.F. Brooker, Jr. Three dimensional imaging in orthopedics: State of the art, *Orthopedics* 11:1021–1026 (1988).
2. Cremin, B.J. The third dimension in imaging — A new role for radiologists, *S. Afr. Med. J.* 74:108–109 (1988).
3. Mayer, J.S., D.J. Wainwright, J.W. Yeakley, K.F. Lee, J.H. Harris, Jr., and M. Kulkarni. The role of three-dimensional computed tomography in the management of maxillofacial trauma, *J. Trauma* 28:1043–1053 (1988).
4. Herman, G.T. Three-dimensional imaging in medicine. A response to a critique by surgeons, *Scand. J. Plast. Reconstr. Surg. Hand Surg.* 22:61–67 (1988).
5. Biondetti, P.R., M.W. Vannier, L.A. Gilula, and R.H. Knapp. Three-dimensional surface reconstruction of the carpal bones from CT scans: Transaxial versus coronal technique, *Comput. Med. Imag. Graph.* 12:75–83 (1988).
6. Adler, S.J., M.W. Vannier, L.A. Gilula, and R.H. Knapp. Three-dimensional computed tomography of the foot: Optimizing the image, *Comput. Med. Imag. Graph.* 12:59–66 (1988).
7. Grivas, A., P.N. Manson, M.W. Vannier, and A. Rosenbaum. Post-traumatic orbit evaluation by three-dimensional surface reconstructions, *Comput. Med. Imag. Graph.* 12:47–57 (1988).
8. Lang, P., H.K. Genant, N. Chaftz, P. Steiger, and J.M. Morris. Three-dimensional computed tomography and multiplanar reformations in the assessment of pseudarthrosis in posterior lumbar fusion patients, *Spine* 13:69–75 (1988).
9. Fishman, E.K., R.A. Drebin, R.H. Hruban, D.R. Ney, and D. Magid. Three-dimensional reconstruction of the human body, *AJR* 150:1419–1420 (1988).
10. Fishman, E.K., D. Magid, D.R. Ney, R.A. Drebin, and J.E. Kuhlman. Three dimensional imaging and display of musculoskeletal anatomy, *J. Comput. Assist. Tomogr.* 12:465–467 (1988).
11. Sartoris, D.J., and D. Resnick. Computed tomography of the spine: An update and review, *CRC Crit. Rev. Diagn. Imag.* 27:271–96 (1987).
12. Hoehne, K.H., R.L. Delapaz, R. Bernstein, and R.C. Taylor. Combined surface display and reformatting for the three-dimensional analysis of tomographic data, *Invest. Radiol.* 22:658–664 (1987).
13. Gillespie, J.E., I. Isherwood, G.R. Barker, and A.A. Quayle. Three-dimensional reformations of computed tomography in the assessment of facial trauma, *Clin. Radiol.* 38:523–526 (1987).

14. Hadley, M.N., V.K. Sonntag, M.R. Amos, J.A. Hodak, and L.J. Lopez. Three-dimensional computed tomography in the diagnosis of vertebral column pathological conditions, *Neurosurgery* 21:186–192 (1987).

15. Wojcik, W.G., B.S. Edeiken-Monroe, and J.H. Harris, Jr. Three-dimensional computed tomography in acute cervical spine trauma; a preliminary report. *Skeletal Radio.* 16:261–269 (1987).

16. Marsh, J.L., and M.W. Vannier. The anatomy of cranio-orbital deformities of craniosynostosis: Insights from 3-D images of CT scans, *Clin. Plast. Surg.* 14:49–60 (1987).

17. Pozzi-Mucelli, R.S., F. Stacul, R.L. Smathers, F. Pozzi-Mucelli, and C. Zuiani. Three-dimensional craniofacial computerized tomography, *Radiol. Med. (Torino)* 72:399–404 (1986).

18. Woolson, S.T., P. Dev, L.L. Fellingham, and A. Vassiliadis. Three-dimensional imaging of the ankle joint, *Comput. Tomogr. Foot Ankle* 6:2 (1985).

19. Weeks, P.M., M.W. Vannier, W.G. Stevens, D. Gayou, and L.A. Gilula. Three-dimensional imaging of the wrist, *J. Hand Surg. [Am.]* 10:32 (1985).

20. Hemmy, D.C., D.J. David, and G.T. Herman. Three-dimensional reconstruction of craniofacial deformity using computed tomography, *Neurosurgery* 13:534 (1983).

21. Artzey, E., G. Frieder, and G.T. Herman. The theory, design, implementation, and evaluation of a three-dimensional surface detection algorithm, *Comput. Graph. Image Proc.* 15:1 (1981).

Three-Dimensional CT for Treatment Planning

1. Burk, D.L., D.C. Mears, L.A. Cooperstein, G.T. Herman, and J.K. Udupa. Acetabular fractures: Three-dimensional computer tomographic imaging and interactive surgical planning, *J. Comput. Assist. Tomogr.* 10:1 (1986).

2. Vannier, M.W., J.L. Marsh, and J.O. Warren. Three-dimensional CT reconstruction images for craniofacial surgical planning and evaluation, *Radiology* 150:179 (1984).

3. Vannier, M.W., G.C. Conroy, J.L. Marsh, and R.H. Knapp. Three-dimensional cranial surface reconstructions using high-resolution computed tomography, *Am. J. Phys. Anthropol.* 67:299 (1985).

4. Kilgus, D.J., G.A. Finerman, and M. Kabo. Three-dimensional imaging techniques for the reconstruction of a complex hip problem. A case report, *Orthopedics* 11:1039–1034 (1988).

5. Murphy, S.B., P.K. Kijewski, M.Bj. Millis, J.E. Hall, S.R. Simon, and H.P. Chandler. The planning of orthopaedic reconstructive surgery using computer-aided simulation and design, *Comput. Med. Imag. Graph.* 12:33–45 (1988).

6. Sartoris, D.J., D. Resnick, D. Bielecki, D. Gershuni, and M. Meyers. Computed tomography with multiplanar reformation and three-dimensional image reconstruction in the preoperative evaluation of adult hip disease, *Int. Orthop.* 12:1–8 (1988).

7. Bannister, C., J. Lendrum, J. Gillespie, and I. Isherwood. Three-dimensional computed tomographic scans in the planning of procedures for reconstructive craniofacial surgery, *Neurol. Res.* 9:236–240 (1987).

8. Lida, H., T. Yamamuro, H. Okumura, T. Ueo, R. Kasai, K. Tada, and T. Tsumi. Socket location in total hip replacement. Preoperative computed tomography and computer simulation, *Acta Orthop. Scand.* 59:1–5 (1988).

9. Engel, J., M. Salai, B. Yaffe, and R. Tadmor. The role of three dimension computerized imaging in hand surgery, *J. Hand Surg. Br.* 12:349–352 (1987).

10. Marentette, L.J., and R.H. Maisel. Three-dimensional CT reconstruction in midfacial surgery, *Otolaryngol. Head Neck Surg.* 98:48–52 (1988).

Chapter 8

THREE-DIMENSIONAL IMAGING: ORTHOPEDIC APPLICATIONS

Elliot K. Fishman, Donna Magid, Derek R. Ney, and Janet E. Kuhlman

TABLE OF CONTENTS

I. INTRODUCTION

The past few years have seen a number of new technologic advances in the field of diagnostic radiology. The introduction of magnetic resonance imaging (MRI), further refinements in computed tomography (CT) with ever-faster scan times and higher resolution, and clinical positron emission tomography (PET scanning) are but a few of the major innovations introduced over the past decade. Another area in radiology which has seen rapid growth and change is three-dimensional imaging (3D). The introduction of new free-standing 3D imaging systems as well as new reconstruction algorithms has provided the impetus for increased interest in this potentially important area.[1,2] This chapter will review some of the basic concepts in 3D imaging as well as some of the clinical applications in which it is valuable. The list of clinical applications is in no way meant to be all inclusive. Rather, it represents a spectrum of common clinical applications in use today.

II. THREE DIMENSIONAL IMAGING: IMAGING SYSTEMS AND TECHNOLOGY

Although a detailed review of available 3D imaging systems as well as a detailed description of available technology is discussed in other chapters of this book, several important factors must be addressed if one is to understand the current state of 3D imaging applications.

1. 3D imaging is in a rapid growth phase with no less than 15 companies offering systems or software at the RSNA meeting in December 1988. Undoubtedly new systems will continue to evolve while consolidation of the market into but a few major players is inevitable. Companies presently offering 3D systems range from the major radiographic manufacturers like Siemens Medical Systems and General Electric, to cutting-edge computer manufacturers like Pixar Inc. and Sun Microsystems, to single-product companies like ISG Technologies, Cemax and Dynamic Digital Displays. Other newer companies like HipGraphics Inc. supply system software for specific hardware configurations.

2. The systems available can be divided into free-standing 3D imaging systems and software packages running on the main computers of the CT scanner. The advantages of the scanner-based systems include lower cost and no additional space or power requirements. The disadvantages generally include a limited user interface, slower reconstruction time, inferior image quality as well as having to compete for computer access with the scanner's primary function: scanning patients. Free-standing systems are more expensive (generally $75,000 to $200,000) but provide greater system flexibility and features. These include user-friendly interfaces, complex reconstruction algorithms for better image quality, near real-time 3D reconstruction, as well as more complex software programs such as disarticulation, volumetrics, and 2D/3D correlation. A free-standing system in our experience tends to enhance radiologist-referring physician interaction by allowing the clinician direct hands-on experience.

3. Although many features are standard on a wide range of 3D imaging systems, other programs will be available only on selected systems. Some examples include real-time 2D imaging, interactive correlation of 2D and 3D images, tumor volumetrics, and the ability to use datasets other than CT (MRI data). These differences will become more noticeable in the future with the introduction of more specific software packages including surgical simulation, prosthetic design, radiation therapy planning, and the introduction of artificial intelligence technology.

III. EXAMINATION TECHNIQUE AND PRINCIPLES

The wide variety of available CT scanners with varying scan parameters (i.e., mAs, collimation, reconstruction algorithm) makes it impossible to define a standard set of scanning protocols for the 3D examination. However, several very basic principles need to be followed in order to optimize the 3D examination.

1. The patient must be made as comfortable as possible prior to the exam to help prevent inadvertent motion during the study. We have found it invaluable to tape the patient when possible to prevent inadvertent motion.[3-5] For example, on a pelvic CT we tape the patient's hips, knees, and ankles together. Although the patient can still move (and break the tape), this tends to reinforce the importance of lying still. A few words of encouragement both prior and during the scan from the technologist will also help keep the patient's mind on the task at hand. A poorly done CT scan will result in an unacceptable 3D image. Following this protocol, we obtain successful studies more than 98% of the time.

2. Scans should be closely spaced with an overlap, to help preserve final image quality. The standard cranial study is often done with 1.5-mm slice thickness at 1.5-mm intervals or 3-mm thick sections at 3-mm intervals. This often results in a stack of 100 or more slices. Most of the other clinical 3D applications are done with a stack of between 20 and 60 CT slices. Our standard protocol at Johns Hopkins Hospital is 4-mm collimation at 3-mm intervals; a 1-mm overlap.[3-5] We are presently using Somatom DR scanners (Siemens Medical Systems, Iselin, New Jersey). If one is using a GE 9800 or 8800 scanner, a protocol of 5-mm thick slices at 3-mm intervals is comparable. The protocol then should obviously conform to the capabilities of the individual scanner used.

3. Scans done for a 3D exam need not be with high mAs. This compromise helps maintain a relatively low dose for the examination. Our standard pelvic 3D CT in the adult involves less radiation than a full series of pelvic films. On a Siemens Somatom DR, the standard scan parameters are either 3 sec, 230 mA, 125 kVp or 4 sec, 310 mA, 125 kVp. We are now routinely scanning pediatric patients at 3 sec, 140 mA, 125 kVp.

4. A 3D examination should be interpreted only in conjunction with the standard transaxial views as well as the multiplanar reconstruction. We routinely generate a full series of coronal, sagittal and/or oblique views in every case using the Orthotool[TM],[6] an interactive real-time display of multiplanar images with single 3D images for orientation.

5. A complete 3D CT exam requires several different projections. We routinely generate spinal axis (rotation of images around the pelvic axis) and z-axis somersaulting rotations (rotation as if the dataset was turning head over heels) of the initial dataset. Additional edited rotations (i.e., disarticulation of femur) are done as needed.[7]

6. Every 3D case should be customized for the specific clinical problem at hand. The decision as to the type of reconstruction chosen (i.e., bone or muscle display, opaque or transparent bone) will be based on the clinical problem. In some cases, multiple series of reconstructions are needed to supply the clinician with the information on the full extent of an abnormality.

IV. THREE-DIMENSIONAL IMAGING: THE RECONSTRUCTION PROCESS

The final image produced on any 3D imaging system is a direct function of the reconstruction algorithm used to generate the image. All commercially available systems, whether they are scanner-based add-on packages or free-standing dedicated 3D systems, use similar reconstruction schemes. Most systems use "thresholding" as a step in the reconstruction process. Thresholding is a binary classification scheme which separates all voxels within a set CT intensity (Hounsfield number) level from those outside of it. Voxels within the window can then be treated as a set of homogeneous cubes which are then used to create the 3D image.[8]

Yet it is widely known that there is often the presence of partial volumes in voxels. The relative sampling rate of CT ensures that many voxels represent a volume which is a combination of at least several different Hounsfield levels. By definition, thresholding is an "all or none" decision and does not permit the representation of tissue mixtures. A voxel must be represented as a single tissue type (100%) and cannot display a combination or summation of tissue values. These limitations are most pronounced on object surfaces, when an object is thinner than a voxel and when two adjacent structures are separated by less than a voxel.

Although with experience and/or trial and error one can choose a "satisfactory" threshold, the selection of an optimal threshold for an entire image is usually impossible. Setting the threshold low enough so that thin bones do not have "holes" will increase the likelihood that minimally displaced fractures, unfused epiphyses, and cranial sutures will appear fused. Conversely, raising the threshold to prevent "false" fusion will increase the likelihood of false holes. This can be illustrated in looking at a 3D reconstruction of the skull. At a low threshold the cranial sutures and fine details of the skull are obscured. Yet by raising the threshold enough to define the sutures there is significant bone dropout in the maxillary and facial region.[9]

These limitations can be overcome by representing voxels as material mixtures instead of cubes. This novel approach was first developed at LucasFilms and later at Pixar, the former research and development group at LucasFilms. In volume rendering,[1,7] each voxel is represented as the percentage of each material present within it. This percentage classification helps prevent the numerous artifacts and problems associated with thresholding. (Figure 1 and Plate 8*).

With the percentage classification scheme a voxel's characteristics are determined by the percentage of each material present within the voxel. The result is that there are no artificially sharp transitions (aliasing), as there are in nonlinear operations like thresholding. Functions applied to the volume are smooth and continuous. This minimizes artifacts and yields a more accurate 3D representation. The sharpness of the final image is then limited only by the resolution of initial CT data.[10]

The volumetric rendering technique with a percentage classification overcomes the three major clinical failings of thresholding:

1. The thinner an object becomes, the more translucent it appears. In this manner the integrity of very thin objects is preserved.
2. Objects of the same material separated by less than a voxel are represented by a drop in the percentage of the material. This results in the ability to define subtle separations between structures. We have shown that fracture gaps less than 0.5 mm are reliably detectable.
3. Material percentage classification maintains the integrity of the initial CT data. Mixtures are represented accurately. Thresholding, on the other hand, by destroying the continuity of the dataset throws away valuable information. Other reconstruction schemes may help overcome some of the limitations of thresholding. Structure-oriented interpolation techniques and normal estimation methods are two methodologies which can greatly compensate for inadequacies of thresholding. Further details on these techniques can be found in Chapter 1 of this book.

* Plate 8 appears following page 294.

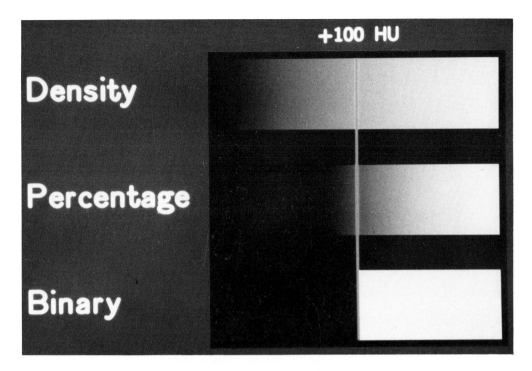

FIGURE 1. Computer-simulated display comparing percentage classification. By looking at a density of 100 Hounsfield units (HU) one can see that a binary classification is an "all or none" phenomenon. The percentage classification allows for voxels both below and above 100 HU to be average included in the final image.

V. CLINICAL APPLICATIONS

For 3D imaging to be clinically useful it needs to be more than just a technique for presenting an assimilation of a series of CT slices. To reach its true potential, 3D imaging must ultimately provide an environment where both the referring physician and the radiologist are able to interact on a patient dataset. This interaction can take many forms, some of which are available today and others which will be available in the immediate future. Yet all of these applications must meet one standard: they must provide definable information that will impact on patient management and decision making. To date, a number of clinical applications have been shown to be successful in a number of medical institutions. Although some of these applications are relatively specific to unusual clinical problems, most can be applied in nearly every hospital or clinic with an emergency room. The following represents the more common applications of 3D imaging to date.[11-13]

A. PELVIC RING AND ACETABULUM (Figures 2 to 7)

Pelvic trauma is high on the list of factors contributing to morbidity and mortality following blunt trauma.[11-16] Pelvic injuries severe enough to cause fractures or dislocations are often associated with significant hemorrhage, neurovascular or compartment syndrome, and concomitant damage to the genitourinary or gastrointestinal tracts.

Successful hemodynamic, neurologic, and orthopedic stabilization requires prompt, accurate, and noninvasive assessment. In the most severe cases, such as would be seen in a shock-trauma center, the principle of "the Golden Hour" serves as a reminder of the positive impact rapid diagnosis and treatment can have on outcome.

The pelvic ring is the bulkiest bony structure in the body and as such can be difficult to

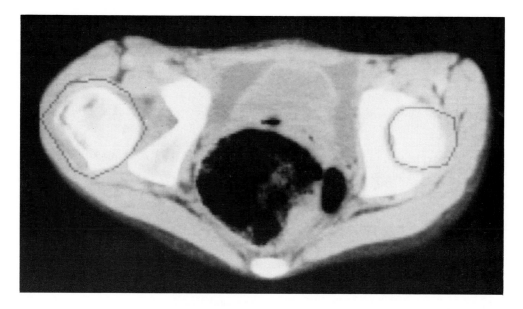

A

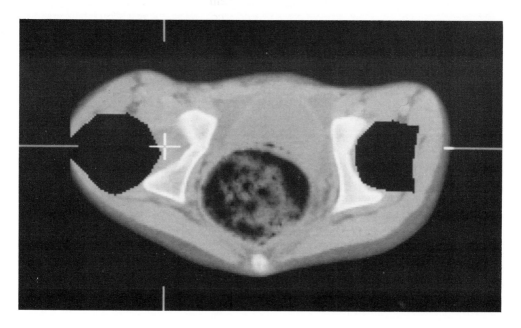

B

FIGURE 2.   The ability to edit the transaxial CT scans is an important feature in 3D imaging. Femoral disarticulation is particularly important in acetabular fractures and congenital hip disease. The editing in this case is done interactively by using the computer mouse as a scalpel.

explore with conventional radiographs alone. We have previously demonstrated,[15] in 20 consecutive pelvic trauma emergency room patients, that animated 3D imaging surpasses the conventional plain film series in providing optimal views of the pelvic ring, providing a large series of obliques and inlet/tangential views at a similar economic and dose cost. Added to the flexibility and detail of Orthotool™ 2D imaging,[6] this forms a comprehensive 2D/3D baseline for diagnosis and therapeutic planning.

The overall integrity of the pelvic ring may be studied in a serial review of axial and coronal 2D images. Since the entire pelvis is presented, comparison with the contralateral side helps assess shift and symmetry. Once the conventional planar studies are reviewed, a new baseline axis can be drawn parallel to the longitudinal axis of the sacrum on a midline sagittal image, which establishes the new x-axis for a new series of orthogonal images generated from the original CT dataset. Within 15 seconds, the three sets of orthogonal 2D views are regenerated with the three orientational 3D views to demonstrate the new simulated patient position. The top left box then demonstrates a series of 2D images approximately parallel to the plane of an inlet view, giving an excellent overview of pelvic ring integrity. The bottom left box approximates the tangential or outlet plane and is useful for evaluating the anterior ring and the sacroiliac (SI) joints. The final step in assessing ring integrity is to review the 3D somersaulting pelvic views, which again demonstrate ring disruptions, anterior or posterior diastases, and fracture fragment position.

The sacrum and SI joints may be the most difficult pelvic components to assess on plain film, being obscured by their own bulk and obliquity, by superimposed intestinal air and feces, and in older patients by osteoporosis. Transaxial and coronal 2D series best document fractures, commonly missed in this area, and subtle diastasis. The sagittal series best demonstrates the central canal and any associated presacral hematoma or soft tissue mass. Realigned orthogonal views as described above will give direct coronals through the obliqued sacrum, allowing best visualization of the paired neural foramina.

The application of this technique to the acetabulum has been described in detail elsewhere.[15,16] This, like the sacrum, is a large bony mass only partially evaluated by conventional

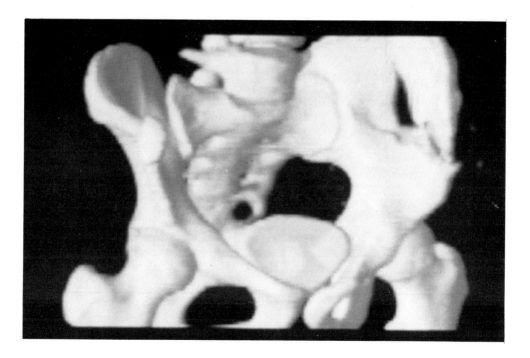

FIGURE 3. CT scan of an 11-year-old female with pelvic trauma. Note the extensive left iliac wing fracture. There is also intravenous contrast in the bladder. A 3D CT scan can be done as part of a routine trauma examination.

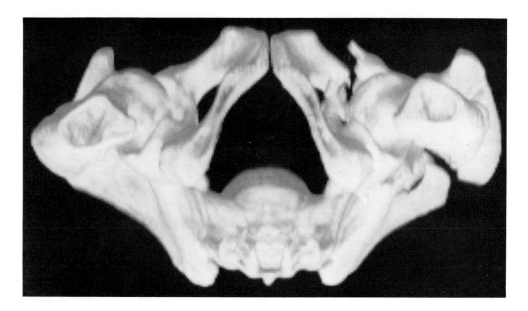

A

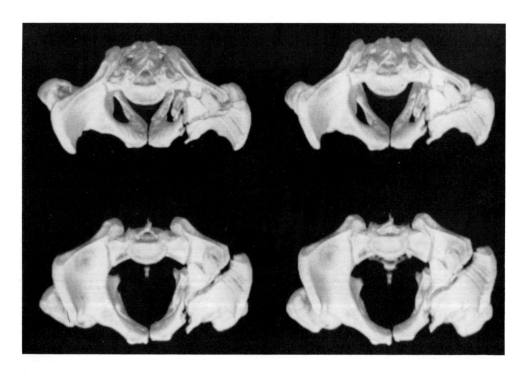

B

FIGURE 4. Complex left acetabular fracture best demonstrated on this pelvic inlet view. Notice disruption of anterior column and displacement of fracture fragments. The views from above show the extent of involvement of the iliac wing.

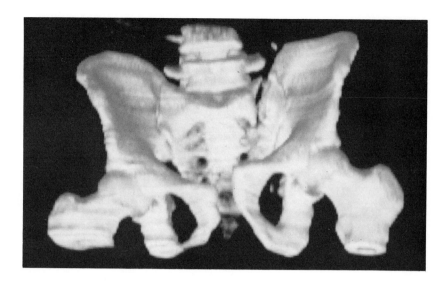

A

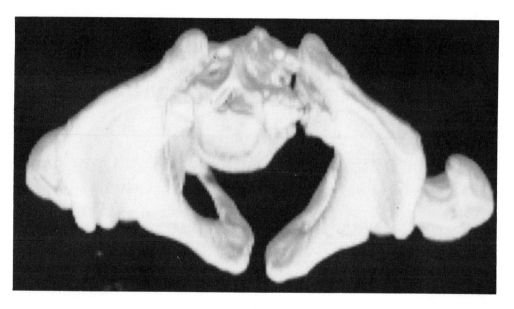

B

FIGURE 5. Vertical strut fracture of left sacrum. Note the fracture and its extension near the sacral foramina. Also, disruption of left hemipelvis with diastasis of left symphysis is seen (A, B). The somersaulting view optimally demonstrates the pelvic ring disruption. The patient was placed in traction and repeat study using Orthotool™ demonstrates early callus formation around fracture site.

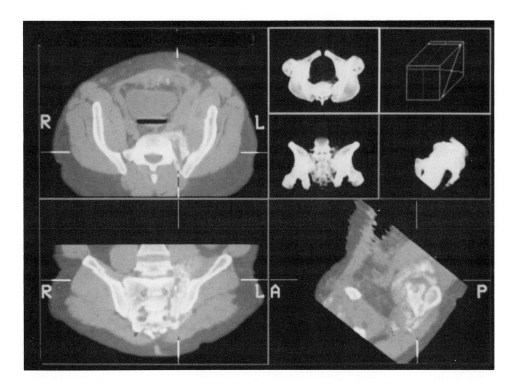

FIGURE 5C

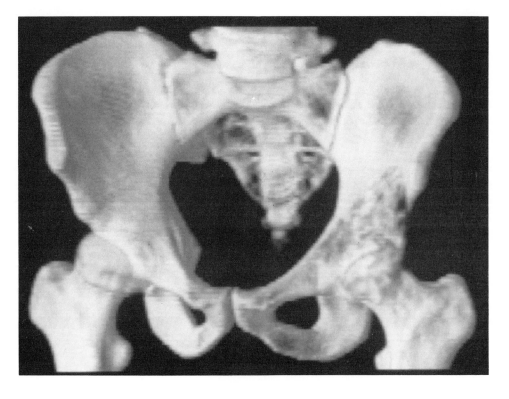

FIGURE 6. Thirty-four-year-old male with left hip pain and a history of testicular tumor. The bone scan was initially negative. The study demonstrates an infiltrating tumor of the dome of the acetabulum.

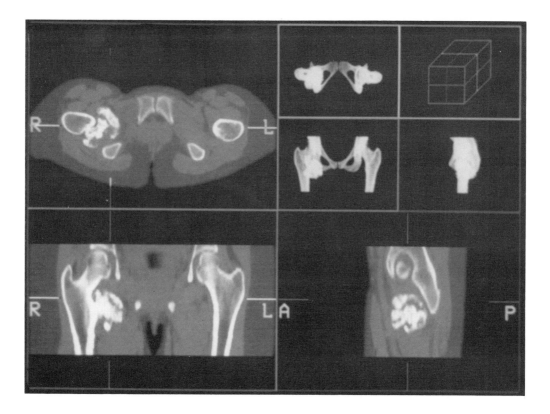

A

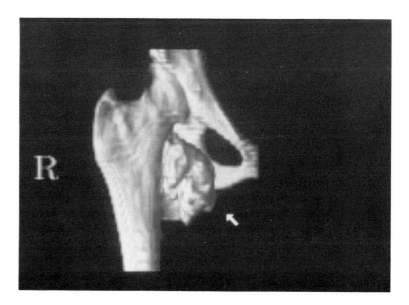

B

FIGURE 7. Chondrosarcoma of the right thigh. (A) Orthotool™ demonstrating chondroid mass arising from lesser trochanter consistent with chondrosarcoma. (B) 3D reconstruction shows full extent of tumor. This study was used for planning successful surgical resection.

plain films. While transaxial CT well-describes fracture lines, anterior and posterior column disruptions, and displacement of the quadrilateral surface, coronal and sagittal series provide more complete visualization of acetabular dome disruptions and of the actual articular surfaces and joint space. The use of orthogonal planes also guards against the potential hazard of missing or underestimating abnormalities running parallel to the initial (transaxial) plane of scanning.

The final step in a sacral or acetabular study, as in overall pelvic ring studies, is to review the animated 3D image series. The surgeon uses this perspective to integrate the 2D data and to finalize surgical plans. Where necessary, extraneous or superimposed structures — the anterior pelvic ring and lateral iliac wings, when looking at the sacrum; the contralateral pelvis or the proximal femur, when looking at the acetabulum — can be edited off for 3D viewing.

B. COMPLEX PELVIC DEFORMITIES (Figures 8 to 12)

3D imaging becomes increasingly important the more complex an anomaly is. Pelvic deformities can be arbitrarily divided into those which are congenital and those which are acquired (i.e., trauma). The congenital deformities include abnormalities of the sacrum (i.e., segmentation defects, duplication, dwarfism), acetabulum (congenital hip dislocation, dwarfism), and pelvic ring. In these cases, 3D CT plays a direct role in arriving at correct patient management decisions. Acquired pelvic deformities are most often posttraumatic in nature, although prior surgery or endocrine problems might be the underlying problem.

The major indication for a 3D CT in this group of patients is an adjunct to surgical planning. Routine plain X-rays are often of limited value in these patients due to the complexity of overlapping structures. This is particularly true in abnormalities involving the sacrum. Figure 12 demonstrates a patient with a congenital duplication of the iliac wing and acetabulum. Associated sacral anomalies are also clearly defined. Albeit a rare entity, the 3D CT study was to be used as a template for surgical resection. However, based on this study the surgeon felt the potential complications of surgery (i.e., location of sacral nerve routes) would make the operation too risky. Continued conservative management was therefore recommended.

Another area where we have used 3D CT has been in the patient with bladder exstrophy prior to primary or secondary repair. The relationship of the pelvic organs, genitalia, and muscles can all be clearly defined using Orthotool™, with the area of the levator muscle best defined on the coronal display. The 3D images are most helpful in understanding the orientation of the pelvis. The z-axis rotation is particularly helpful in this regard. The major decisions to be made include the position of the osteotomy and the degree of rotation needed to create an optimal repair. Two examples of such a study are seen in Figure 11.

C. THE KNEE (Figure 13)

The knee is another joint where 3D imaging has been of value in our experience. Although the knee is involved in a wide range of pathological processes, the indications for 3D CT have mainly been for evaluation of fractures or determining the extent of bony tumors.[17,18] A key component of any exam of the knee is a careful review of the transaxial slices and the coronal and sagittal 2D images. These images are particularly helpful in the detection of joint fragments or soft tissue involvement of the knee joint. The sagittal image is particularly helpful in these cases. We feel that the 2D/3D exam is the study of choice for the evaluation of the knee joint, obviating the need for standard tomography. Other uses of 3D CT in the knee include evaluation of the extent of bony tumors including the determination of whether or not they extend into or preserve the joint space.

The primary use of 3D CT of the knee in our experience has been for evaluation of the patient following trauma. Standard transaxial CT has proven valuable in the detection of nondisplaced or minimally displaced fractures. The position and orientation of the fracture

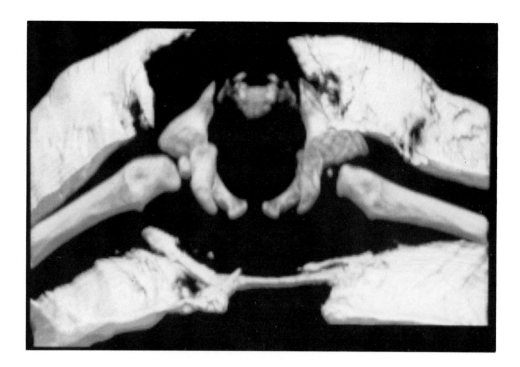

FIGURE 8. CT scan of 1-year-old with reduction of dislocated left hip. Notice the small left femoral epiphysis. The study was done with the patient in cast and only consisted of eight CT slices. The study demonstrated satisfactory reduction.

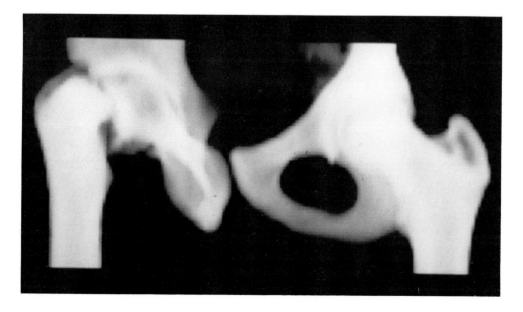

FIGURE 9. 3D reconstructions of patient with a right slipped capital femoral epiphysis. Notice rotation of femoral head downward.

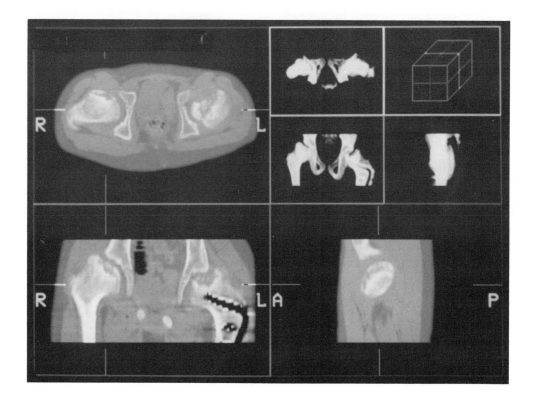

A

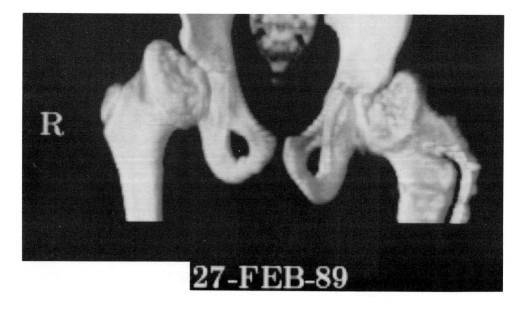

B

FIGURE 10.   Deformed bilateral femoral heads in patient with prior osteotomy on the left. The deformities are due to a slipped capital femoral epiphysis with avascular necrosis of the femoral heads. This study shows the advantage of combining the information from 2D and 3D image sets.

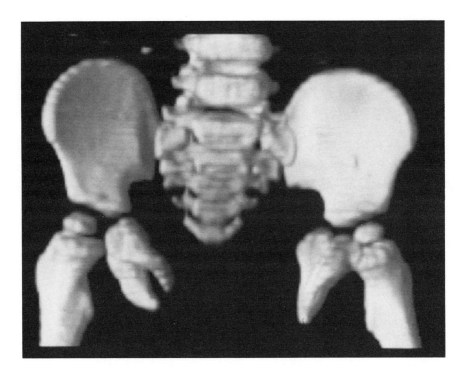

A

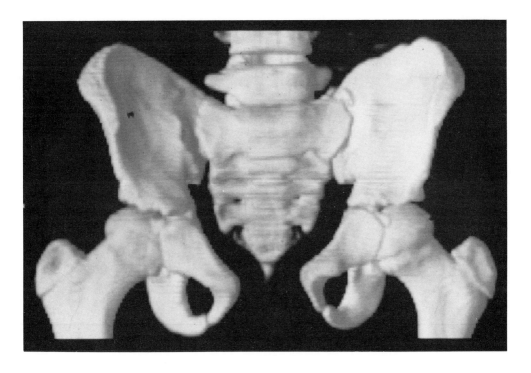

B

FIGURE 11.   Two examples of patients with exstrophy of the bladder. The patient in Figure A has not had an attempt at closure. Patient in Figure B has had one surgical attempt. These images are used as templates for designing where the osteotomy is to be performed.

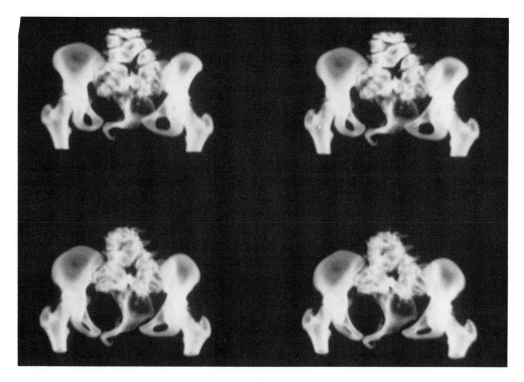

A

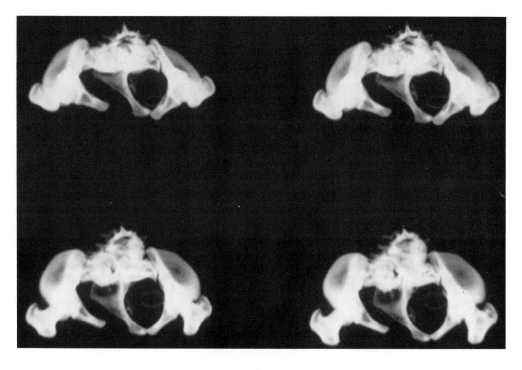

B

FIGURE 12.   Twelve-year-old male with duplication of iliac bone and acetabulum. Also present are multiple segmentation defects in lumbar spine and sacrum. Due to the complexity of the fusion of the duplications, surgery was not attempted based on review of the 3D images.

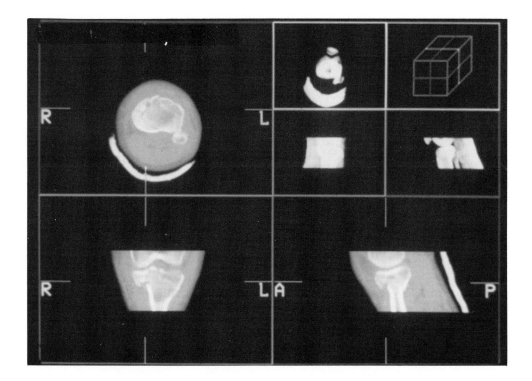

A

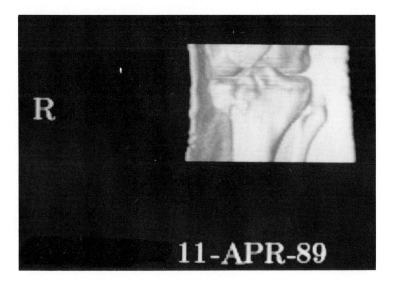

B

FIGURE 13. 2D and 3D reconstructions of a tibial plateau fracture. The study demonstrates the medial plateau to be depressed downward, indicating surgical intervention would be needed. The study was done with the patient in cast. The front part of the cast was cut away.

fragments can all be resolved with standard techniques. The addition of 2D/3D imaging to the exam is invaluable in the detection and definition of several other factors critical in patient management in fractures of the distal femur or proximal tibia.

Although the presence of a fracture is often detected with properly positioned and exposed X-rays, in many cases there is still uncertainty. The presence of a fracture is often obvious but its extent and area of involvement uncertain. The extent of tibial plateau displacement is critical in determining whether the patient gets surgical intervention or is treated conservatively.

D. ANKLE FRACTURES (Figure 14)

The ankle, or the talocrural joint, is often involved in trauma. Most fractures are produced by combinations of rotation with supination or pronation, with a minority produced by axial forces. Complete imaging assessment of the traumatized ankle must include evaluation of bony integrity, soft tissue and ligamentous structures, and the preservation or loss of normal spatial relationships between component parts. Even small tibio-talar shifts or alterations can significantly reduce the normal articular contact surface.

In evaluating pre- and postreduction plain films of the ankle, clues to ligamentous integrity are sought in the configuration of the mortise and the extent of displacement at fracture sites.

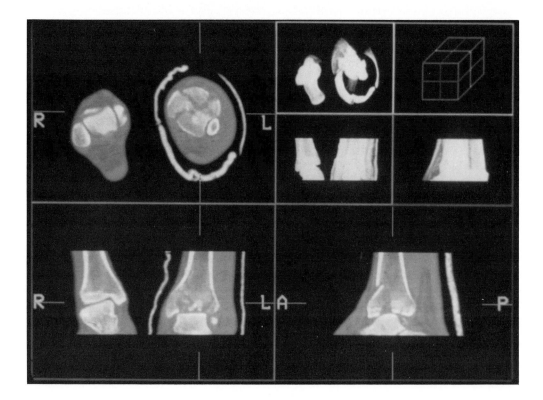

A

FIGURE 14.   2D and 3D reconstruction of ankle fracture. The study is done with the patient in cast. (A) The disruption of the ankle mortise is best seen on the 2D images in coronal and sagittal plane. (B) The spinal axis 3D image vividly demonstrates the extensive nature of the fracture. Notice how the cast has been removed. (C) Image reconstruction from above ankle joint with cast still present.

FIGURE 14B

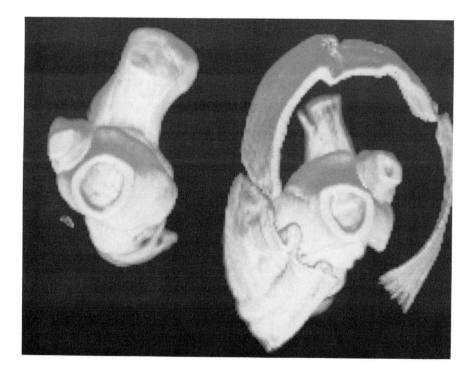

FIGURE 14C

Transaxial CT has proven its superiority over plain film in demonstrating bone and soft tissue trauma. CT, removing the element of superimposition limiting plain films, often shows fractures either missed or underestimated by plain film, and more precisely maps fragment displacement and rotation. While CT cannot separate ligaments from adjacent soft tissue to the extent possible with MRI, it surpasses plain films at defining joint effusion, soft tissue swelling, and tissue plane effacement which may suggest the extent of trauma. It is also more sensitive to cortical disruptions or avulsions resulting from ligamentous or tendinous forces.

In the ankle, as in the tibial plateau or the vertebral disc and end plate, some of the primary structures of interest — the tibial plafond, the central portion of the mortise, and the talar dome — are horizontally oriented and therefore parallel the conventional plane of scanning. Therefore, while transaxial CT gives better bone and tissue detail than plain film, the addition of multiplanar reformatting (sagittal and coronal images) allows more precise demarcation of the anatomy in question.[19]

The Orthotool™[6] workstation integrates all three sets of planar or 2D images — transaxial, coronal, and sagittal — on one screen, which, with the traveling cross-haired reference axes, allows precise localization and confirmation of findings from three perspectives. The Orthotool™ software also allows simple measurement of distance, size, and angles. Preservation of the entire CT dataset usually means having the contralateral normal ankle for comparison. While cast material does not impede 2D imaging, the 3D Edit feature can rapidly remove cast and, if necessary, the contralateral extremity prior to simulated 3D image reformatting.

In comparing CT to existing plain film parameters commonly used to assess fractures, we find that the plain film criteria do not necessarily translate directly to CT. We believe that the CT images are giving a more precise impression of displacement and rotation. The degree of apparent plain film displacement and distraction at a distal fibular fracture, for instance, is felt to be a good indicator of more distal tibio-fibular and fibular-talar disruption. Our CT experience has not supported this; we find the more distal joints often to be more stable than would be expected given the degree of more proximal loss of alignment.

Preserving both sides for comparison allows us to evaluate an absolute talo-crural angle measurement in terms of the presumed change from the normal baseline, i.e., in comparison to the angle on the normal side. In our experience, the normal talo-crural angle may be slightly lower than the average angle ($84° \pm 5°$) given in the literature. This angle, when significantly different from the normal side — i.e., *relatively* altered — may be of significance, but the significance of *absolute* angle measurements remains to be seen.

The fractured posterior tibial lip must also be assessed for percent involvement of the articular surface, best done by reviewing sagittal and transaxial series together, and for determining the degree to which fracture distraction or displacement has created a gap or step-off in the articular surface.

Review of the coronal and sagittal series best identifies osteochondral defects or fractures involving the talar dome; the cross-hair axes can pinpoint any such fragments prior to surgery or arthroscopy.

While radiologists prefer the detail and measurement precision of the animated interactive Orthotool™ 2D images, the final study is to edit off the cast, if necessary, and to create animated rotating volumetric 3D views. This is the perspective requested by the surgeon and used in the final steps of assessment to integrate 2D information and to complete surgical or therapeutic planning.

E. THE SHOULDER AND SHOULDER JOINT (Figures 15 to 17)

Routine radiographic evaluation of the shoulder and shoulder girdle is often limited by a number of technical factors including poor patient positioning, improperly exposed radiographs, or overlapping shadows on the standard radiographic views. Routine transaxial CT scanning overcomes many of these limitations by requiring less patient cooperation, providing a more optimal view without overlapping radiographic lines, and the ability to visualize soft tissue and bone. 3D CT provides an even more comprehensive visualization of this area.[20]

The clinical problems in which 3D CT of the shoulder has proven valuable include acute trauma, acute or chronic dislocations, suspected scapular injuries, and in the evaluation of axillary masses. In all cases, the standard CT imaging technique is used. One potential

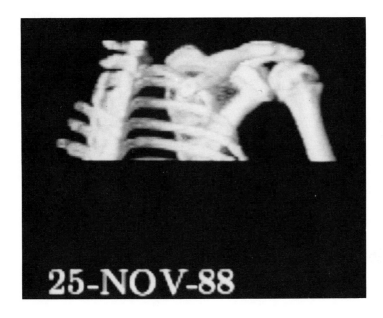

FIGURE 15 . Fracture of humerus. (A) Edited left hemithorax demonstrates shear fracture along neck of humerus. No evidence of dislocation was seen.

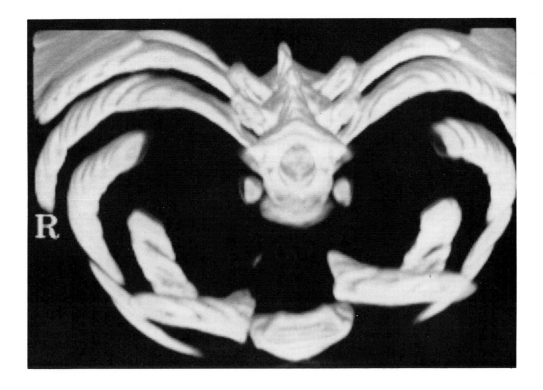

FIGURE 16. Following injury to chest in automobile accident, the patient experienced increasing chest pain. 3D study demonstrates dislocation of left sternoclavicular joint posteriorly. The somersaulting view was the optimal view in this case.

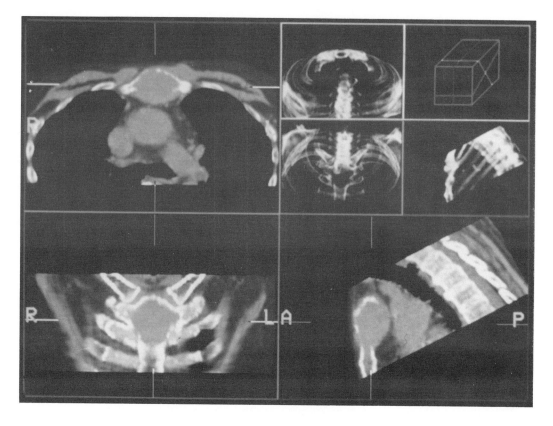

A

FIGURE 17.   Metastatic hepatoma to sternum. (A) 2D reconstruction demonstrates a large lytic lesion in manubrium compatible on biopsy with metastatic hepatoma. 3D representation of the same study is seen in Figures B and C. Figure C is imaged following editing of ribs and spine for a clearer look at tumor.

problem in imaging the chest or shoulder is respiratory motion. To prevent any inadvertent motion we instruct the patient to breathe with shallow breaths during the study. This is effective in the majority of cases.

The ability to disarticulate the scapula, sternum, or humerus in 3D CT images helps to optimize the information generated in any individual case. We have found it particularly valuable to edit the scapula off the chest wall and visualize it directly. Scapula fractures are easily missed on routine radiographs, because of overlapping structures, which can be eliminated from view with edited 3D views. The value of volume rendering is particularly well demonstrated in imaging the scapula. The body of the scapula is often paper-thin, yet the acromion and glenoid are of thicker bone. The 3D images generated are able to simultaneously show features of both of these areas.

The major clinical use of 3D of the shoulder has been in trauma. The presence and extent of a fracture, as well as the presence or absence of an associated dislocation, are best demonstrated on 3D images. Associated injuries to the sternum and clavicle or sternoclavicular joint can be imaged at the same time. Sternoclavicular injuries are well demonstrated on z-axis rotations. This display projects the sternum/clavicle relationship from directly above and below, optimizing the detection of even the most subtle dislocation. Edited x-axis views are obtained to define the extent of humeral injury. Usually the humeral head need not be disarticulated from the glenoid cavity for an adequate evaluation. However, when joint

245

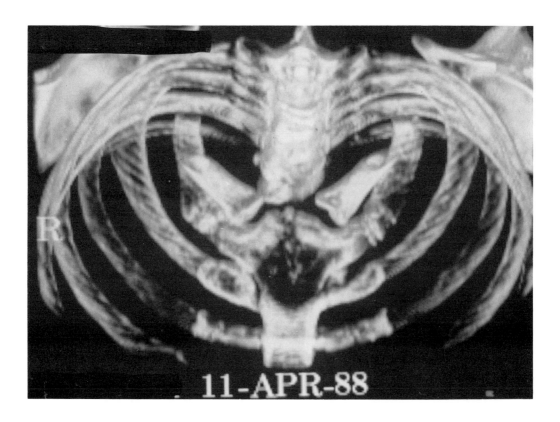

FIGURE 17B

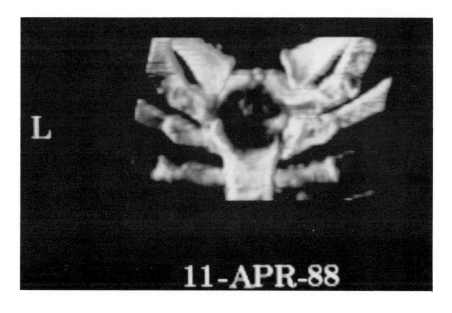

FIGURE 17C

fragments are suspected from the transaxial images or the fracture appears to extend medially, then disarticulation is advised. 3D reconstructions with a high degree of transparency is particularly helpful in these cases. The presence or absence of a dislocation is best seen on the z-axis views.

Although most trauma cases require bone reconstructions only, 3D reconstructions of the soft tissues and muscle are helpful in other cases. Evaluation of a clinically palpable mass or patients with symptoms related to a suspected mass in the region of the neurovascular bundle are ideal candidates for 3D imaging. The presence and extent of a mass can be well documented for further therapeutic intervention. As 3D images of the soft tissues and internal organs improve, evaluation of the axilla and brachial plexus may become an important clinical indication for the study.

F. SPINE (Figures 18 and 19)

3D reconstruction has found a number of successful applications in the evaluation of the spinal column. The major areas of use include trauma, oncology, infectious etiologies, and degenerative disease. Although any area of the spine can be evaluated, the lumbar spine has been the major area of interest.[12,21,22]

In the traumatized patient the presence and extent of the fracture is important. Although in most cases, the routine transaxial views are sufficient for the detection of a fracture, its true extent is best documented on a 2D or 3D display. The 2D display is especially helpful in the detection of and definition of small bone fragments which may extend into the spinal canal. Associated hematomas either in an extradural location or in the paraspinal zones are best seen on the transaxial CT slices or on reformatted coronal/sagittal planes. An interactive 2D display

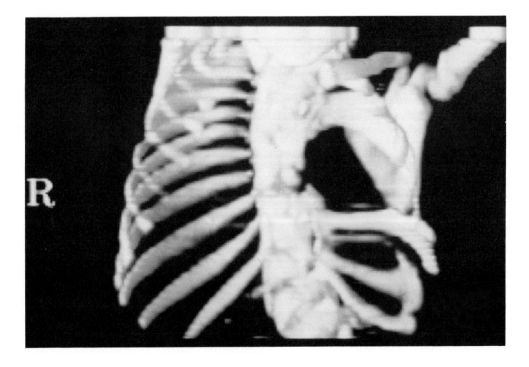

FIGURE 18.   Three-year-old female with complex deformities of rib and spine. Notice multiple segmentation defects in thoracic spine as well as deformed ribs, including fusion of several upper ribs. At surgery, the ribs were resected in order to allow adequate growth of left hemithorax.

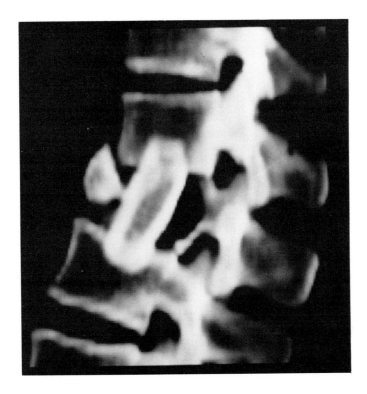

FIGURE 19. 3D reconstruction following spinal fusion for collapse of vertebral body. The area of fusion is clearly seen, as well as the foramen both above and below.

is especially useful in the sacrum where an oblique reconstruction along the major plane of the sacrum presents the foramen in an end-on appearance. This allows one to evaluate foraminal involvement in trauma or neoplastic involvement.

The 3D images can be useful in the spine by providing additional visualization to the orthopedic surgeon, neurologist, or neurosurgeon. The clinical applications include the following.

1. Trauma

Although most instances of spinal trauma are vividly displayed on transaxial CT, in other instances the use of coronal, sagittal, or oblique imaging can help with the more difficult case. Unusual lines created by congenital variations, poor patient positioning, or unusual fractures are often best explained on the reformatted views. An example of this additional information is illustrated in Figure 5 where the true extent of a vertical strut fracture is defined. In cases of suspected failed back syndrome the presence or absence of fusion may be defined with this presentation.

The 3D presentation is especially helpful in evaluating encroachment on the spinal canal, neural foramina, or adjacent joint spaces. To optimize these findings the use of edited 3D views is encouraged. The neural foramina are particularly well seen when the spine is cut in half along a sagittal plane (Figure 19). This view also allows for a full evaluation of the texture of the vertebral bodies if 3D reconstruction is done with percentage classification. Several authors have suggested that 3D reconstruction can determine whether there is proper union following spinal fusion or detect unsuspected fractures where plain CT is negative. However, these ideas should be viewed with some skepticism on several grounds. There has been no proof that thresholded images can accurately detect subtle fractures, define fusion, or detect the accurate status of cranial sutures. In fact, information to the contrary exists. Drebin et al.[9] has shown that thresholded images cannot accurately detect small fractures or define cranial sutures. In fact, observer bias with subsequent choices of threshold level can produce the wrong results. Several papers have suggested that 3D images created with percentage classification can be correct in these areas.[23,24] However, further studies with phantoms are needed to scientifically prove the accuracy of these studies.

2. Tumors

A combination of 2D and 3D images is helpful in determining the extent of primary and malignant tumors. Transaxial CTs are often obtained to help in difficult cases where either clinical exam, plain X-rays, or bone scan are equivocal or contradictory. Detection of the extent of a destructive lesion as well as the presence or absence of an associated mass can be shown on the routine scans. The 3D image adds a better understanding of the extent of a lesion especially when 2 or more vertebrae segments are involved. The full 3D display can also prove useful in the construction of radiation therapy portals.

3. Infection

Infection of the spine or disc space due to hematogenous or direct spread of infection is an uncommon phenomenon. Tuberculous involvement of the spine (Pott's disease) is now a rare presentation. However, in those cases with osteomyelitis 3D display can help define the extent of involvement and be used for preoperative planning.

G. CRANIOFACIAL (Figures 20 to 22)

One of the first areas analyzed in detail with 3D imaging was the craniofacial region and skull. Mike Vannier and Jeffrey Marsh published a series of articles on the use of 3D imaging in craniofacial surgery. Some of the more common uses include:[23,24]

1.   Evaluation of craniofacial trauma
2.   Preoperative evaluation of craniofacial anomalies
3.   Plastic surgery
4.   Orbital tumors
5.   Tumors and inflammatory disease of the sinuses
6.   Temporomandibular joint abnormalities
7.   Evaluation of postoperative results

The images were a valuable adjunct to the surgeon by allowing him to precisely visualize the abnormality to be corrected at operation. The surgeon could use the images created as a template for surgical planning and rehearsal. The 3D CT was most helpful in complex congenital anomalies such as Apert's syndrome and Crouzon's syndrome.

Images of the skull can be generated with either opaque or transparent bone, depending on the clinical problem being evaluated. We have found transparent bone most valuable in visualizing pathology of the sinuses, mastoids, and orbits. An opaque bone reconstruction is most helpful around the zygomatic arches and the mandible.

H. MUSCLE AND SOFT TISSUES (Figures 23 and 24 and Plate 9*)

Although most of the initial work on 3D imaging has been on reconstruction of bony structures, advances in computer software and hardware now provide the capabilities of high-

* Plate 9 appears following page 294.

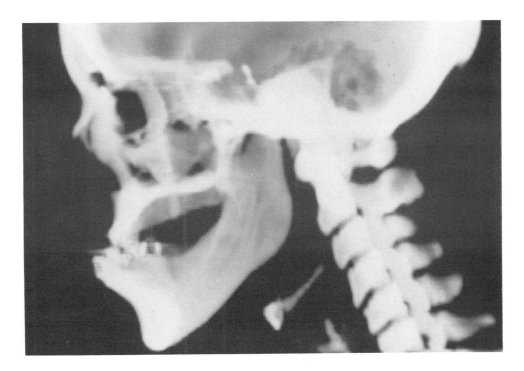

A

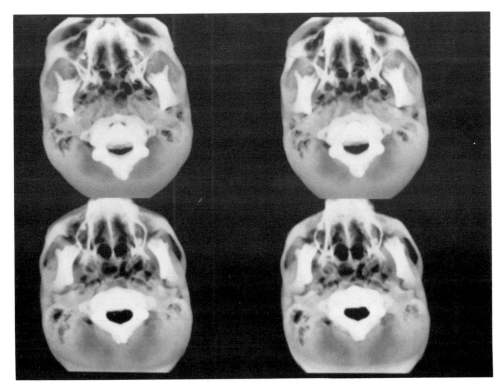

B

FIGURE 20. Transparent view of edited skull clearly allows one to see the mastoid air cells. The vascular groove in the mandible is also seen.

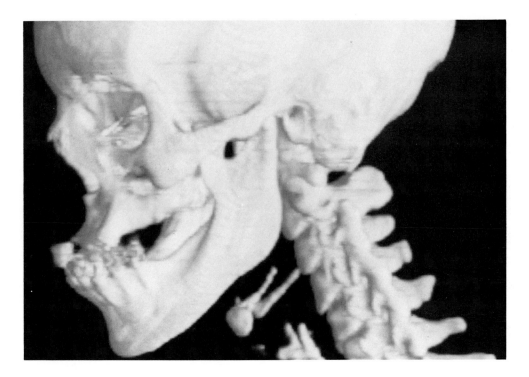

A

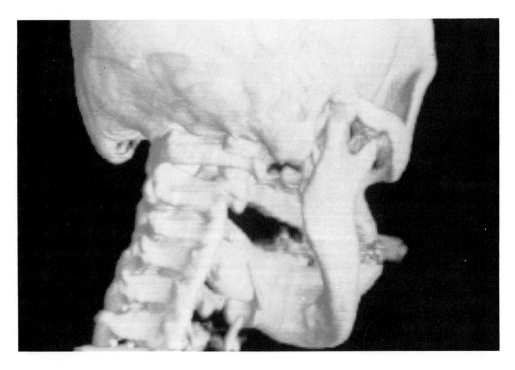

B

FIGURE 21. The opaque view of the skull does not allow one to see the sinuses clearly. However, the zygomatic arch and the mandible are much better defined. Optimization of 3D reconstruction means choosing a particular algorithm depending on the clinical dilemma.

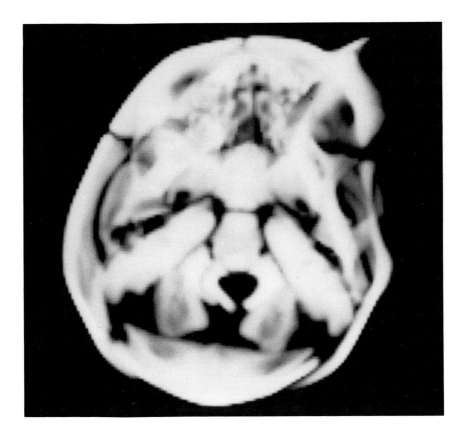

FIGURE 22. 3D reconstruction of base of the skull in patient with meningocele and resultant deformity of left side of the skull. Notice detail of sutures in the base of skull.

quality displays of soft tissue structures. This ability opens up a wide range of new potential applications of 3D imaging in such varied areas as evaluation of soft tissue tumors, plastic surgery, radiation therapy planning, surgical simulation, and correlative imaging (simultaneous data display of CT and/or MRI and/or PET data).[25,26]

The detection, definition, and display of a soft tissue or muscular mass is important in determining the optimal management of the individual patient. The extent of a tumor, its relationship to adjacent bone or muscle, as well as its vascularity are all important pieces of information. Combined with a review of the transaxial and 2D data, we have used 3D imaging for defining the exent of a planned surgery.

Evaluation of the chest wall and axilla is especially important in the patient with a history of breast cancer. Tumor recurrence is common in the axilla, occurring most commonly around the region of the axillary artery and vein (near neuromuscular bundle). Other common areas of recurrence are in the chest wall, internal mammary nodal region, and the supraclavicular nodes.

The detection of every recurrence may be difficult to distinguish from postsurgical scarring or fibrosis, particularly when only viewing transaxial images. A multidimensional display with 2D and 3D images can better define the normal anatomy and extent of disease. These images can then be used as a basis for surgical resection or radiation therapy planning.

The evaluation of the location, extent, and relationship of soft tissue tumors as an adjunct to surgical planning is illustrated in Plate 9. The patient had a large tumor in the right mid-thigh which was to be surgically resected. Using an infusion of iodinated contrast (Hypaque-

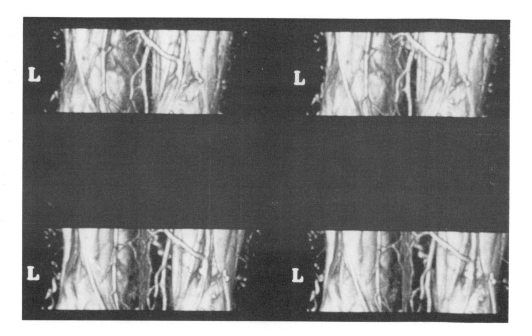

A

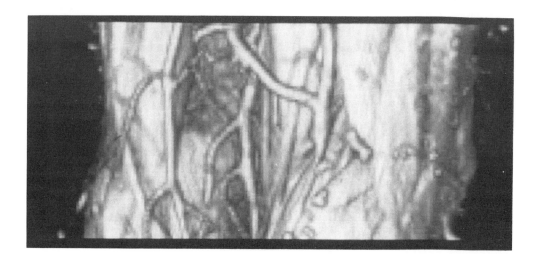

B

FIGURE 23.   Scans of the mid-thigh demonstrate in detail the collateral vessels in this patient with a history of deep venous thrombosis. Notice how smaller vessels are clearly defined on the 3D CT scans.

60), one can clearly define the increased vascularity of the lesion. This was important in planning the subsequent tumor resection. In addition, an unsuspected second nodule (Plate 9) was also present.

One of the areas where soft tissue imaging will be particularly important is in the definition of tumors of organs like the liver, kidneys, and pancreas. To date, however, there has been little success in this area. The potential problems to overcome include motion artifacts, inability to separate various organs due to similarity in CT attenuation, and difficulty in

253

FIGURE 24. Axillary node due to breast cancer. 3D reconstruction of upper chest and axilla demonstrates 3-cm node. Nodal biopsy revealed metastatic breast cancer.

separating normal tissue and tumor. These problems however should be solvable by various hardware and software improvements. For example, at present we routinely obtain our CT scans at 3-second scan times. With newer scanners like the Imatron and Siemens Somatom Plus, subsecond scan times are now possible.

In the future, 3D displays of MRI datasets will be especially valuable in the evaluation of musculoskeletal pathology. With its inherent superior contrast resolution and ability to image in any place, as well as the ability to acquire large datasets, 3D MRI should be very useful clinically. Work is presently being done in this area in several research labs across the country.

VI. CONCLUSION

This chapter has attempted to review some of the more common clinical applications of 3D imaging. With increased user sophistication, improvements in computer hardware and software, as well as a close collaboration with our clinical colleagues, new and more sophisticated applications will develop. The use of 3D CT as both a diagnostic and therapeutic imaging study will undoubtedly prosper in the coming years.

REFERENCES

1. Drebin, R.A., L. Carpenter, and P. Hanrahan. Volume rendering, *Comput. Graph.* 22(4):65–74 (1988).
2. Lorensen, W.E., and H.E. Cline. Marching cubes: A high resolution 3D surface reconstruction algorithm, *Comput. Graph.* 21(4):163–169 (1987).
3. Fishman, E.K., D. Magid, B.R. Mandelbaum, et al. Multiplanar (MPR) imaging of the hip, *RadioGraphics* 6:7–54 (1986).

4. Magid, D., E.K. Fishman, W.W. Scott, Jr., et al. Femoral head avascular necrosis: CT assessment with multiplanar reconstruction, *Radiology* 157:751–756 (1985).
5. Fishman, E.K., D. Magid, D.D. Robertson, A.F. Brooker, Jr., P.J. Weiss, and S.S. Siegelman. Metallic hip implants: CT with multiplanar reconstruction, *Radiology* 160:675–681 (1986).
6. Ney, D.R., E.K. Fishman, D. Magid, and J.E. Kuhlman. Interactive real-time multiplanar CT imaging, *Radiology* 170:275–276 (1989).
7. Fishman, E.K., R.A. Drebin, D. Magid, W.W. Scott, Jr., D.R. Ney, A.F. Brooker, Jr., L.H. Riley, Jr., J.A. St. Ville, E.A. Zerhouni, and S.S. Siegelman. Volumetric rendering techniques: Applications for 3-dimensional imaging of the hip, *Radiology* 16(7):56–61 (1987).
8. Marsh, J.L., and M.W. Vannier. Surface imaging from computerized tomographic scans, *Surgery* 94(2):159–165 (1983).
9. Drebin, R.A., D. Magid, D.D. Robertson, and E.K. Fishman. Sensitivity of three-dimensional imaging for detecting fracture gaps, *J. Comput. Assist. Tomogr.* 13(3):487–489 (1989).
10. Ney, D.R., R.A. Drebin, E.K. Fishman, and D. Magid. Volumetric rendering computed tomography data: Principles and techniques, *Comput. Graph. Appl.* 10(2):24–32 (1990).
11. Fishman, E.K., D. Magid, R.A. Drebin, A.F. Brooker, Jr., W.W. Scott, Jr., and L.H. Riley, Jr. Advanced three-dimensional evaluation of acetabular trauma: Volumetric image processing, *J. Trauma* 29(2):214–218 (1989).
12. Totty, W.G., and M.W. Vannier. Complex musculoskeletal anatomy: Analysis using three-dimensional surface reconstruction, *Radiology* 150:173–177 (1984).
13. Pate, D., D. Resnick, M. Andre, et al. Perspective: Three-dimensional imaging of the musculoskeletal system, *AJR* 147(3):545–551 (1986).
14. Burk, D.L., Jr., D.C. Mears, W.H. Kennedy, L.A. Cooperstein, and D.L. Herbert. Three-dimensional computed tomography of acetabular fractures, *Radiology* 155:183–186 (1985).
15. Scott, W.W., Jr., E.K. Fishman, and D. Magid. Actabular fractures: Optimal imaging, *Radiology* 165:537–539 (1987).
16. Magid, D., E.K. Fishman, A.F. Brooker, Jr., and L.H. Riley, Jr. Volumetric three-dimensional image processing: An introduction to orthopaedic applications, *Contemp. Orthop.* 16(2):29–34 (1988).
17. Vannier, M.W., W.G. Totty, G.W. Steven, et al. Musculoskeletal applications of three-dimensional surface reconstruction, *Orthop. Clin. NA* 16(3):543–555 (1985).
18. Sunburg, S.B., B. Clark, and B.K. Foster. Three-dimensional reformation of skeletal abnormalities using computed tomography, *J. Pediatr. Orthop.* 6:416–420 (1986).
19. Magid, D., E.K. Fishman, D.R. Ney, and J.E. Kuhlman. Two and three-dimensional CT analysis of ankle fractures, *Radiology* 169(P):265 (1988).
20. Kuhlman, J.E., E.K. Fishman, D.R. Ney, and D. Magid. Complex shoulder trauma: Three-dimensional CT imaging, *Orthopedics* 11(2):1561–1563 (1988).
21. Wojcik, W.G., B.S. Edeiken-Monroe, and J.H. Harris, Jr. Three-dimensional computed tomography in acute cervical spine trauma: A preliminary report, *Skel. Radiol.* 16(4):261–269 (1987).
22. Kilocyne, R.F., and L.A. Mack. Computed tomography of spinal fractures, *Appl. Radiol.* p. 40–54 (1987).
23. Vannier, M.W., J.L. Marsh, and J.O. Warren. Three-dimensional CT reconstruction images for craniofacial surgical planning and evaluation, *Radiology* 150:173–177 (1984).
24. Hemmy, D.C., D.J. David, and G.T. Herman. Three-dimensional reconstruction of craniofacial deformity using computed tomography, *Neurosurgery* 13(5):534–541 (1983).
25. Fishman, E.K., R.A. Drebin, R.H. Hruban, D.R. Ney, and D. Magid. Three-dimensional reconstruction of the human body, *AJR* 18(1):53–59 (1988).
26. Fishman, E.K., D.R. Ney, D. Magid, and J.E. Kuhlman. Three-dimensional imaging of the vascular tree, *Dyn. Cardio. Imag.* 2(1):55–60 (1989).

Chapter 9

THREE-DIMENSIONAL MODELING IN THE DESIGN AND EVALUATION OF HIP AND KNEE PROSTHESES

Douglas D. Robertson and Peter S. Walker

TABLE OF CONTENTS

I. INTRODUCTION

There is great interest in utilizing three-dimensional (3D) display for diagnosis, operative planning, prosthesis design, and biomechanical modelling.[1-11] In orthopedics 3D methods improve fracture definition, improve the assessment of the postoperative pelvis, improve the staging and treatment of femoral head avascular necrosis, and provide surgically and mechanically relevant images and models of joints.[6-11]

3D numerical models have also been used to design custom and standard (off-the-shelf) implants for the hip and knee.[9-15] These 3D models were created from stacked image slices obtained using computed tomography (CT). Thresholding was used to detect the boundary coordinates of the object of interest. The resultant boundary contours were used to recreate the surfaces of the object or a solid representation of the object. The volume elements of the solid were assigned material properties (density, stiffness, etc.) related to the attenuation coefficients of that volume or were given values from the literature.

This chapter presents an overview of the Brigham and Women's Hospital's Orthopedic Biomechanics Laboratory's experience using 3D models to design and evaluate hip and knee joint replacements.

II. HIP STEMS

Improved fit appears to be critical to the clinical success of noncemented and cemented total hips. Experimental and clinical evidence demonstrate that for noncemented hip stems maximal stem-cortical bone contact, especially proximally, produces more normal strain values, reduces micromotion and sinkage, and improves clinical results. Improved fit has also been shown to be important in cemented hip stems, where uneven cement mantles have been shown to be more prone to cracking and failure. In response to these problems standard (off-the-shelf) systems have added more sizes and instrumentation.

A properly designed custom implant can produce an optimal fit for a given individual. Custom implant fit has been improved by using two orthogonal planar radiographs and cross-sectional data extracted from several axial CT images. However, while physical bone models have been made for several years from 3D computer model surfaces generated from CT data, custom designs have not. Custom implant designs utilizing 3D modeling are still in the development phase commercially.

A. OPTIMAL FIT DESIGN

The design of an optimal fit implant requires extraction of the accurate 3D geometry of the canal and the incorporation of that data with design features that produce an optimal strain environment for an insertable stem[12-14] (Figure 1). Accurate canal geometry cannot be recreated using two orthogonal radiographic views. These two projections alone do not adequately define the canal's complex geometry.

Recreation of the exact canal geometry may be obtained using CT. However, there are dimensional errors. Beam hardening, partial volumes, and scatter introduce errors in single-energy X-ray CT-generated sizing measurements.[16-21] Beam hardening has been shown to increase the overall diameter and decrease the inner diameter of cortical bone rings.[16] Scanning techniques may be used to reduce partial volumes and scatter and beam hardening may be reduced using dual-energy scanning or postprocessing corrective procedures.[16,17,22,23] However the easiest method to quantify and correct dimensional errors is to include a bone sizing reference phantom in the scan field (Figure 2). This method also quantifies dimensional errors created by the use of various image convolvers and digital filters.

To extract the boundary contours from the axial slices we used a combination of digital

OPTIMAL FIT STEM DESIGN

Extraction of Accurate Canal Geometry from Axial CT images
- digital filtering and boundary detection
- bone sizing error quantification and correction

Canal Surface Model Generation
- spline & stack axial contours then tile surface
- "surgical" preparation of canal model
- assignment of priority regions

Optimal Fit Design
- maximum stem-canal contact (in priority regions) for an insertable stem

Stem Verification, Refinement, and Manufacture
- examine stem-canal contact
- adapt parts of stem for easy machining
- stress analysis of stem
- translate stem surface into tool paths
- CNC machine stem

FIGURE 1. Outline of the optimal fit stem design process.

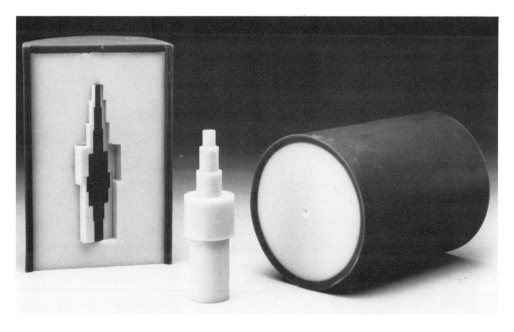

FIGURE 2. Bone sizing reference phantom. The phantom is composed of cortical bone, marrow, soft tissue, and fat equivalent materials. The cortical bone has sections with constant inner to outer ratios and sections with constant outer diameter and varying inner diameters.

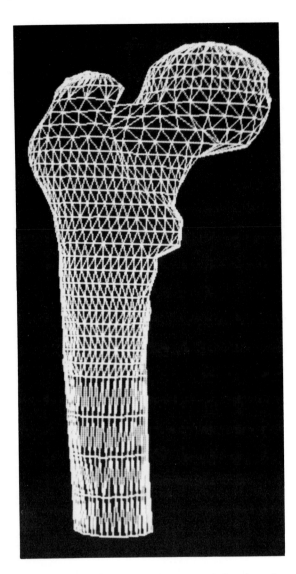

FIGURE 3. Posterior view of the outer cortical surface of a femur. The axial contours were generated from CT scans, which were then stacked and a polygonal surface mesh added. The canal is not displayed in this rendering.

filtering (low pass) and boundary detection based on a CT number threshold (half mean of cortical bone and the canal). The bone's closed canal and outer cortical contours were extracted from each axial slice, the contours splined, and the coordinates equally spaced. The bony contours were then stacked and either the surface tiled or a solid defined (Figure 3).

Dimensional error checking and correction was performed using the sizing reference phantom. All CT scanning and digital image procedures that were performed on the bone were also performed on the included phantom. Automatic measurements were made of the phantom's radii and correction curves generated. If needed, the corrections were applied to the bony contours.

Even with the bony geometry accurately defined in three dimensions the design of an optimal fit stem was incomplete. A maximum fit custom implant would be an exact canal-shaped stem. However, this stem would be impossible to insert due to the complex geometry (three-plane curves) of the canal. Additionally, such a stem would not produce an optimal bone strain environment for stems made out of today's alloys. Thus some stem material must be removed in an intelligent manner. Optimal fit design maintains contact in regions of load transfer and regions important to stem fixation, but removes material in order to make the stem insertable. The important or priority regions for stem-bone contact were the proximal medial wall and calcar for transfer of axial forces, sagittal bending moments, and resistance to torsional loads and the distal lateral wall for prevention of medial-lateral rocking.

Prior to designing the stem, the 3D bone surface model was interactively "surgically" prepared using the keyboard and trackball. A femoral neck osteotomy was made by cutting across the model's tiled surface. A greater trochanteric osteotomy was also performed if desired. Reamimg and rasping of the canal was performed by point editing or surface resplining. During this session the stem length was selected and the priority regions for stem-bone contact assigned. The priority score for each axial contour point was automatically assigned by placement of a predefined circular template. The contour point was assigned the score of the template sector that overlaid it (Figure 4).

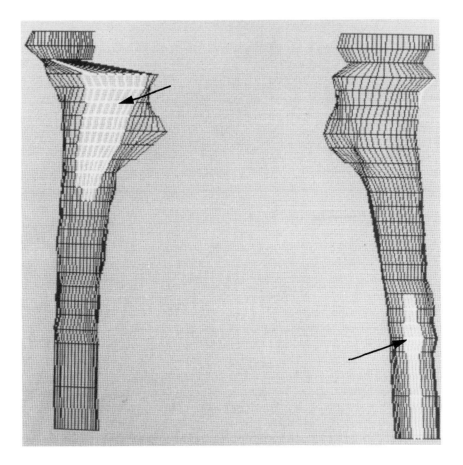

FIGURE 4. A medial-oblique (left) and lateral (right) view of the prepared canal. The femoral neck osteotomy has been made. The white surfaces depict the proximal medial and distal lateral priority regions (arrows). (From Robertson, D.D., P.S. Walker, S.K. Hirano, X.M. Zhou, J.W. Granholm, and R. Poss. Improving the fit of press-fit hip stems, *Clin. Orthop. Rel. Res.* 228:134–144 (1988). With permission.)

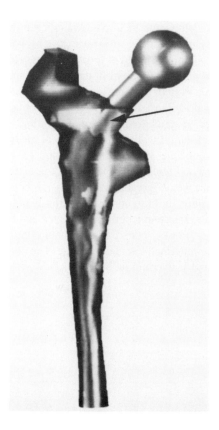

FIGURE 5. Graphic display of stem-inner cortical bone contact. In regions of contact the lighter shaded stem shows through (arrow) the darker shaded canal.

The prepared 3D bone model was input into the optimal fit stem design software. This process began by setting the initial stem design equal to the shape of the prepared canal model. This exact fit stem was then placed fully within the canal. The stem was moved vertically an incremental amount. Its lateral tip was moved into contact with the canal and rotations about the tip's three orthogonal axes performed. At each position a stem-canal surface overlap score was calculated by summing over all the stem points the distance from the point to the canal surface. This result was weighted by the priority score of each stem point. The orientation with the minimum overlap score was selected. The stem was moved to that position and if any of its surface was outside the canal, those portions redefined at the intersection with the canal. Thus the stem was once again within the canal. The stem was again elevated and the process repeated until the stem had passed through the femoral neck osteotomy. The resultant stem was insertable but had maintained as much stem-canal priority contact as possible.

Following design, the 3D stem model was assessed and refined. The design was examined both as axial contours and in three dimensions as a wire-frame and shaded surface. A cutting option sectioned and displayed the stem-bone model at any level in any plane. Stem-canal contact was visually assessed in plots of the axial cross-sections and using MOVIE.BYU (Brigham Young University, Provo UT) (Figures 5 and 6). A program also quantified the contact (fit) and canal fill of the stem. Contact was calculated from the axial sections by determining the distance between the stem and the canal along lines perpendicular to the stem's surface.

Point editing and surface splining may be performed post design. The stem's surface may be redefined in nonpriority regions to make machining of the stem easier and faster. A maximum stress analysis of the final stem was performed and, if acceptable, the stem's surface was translated into computerized numerically controlled (CNC) machine tool paths and the stem produced.

In addition to custom implants, optimal fit design has been used to produce standard (off-the-shelf) implants from a 3D "average femur". The "average right femur" was generated from 26 cadaveric femurs.[14] The 26 femurs were placed in a jig and aligned to a reproducible axis system (based on the femoral head and long axis of the femur). The femurs were then embedded and axially sectioned, with the spacing based on femoral length. The outer and inner cortical contours were digitized and the boundary coordinates triaxially scaled according to equations derived from our studies of the Smithsonian Institute's Terry Collection (skeleton bank). The average contour for each level was calculated and the final averaged contours stacked to construct the average femur.

The average femur was input into the optimal design program and a stem designed. Each axial section of the stem was mirrored (rotated 180 degrees about the anterior-posterior axis) and a left stem created. The average right and left stems were scaled to create six sizes (Profile Hip, DePuy Inc., Warsaw, IN). The range of sizes and change in size per step were based on

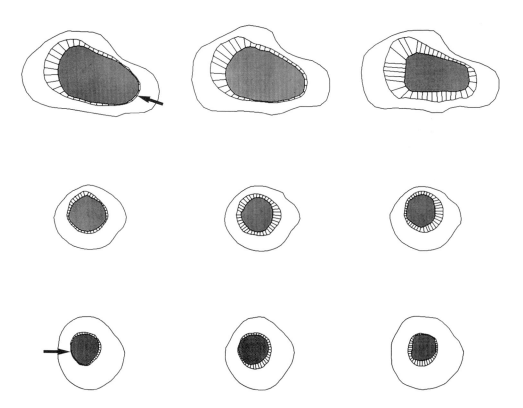

FIGURE 6. Representative proximal (top row), midstem (middle row), and distal (bottom row) axial sections showing the fit of the custom optimal fit (left), standard optimal fit (center), and standard symmetric (right) stems. Note the fit in the proximal medial and distal lateral regions (arrows). (From Poss, R., D.D. Robertson, P.S. Walker, D.T. Reilly, F.C. Ewald, W.H. Thomas, and C.B. Sledge. Anatomic stem design for press-fit and cemented application, in *Non-Cemented Total Hip Arthroplasty*, R.H. Fitzgerald (Ed.) (New York, NY: Raven Press, 1988), pp. 343–363. With permission.)

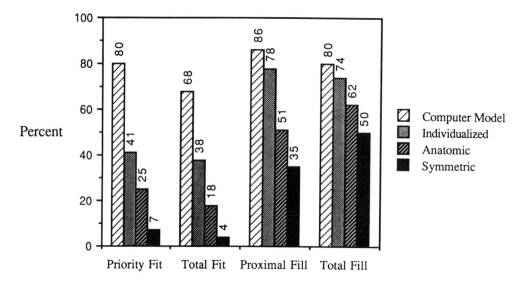

FIGURE 7. Fit (contact) and fill results of the custom stem computer model (computer model), a custom stem inserted into an actual bone (individualized), a standard optimal fit stem (anatomic), and a standard symmetric stem (symmetric). (From Poss, R., D.D. Robertson, P.S. Walker, D.T. Reilly, F.C. Ewald, W.H. Thomas, and C.B. Sledge. Anatomic stem design for press-fit and cemented application, in *Non-Cemented Total Hip Arthroplasty,* R.H. Fitzgerald (Ed.) (New York, NY: Raven Press, 1988), pp. 343–363. With permission.)

the sizes of femurs obtained from the literature and clinical experience. By using the average femur we hoped to minimize the average priority region misfit of the stem over a large number of "routine" clinical cases. The concept of an average femur may also be used for various subpopulations of total hip replacement recipients.

B. HIP STEM EVALUATION

Two studies using a total of 15 cadaver femurs were performed to compare the fit of three noncemented press-fit stem designs.[13] The first study compared a standard symmetrically shaped stem and the standard anatomically shaped stem (designed using the optimal fit design software on the "average" femur model). Five right and left pairs of femurs had appropriately sized implants inserted using each system's respective instrumentation.

The second study compared the above two implant types to a custom optimal fit implant. Custom optimal fit implants were made for five femurs. The standard symmetrically shaped stem's instrumentation was used to prepare the canal and determine the best fitting symmetric stem. Further canal preparation (removal of additional cancellous bone) was performed using the standard anatomically shaped stem's instrumentation. The best fitting anatomic stem was selected and all remaining soft cancellous bone removed. Exact plastic models of the canals were produced and a rigid foam placed into the plastic canals to simulate cancellous bone. Three models were made for each of the five femurs. Each standard stem was inserted using its respective instrumentation. The custom stems were inserted using the standard anatomically shaped stem's instruments and curettes.

All stem-bone specimens, for both studies, were embedded in acrylic and axially sectioned every 2 mm. The stem and canal contours were digitized and submitted to the contact (fit) and fill analysis program. Contact was defined as a distance of 1 mm or less. This analysis of contact showed that along the total length of the stem as well as in the priority regions standard anatomic stems fit better than symmetric stems, but that custom optimal fit stems fit the best (Figures 6 and 7).

Another study, using cadaver femurs and rigid foam bone models, was performed to examine the cement mantle of down-sized standard anatomically shaped stems.[14] Results showed that a symmetric and uniform cement mantle could be achieved by down-sizing the standard stems.

Finite element analyses (FEA) have been used to analyze the mechanical environment of components and bone in total hip replacements.[14,24–27] We have used 3D FEA to study noncemented and cemented optimal fit standard stems.[14] CT imaging of a plastic optimal fit stem implanted into a femur was used to determine the stem, cortical bone, and cancellous bone geometries. The average cortical and cancellous bone mineral densities for each axial section were also obtained from the CT scans. The material properties of the cortical bone, cancellous bone, stem's titanium alloy, and acrylic bone cement were taken from the literature. The final model had almost 2000 3D orthotropic bone elements, 1000 3D interface elements (gaps or cement), and more than 1000 isoparametric stem elements (Figure 8).

The numerical stress analysis was performed using Ansys (Swanson Analysis Systems, Inc., Houston, PA). A physiologic load for one-legged stance was applied to the stem's head. Results showed that the noncemented (press-fit) stem's overall bone stress pattern was closer to that of intact femurs than were the stresses for the cemented simulation. With a press-fit the load transfer occurred primarily in the proximal medial femur, while in the cemented stem transfer occurred primarily in the distal femur (Figure 9).

III. KNEE IMPLANTS

Successful total knee replacement (TKR) designs must balance the requirements of having enough laxity to allow normal knee motion without being unstable, provide a favorable mechanical environment to prevent loosening or subsidence, and be able to be properly positioned and aligned by surgeons. With these in mind, 3D computer modeling of knees and knee implants has been used to improve the design of TKR.[15,28–36]

3D numerical models of 35 knees were made to improve the design and to evaluate femoral and tibial implants.[15,32–36] The knee models each consisted of femoral, tibial, patellar, meniscal, and fibular surface geometry, the origins and insertions of the cruciate and collateral ligaments, and motion data (Plate 10*).

The geometry was obtained by placing cadaver knees in extension and aligning them to a reference axis. The reference axis was a line connecting the centers of the spherical surfaces of the posterior femoral condyles. The longitudinal axis was a line intersecting the midpoint of the reference line and parallel to the long axis of the tibia. The anterio-posterior axis was mutually perpendicular to these two axes. Once fixed in the jig, a 2-mm diameter plastic rod was placed through the reference axis, the knees were embedded, and the knees sagitally sectioned into 24 sections. The contours of interest were digitized, splined, and surfaces produced for the femur, tibia, patella, menisci, and fibula.

After determination of the reference axis but prior to embedding, the origins and insertions of the collateral ligaments (medial-3 bands, lateral-1 band), cruciate ligaments (2 bands each), and patella tendon (1 band) were marked with pins. The knees were placed in extension and aligned in a second jig. Anterior to posterior and medial to lateral radiographs were taken, the centers of the pin head spheres digitized, and the coordinates of the origins and insertions calculated (Figure 10).

The motion data was either assigned the average knee motion of Kurasawa et al.[37] for 23 knees or individually determined for 12 knees. The individual motion data was produced by placing the 12 fresh cadaveric knees in a rig, dynamically flexing them, and continuously measuring the 3D motion.[38] Flexion was produced by using a step motor to pull on the

* Plate 10 appears following page 294.

quadriceps tendon. Six displacement tranducers attached to a yoke about the reference axis were used to measure the Euler rotations and translations of the reference axis.

A. IMPLANT DESIGN

A new femoral implant surface was generated by scaling all the distal femoral models to a given medial-lateral width and averaging identical sagittal sections. Geometric analogs were then used to describe the bearing surfaces of the averaged distal femur. The posterior condyles

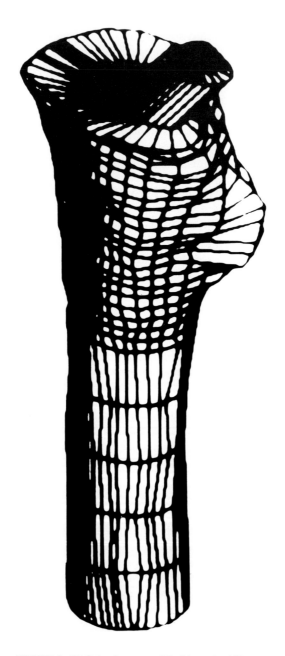

FIGURE 8.   3D finite element model of the optimal fit stem in a bone.

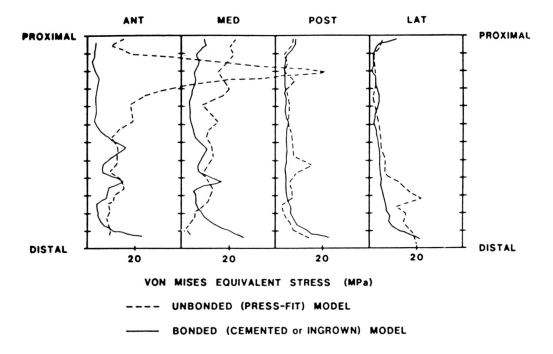

FIGURE 9. Stress analysis of the noncemented (dotted lines) and cemented (solid lines) models. The high-stress peaks correspond to local stem-bone contact. (From Poss, R., D.D. Robertson, P.S. Walker, D.T. Reilly, F.C. Ewald, W.H. Thomas, and C.B. Sledge. Anatomic stem design for press-fit and cemented application, in *Non-Cemented Total Hip Arthroplasty*, R.H. Fitzgerald (Ed.) (New York, NY: Raven Press, 1988), pp. 343–363. With permission.)

were modeled as spheres, the distal condyles as toroids, and the patellar groove as a cone with a toroidal base and sides (Figure 12).

An optimal tibial implant surface was also designed. The criteria were a surface which permitted normal knee motion, was stable, had decreased shear and torque at the implant-bone interface, and had decreased contact stresses on the tibial bearing surface. Such a surface would be a compromise between a low-constraint surface and a high-constraint surface. The low-constraint surface permits most motions and has decreased shear and torque but is unstable and has increased contact stresses, while the high-constraint surface is stable and has decreased contact stresses but has limited motion and increased shear and torque.

To create an optimal tibial surface, the femoral implant was positioned atop a flat tibial surface. The femoral implant was then placed through a series of different mathematically described motions. The motion ranges from normal knee motion to uniaxial motion. The degree of laxity (high, normal, or low) was also a determinant of the described motion. For each described motion the femoral component would "dish out" a surface on the initially flat tibial surface (Plate 11*). The theory of elastic bodies in contact was used to calculate the contact area and maximum contact stress for the generated surfaces.

The best overall compromise between motion, stability, shear and torque, and contact stress was obtained from a tibial surface created using low-constraint normal knee motion and normal laxity. This computer-generated tibial component, as well as the previously described femoral component, are currently being manufactured and implanted (Kinemax Knee System, Howmedica, Rutherford, NJ) (Plate 12*).

* Plates 11 and 12 appear following page 294.

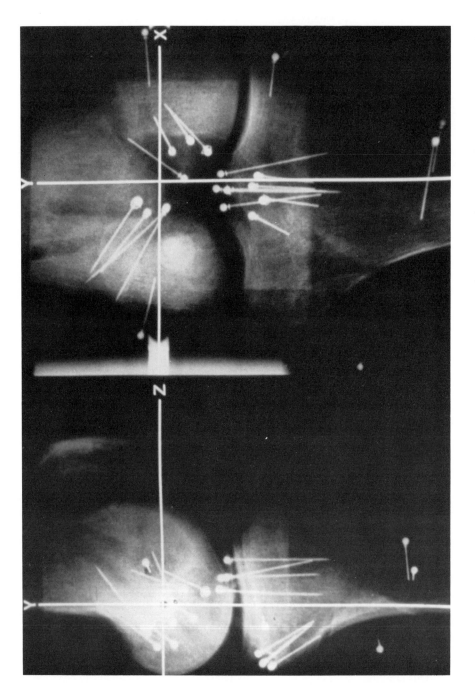

FIGURE 10.   Medial to lateral (a) and anterior to posterior (b) radiographs of a knee specimen with the pins marking the origins and insertions of the ligaments. The x-rays were centered on the reference axis. (From Walker, P.S., J. Rovick, and D. Robertson. The effects of knee brace hinge design and placement on joint mechanics. *J. Biomech.* 21(11):965–974 (1988). With permission.)

267

B. KNEE IMPLANT EVALUATION

The 3D numerical knee models have also been used to evaluate the mechanical behavior of the knee after joint replacement. A dynamic simulation was performed to predict the kinematic and dynamic behavior following resection of the femoral and tibial bearing surfaces and placement of implants.[36] The patella was not resurfaced. Material properties for the model were taken from the literature. Solutions were determined using the principle of total energy minimization, knowing the structural properties of the bearing surfaces and the body weight. Thus the joint motion, contact areas and stresses, quadriceps forces, and ligament lengths and forces were calculated at 5-degree increments between 0 to 120 degrees of flexion. These simulations produced motion and ligament length patterns which were in general agreement with *in vitro* bench testing of knees with joint replacements.

Simulations have also been performed to predict the effect of tibial design and the effect of surgical placement of femoral and tibial components on the range of motion of the knee.[35] Results demonstrated that the greatest flexion was obtained with a flat tibial surface. However this additional 15 degrees of flexion did not outweigh the instability and increased contact stresses of such a design. The simulations also showed that slight anterior displacement of the femoral component improved flexion, while posterior displacement significantly decreased

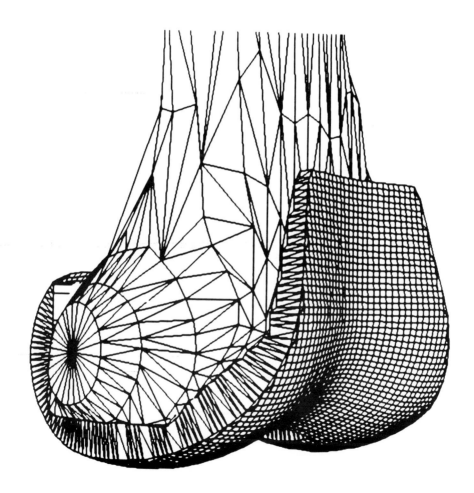

FIGURE 11.   Wire-frame rendering of the computer-generated femoral component implanted on the femur.

the flexion. Tibial component displacement had a negligible effect on motion. However posterior tilting of the tibial component improved flexion and posterior tilting decreased flexion.

IV. CONCLUSIONS

3D modeling and simulations using these models have proven to be valuable experimental tools for designing and studying hip and knee joint replacements. Custom hips can now be made to have an opitmal fit and improved stem-bone stress environments. They are also competitive in cost as a result of computer designing and manufacturing. Modeling has improved the fit of standard stems for noncemented applications and improved the symmetry of cement mantles for cemented applications. Modeling has improved the design of knee implants, providing increased range of motion and decreased contact stresses. Further advances can be expected as the accuracy of our 3D models and simulations continues to improve.

ACKNOWLEDGMENTS

This chapter is a composite of many studies carried out in the Orthopedic Biomechanics Laboratory, for which specific references are cited in the text. Special acknowledgments are made to Robert Poss, M.D., Frederick C. Ewald, M.D., Ashima Garg, Ph.D., Phillip Nelson, Elliot K. Fishman, M.D., Jacques R. Essinger, Ph.D., X. M. Zhou, M.D., Donna Magid, M.D., and Clement B. Sledge, M.D. The bone sizing reference phantom was manufactured courtesy of Computerized Imaging Reference Systems, Inc., Norfolk, VA.

REFERENCES

1. Marsh, J.L., and M.W. Vannier. The "third" dimension in cranio-facial surgery, *Plast. Reconstr. Surg.* 71(6):759–767 (1983).
2. Hemmy, D.C. , D.J. David, and G.T. Herman. Three-dimensional reconstruction of craniofacial deformity using computed tomography, *Neurosurgery* 13:534–541 (1983).
3. Mankovich, N., and J. Beumer. The use of CT scans for the fabrication of cranial prostheses, *Radiology* 153(P):321–322 (1984).
4. Rothman, S.L., W. Glenn, M. Rhodes, R. Bruce, and C. Pratt. Individualized prothesis production from routine CT data, *Radiology* 157(P):177 (1985).
5. Burk, D., D. Mears, W. Kennedy, L. Cooperstein, and D. Herbert. Three-dimensional computed tomography of acetabular fractures, *Radiology* 155:183–186 (1985).
6. Fishman, E., D. Magid, D. Robertson, A. Brooker, P. Weiss, and S. Seigelman. CT/MPR imaging of metallic hip implants, *Radiology* 160:675–681 (1986).
7. Fishman, E.K., D. Magid, B.R. Mandelbaum, W.W. Scott, Jr., P. Weiss, R. Hadfield, B. Mudge, S.E. Kopits, A.F. Brooker, and S.S. Siegelman. Multiplanar (MPR) imaging of the hip, *Radiographics* 6(1):7–53 (1986).
8. Fishman, E., B. Drebin, D. Magid, W. Scott, A. Brooker, L. Riley, E. Zerhouni, and S. Siegelman. Volumetric rendering techniques: Applications, *Radiology* 163:737–738 (1987).
9. Nerubay, J., Z. Rubinstein, and A. Katznelsons. Technique of building a hemipelvic prothesis using computer tomography, *Prog. Clin. Biol. Res.* 99:147–152(1981).
10. Rhodes, M.L., Y. Azzawi, E. Chu, W. Glenn, and S. Rothman. Anatomic model and prostheses manufacturing using CT data, in *Proceedings Conference and Exposition of the NCGA 1985*, pp. 110–124.
11. Woolsen, S.T., P. Dev, L. Fellingham, and A. Vassiliadis. Three-dimensional imaging of bone from computerized tomography, *Clin. Orthop. Rel. Res.* 202:239–248 (1986).
12. Robertson, D.D., P.S. Walker, J.W. Granholm, P.C. Nelson, P.J. Weiss, E.K. Fishman, and D. Magid. Design of custom hip stem prostheses using three-dimensional CT modeling, *J. Comput. Assist. Tomogr.* 11(5):804–809 (1987).

13. Robertson, D.D., P.S. Walker, S.K. Hirano, X.M. Zhou, J.W. Granholm, and R. Poss. Improving the fit of press-fit hip stems, *Clin. Orthop. Rel. Res.* 228:134–14 (1988).

14. Poss, R., D.D. Robertson, P.S. Walker, D.T. Reilly, F.C. Ewald, W.H. Thomas, and C.B. Sledge, Anatomic stem design for press-fit and cemented application, in *Non-Cemented Total Hip Arthroplasty*, R.H. Fitzgerald (Ed.) (New York, NY: Raven Press, 1988), pp. 343–363.

15. Walker, P.S., and X-M. Zhou. The dilemma of surface design in total knee replacement, *Trans. Orthop. Res. Soc.* 12:291 (1987).

16. Robertson, D.D., and H.K. Huang. Quantitative bone measurements using x-ray computed tomography with second-order correction, *Med. Phys.* 13(4):474–479 (1986).

17. Joseph, P.M., and R.D. Spital. A method for correcting bone induced artifacts in computed tomography scanners, *J. Comput. Assist. Tomgr.* 2:100–108 (1978).

18. Zatz, L., and R. Alvarez. An inaccuracy in computed tomography: The energy dependence of CT values, *Radiology* 24:91–97 (1977).

19. Joseph, P.M., and R.D. Spital. The exponential edge-gradient effect in x-ray computed tomography, *Phys. Med. Biol.* 26(3):473 (1981).

20. Glover, G.H., and N.J. Pelc. Non-linear partial volume aritfacts in x-ray computed tomography, *Med. Phys.* 7(3):238 (1980).

21. Joseph, P.M., and R.D. Spital. The effects of scatter in x-ray computed tomography, *Phys. Med. Biol.* 9(4):464 (1982).

22. Alvarez, R.E., and A. Macovski. Energy-selective reconstructions in x-ray computed tomography, *Phys. Med. Biol.* 21(5):733–744 (1976).

23. Kalender, W., W. Perman, J. Vetter, and E. Klotz. Evaluation of a prototype dual-energy computed tomographic apparatus. I. Phantom studies, *Med. Phys.* 13(3):334–339 (1986).

24. Huiskes, R., and E. Chao. A survey of finite element analysis in orthopedic biomechanics, *J. Biomech.* 16:385–409 (1983).

25. Crowninshield, R.D., R. Brand, R. Johnston, and J. Milroy. An analysis of femoral component stem design in total hip arthoplasty, *J. Bone Joint Surg.* 62(1):68–78 (1980).

26. Huiskes R., and W. Vroeman. A standardized finite element model for routine comparative evaluations of femoral hip prostheses, *Acta Orthop. Belg.* 52(3):258–261 (1986).

27. Rapperport, D., D. Carter, and D. Schurman. Contact finite element analysis of porous ingrowth acetabular cup implantatio, ingrowth, and loosening, *J. Orthop. Res.* 5:548–561 (1987).

28. Wismans, J., F. Velpaus, J. Janssen, A. Huson, and P. Struber. A three-dimensional mathematical model of the knee joint, *J. Biomech.* 13:677–685 (1980)

29. Andriacchi, T., R. Mikosz, Hampton, and J. Galante. Model studies of the stiffness characteristics of the human knee joint, *J. Biomech.* 16(1):23–29 (1983).

30. Garg, A., and P.S. Walker. The effect of the interface on the bone stresses beneath the tibial components, *J. Biomech.* 19(12):957–967 (1986).

31. Sidles, J.A., R. Larson, J. Garbini, D. Downey, and F. Matsen. Ligament length relationships in the moving knee, *J. Orthop. Res.* 6:593–610 (1988).

32. Walker, P.S. Joints to spare, *Science* 85:56–57 (1987).

33. Walker, P.S., J. Rovick, and D. Robertson. The effects of knee brace hinge design and placement on joint mechanics, *J. Biomech.* 21(11):965–974 (1988).

34. Garg, A., P.S. Walker, D.D. and Robertson. Theoretical model of knee kinematics using computer simulation, in *Proceedings of the 11th International Congress of Biomechanics 1987*, p. 151.

35. Garg, A. Effect of Component Design and Surgical Placement on the Mechanics of Total Knee Replacement, Ph.D. thesis, Massachusetts Institute of Technology, Cambridge, MA (1988).

36. Essinger, J.R., P.F. Layvarz, and D. Robertson. A mathematical model for the evaluation of the behavior during flexion of condylar-type knee prostheses, *J. Biomech.* (in press).

37. Kurosawa, H., P. Walker, S. Abe, A. Gara, and T. Hunter. Geometry and motion of the knee for implant and orthotic design, *J. Biomech.* 18(7):487–499 (1985).

38. Reuben, J., J. Rovick, P. Walker, and R. Schrager. Three-dimensional kinematics of normal and cruciate deficient knees — a dynamic *in-vitro* experiment, *Trans. Orthop. Res. Soc.* II:385 (1986).

Chapter 10

INTEGRATED 3D DISPLAY OF MR, CT, AND PET IMAGES OF THE BRAIN*

David N. Levin, Xiaoping Hu, Kim K. Tan, Simranjit Galhotra,
Andreas Herrmann, Charles A. Pelizzari, George T. Y. Chen,
Robert N. Beck, Chin-Tu Chen, and Malcolm Cooper

TABLE OF CONTENTS

*   Reprinted in part from *Proc. Natl. Comput. Graph. Assoc.* 1:179–186 (1989). With permission.

I. ABSTRACT

This paper describes a new method for using magnetic resonance (MR) images to create 3D views of the surface and internal structure of the brain. The 3D images depict important landmarks on the surface of the brain, which cannot be clearly identified by inspection of the usual cross-sectional views. Clinical cases illustrate how this new information can be used for neurosurgical planning. A novel technique for producing integrated 3D displays of data from multiple scanning modalities is also discussed. Specifically, MR data are used to create a 3D model of each patient's brain anatomy for the purpose of localizing measurements of brain function, derived from positron emission tomography (PET).

II. INTRODUCTION

The brain is an unusual organ in that many of its most important functional centers are located on its surface. Specifically, certain ridges (or gyral convolutions) on the brain surface are known to be associated with control of movement, sensation, hearing, and speech. Neurosurgeons, who are preparing to resect a tumor, would like to know its location relative to these areas so that they can plan the operation to minimize disruption of these vital functions. Unfortunately, these surface structures cannot be identified with certainty by inspection of ordinary cross-sectional images obtained with X-ray computed tomography (CT) or MR scanners. This is because the characteristic shapes of these convolutions are not evident on individual slice images. It is necessary to examine a 3D view of the brain surface in order to recognize specific gyral landmarks.

In order to create an accurate 3D rendition of an anatomical structure, it is necessary to choose the appropriate combination of scanning modality (e.g., CT vs. MR) and reconstruction algorithm (e.g., surface rendering[1] vs. volume rendering).[2,3] Most investigators believe that skeletal structures are best reconstructed from CT images since bones are depicted with high contrast-to-noise levels. On the other hand, it is reasonable to use MR data to render the brain in three dimensions since MR images are characterized by high soft tissue contrast. There is considerable controversy about which reconstruction algorithm is most appropriate in each of these cases. Surface rendering is the most popular method for creating 3D views of the bones although volume rendering has also been advocated for this purpose.[4] Investigators[5–9] have had varying success using surface rendering to create 3D views of the brain surface. Two difficulties must be overcome in order to obtain a satisfactory result. First, it is necessary to produce a precise digital description of the complex surface of the brain. Second, the reconstruction algorithm must include a routine to account for voxels which straddle the brain surface and contain varying mixtures of brain and adjacent tissues. The discrete tissue characterization step, which is part of most surface-rendering algorithms, makes it difficult to handle these partial volume averaging effects without creating jagged artifacts or "filling in" the grooves (sulci) between convolutions. Nevertheless, the literature [6–9] contains reasonable 3D renditions of the brain, derived by surface rendering. In the third section of this paper we show how to use volume rendering[10,11] to create 3D views of the brain surface. Since our algorithm is based on continuous tissue characterization, it leads to 3D images with minimal artifacts (fuzziness) from partial volume averaging. Furthermore, these images can be produced without specifying an explicit digital description of the convoluted surface of the brain; it is only necessary to subject the original cross-sectional views to a simple slice-editing procedure which draws any contour between the brain surface and overlying skull-scalp tissues.

Surgeons would also like to have 3D views of the internal structure of the brain so that they can assess the depth and shape of a lesion as well as its geometrical relationship to other internal structures. In order to create such views, it is first necessary to characterize the structures to be rendered so that the computer can identify them. As discussed in the fourth section of this chapter, we have done this in individual cases by means of an interactive image-editing operation. The same section also describes a more comprehensive approach to the segmentation of MR images by means of a continuous tissue characterization program based on multispectral analysis.[8,11,13,14]

Some patients are imaged with two or more cross-sectional imaging techniques which reveal complementary information. For example, MR may be used to depict brain anatomy but provides little insight into brain function; on the other hand, PET delineates brain function (e.g., metabolism, blood flow) but provides a poor description of brain antomy. In these situations, an integrated display of information from both modalities can be more useful than either set of single modality images. Some of us[15,16] have recently developed a retrospective image registration program which makes it possible to fuse images from different modalities without prospective manipulations of the patient or images (e.g., without external or internal landmarking). In the fifth section of this chapter, we describe how this technique can be used to create *integrated 3D* displays of MR, PET, and CT images of the brain.

Our group has also written software for interactive manipulation of the aforementioned 3D views of the structure or function of the brain. Specifically, the sixth section of this chapter describes a neurosurgery simulation program which we have used for surgical planning in 20 clinical cases. The same section discusses a "plane-picker" program for displaying cross-sectional views which are selected interactively by positioning a cursor on a 3D image of the brain.

III. 3D VIEWS OF THE BRAIN SURFACE

The surfaces of the brain and of the skin can be reconstructed from data acquired during a single 10-minute volumetric pulse sequence on our 1.5-Tesla unit (Magnetom, Siemens Medical Systems, Inc.; Iselin, New Jersey). This exam produces images of 63 contiguous slices with 3-mm thickness; each image is a 256×256 matrix of square pixels with 1.2-mm sides. This pulse sequence can easily be added to a routine clinical MR study of the brain.

Each cross-sectional image is subjected to an interactive editing program of our own design, which serves to isolate the brain and surrounding cerebrospinal fluid (CSF). This software uses a threshold tracking algorithm to draw any contour between the surface of the brain and the inner surface of the skull on each slice image. The operator specifies a "seed" point to initiate the contour generation on each image, and he may have to perform some manual operations to correct contouring errors which occur on approximately 15% of the slices. Notice that this contour need not follow the complex convolutions of the brain surface. As described below, the surface of the brain will be delineated automatically by a volume-rendering process in which CSF is made invisible (transparent). It takes approximately 45 minutes to edit all 63 slice images with this software which is currently running in C on a Sun SPARC station. It should be possible to reduce this time significantly by using faster hardware and by increasing the automation of the algorithm.

The processed images are then subjected to 3D volume rendering[2,10,11] by means of the ChapVolumes program on a Pixar Image Computer (Vicom Systems, Inc., Fremont, California). A continuous lookup table is applied to the grey scale value of every voxel in order to estimate the fractions of brain and CSF within it. Each voxel is then assigned color and transparency values, which depend on these fractions and on the attributes of "pure" brain (white, opaque) and CSF (colorless, transparent). 3D views are generated by tracing rays through this volume. The ChapVolumes program also provides for the use of spatial gradients to shade the brain-CSF interface.

It takes approximately 10 minutes to devise an appropriate lookup table for each subject; each 3D view is then generated in 45 to 60 seconds. Therefore, almost 1 hour is required to

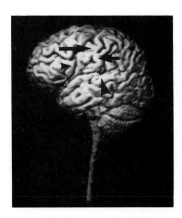

FIGURE 1. MR-derived 3D view of a
patient's brain, showing areas controlling
movement (long arrow), sensation (short
arrow), hearing (large arrowhead), and
speech (small arrowhead).

create a 64-frame "movie" of the rotating (or tumbling) brain surface. 3D views of the
subject's skin surface can be generated by using the unprocessed MR images in conjunction
with a lookup table which characterizes the air-skin interface.

We have applied these techniques to more than 25 subjects, including 20 patients. In almost
every case (Figure 1) it was possible to recognize important landmarks of the brain surface
(e.g., the movement, sensory, auditory, and speech areas) which could not be identified with
certainty on the original cross-sectional images. The 3D views depicted the anatomical
relationship of these critical areas to superficial brain lesions which were visualized as
deformities of the brain surface (Figure 2). This information, which could not be obtained by
any other noninvasive means, had significant diagnostic and therapeutic value.

IV. 3D VIEWS OF THE INTERNAL ANATOMY OF THE BRAIN

Some brain tumors have well-defined margins on the usual cross-sectional MR images. In
such cases, it is possible to isolate the lesion by means of the interactive contour generation
program described in the previous section. If the tumor voxels are given distinctive attributes
(e.g., red color and complete opacity) and brain voxels are made sufficiently translucent,
volume rendering will result in a clear 3D view of the location of the lesion beneath the surface
of the brain (Figure 3).[11,12] The relationship of the tumor to overlying cortical structures can
then be visualized by interactively "toggling" between these translucent 3D views and 3D
images of the surface of the opaque brain. This type of display makes it easier for a
neurosurgeon to appreciate the location of a tumor as well as irregularities of its shape, which
are not as evident on cross-sectional images.

We have also experimented with a more comprehensive approach to image segmentation,
which uses supervised multispectral analysis.[8,11,13,14] Every slice of the brain is scanned with
two pulse sequences (e.g., T1-weighted and T2-weighted), each of which produces an image
with characteristic tissue intensities. Regions of interest (ROIs) are interactively placed on a
representative portion of every important tissue of the imaged volume. The pair of signal
intensities is measured in each voxel in these ROIs. The measurements within each tissue tend
to cluster at characteristic locations in the two-dimensional space of signal intensities. These
signal intensity distributions are fitted with bivariate Gaussian functions, and qualitative
estimates of the relative volumes of different tissues are made by visual inspection of the

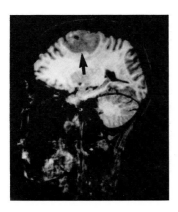

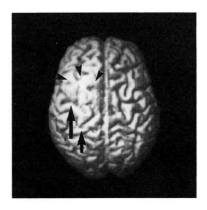

FIGURE 2A. Cross-sectional MR image of a patient with a tumor (arrow) which has effaced the overlying convolutions of the surface of the brain. The specific convolution, which controls movement, cannot be identified with certainty.

FIGURE 2B. MR-derived 3D view of the brain in Figure 2A, showing the flattening of the surface by the tumor (arrowheads), just in front of the left movement and sensory strips (long and short arrows, respectively).

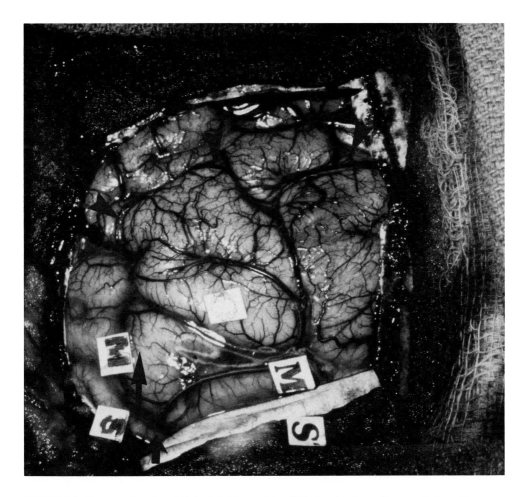

FIGURE 2C. Intraoperative view of the brain in Figures 2A and 2B, showing the swollen convolutions of the tumor (arrowheads), just in front of the motor strip ("M" and long arrow) and sensory strip ("S" and short arrow). The positions of the arrows and arrowheads are approximately the same as in Figure 2B.

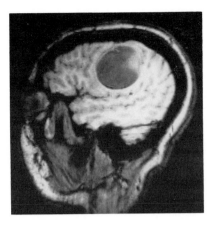

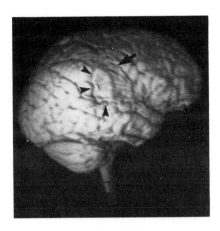

FIGURE 3A. Cross-sectional MR image of a patient with a cystic brain tumor containing a mural nodule. The movement and sensory areas on the brain surface cannot be identified.

FIGURE 3B. MR-derived 3D view of the brain in Figure 3A, showing the tumor (small arrowheads) as an area of distortion behind the right sensory strip (short arrow).

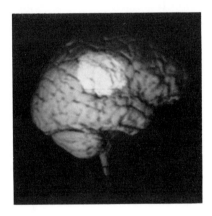

FIGURE 3C. MR-derived translucent model of the brain in Figures 3A and 3B, showing the tumor mass undermining the right sensory strip.

images. These numbers are used in Bayes formula for estimating the fractional amount of each tissue type in every voxel from the signal intensities of that voxel. The color and opacity attributes of each voxel are then calculated as fractional combinations of the colors and opacities selected for each "pure" tissue. In an alternative scheme voxels, which are closer than a critical "distance" from a cluster in intensity space, are given attributes of the corresponsing "pure" tissue while all other voxels are considered to be unclassified. Finally, 3D views are created by using the ray-tracing and shading algorithms of ChapVolumes.

Although this tissue characterization algorithm leads to correct classification of most voxels, a noticeable number of voxels are usually misclassified or unclassified (Figure 4). These classification errors can be traced to several sources: noisy data, overlapping of the intensity distributions of the various tissue types, and inhomogeneity of the scanner's sensitivity. We have developed semi-automatic algorithms for "correcting" these classification errors prior to the 3D rendering step. For example, contiguous misclassified voxels can be collected by 3D region-growing, and then the tissue type of the entire collection can be reassigned with a single command. Also, isolated misclassified voxels can be reassigned automatically to adjacent tissues.

We tested this technique in six clinical cases. Neurosurgeons felt that these renditions of internal structures (e.g., tumor, edema, normal brain tissue, CSF) helped them visualize abnormal brain anatomy. However, certain parts of the reconstruction process (e.g., the error correction procedure) required too much time and interaction to be practical in most clinical situations.

V. INTEGRATED 3D DISPLAY OF DATA FROM MULTIPLE MODALITIES

In order to merge images from different modalities, it is necessary to know the coordinate transformation which maps each voxel of one data set onto the corresponding voxel of the other. Some of the authors have recently reported a retrospective method of performing this image registration operation[15,16] in which the required coordinate transformation was derived by matching models of the brain or skin surfaces, detected in the two data sets. These surface models could be rather crude since they only needed to represent the orientation and ellipsoidal shape of the head or brain. It was sufficient to use simple "prism" models comprised of brain or skin contours, derived by applying the threshold-tracking program (see first section) to MR, CT, and PET (emission or transmission) cross-sectional images. The interscan coordinate transformation was then used to create integrated cross-sectional views by fusing complementary features of images from different modalities. We have now taken this process one step further by forming *integrated 3D* displays of data from MR, PET, and CT scans.[11,18-20]

The average PET measurements in the superficial layer (5 mm) of brain were used to color

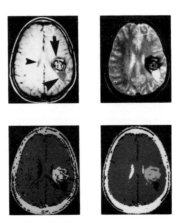

FIGURE 4.   Matched T1-weighted (upper left) and T2-weighted (upper right) MR images of a patient with a vascular mass (arrow), postsurgical cyst (large arrowhead), and normal ventricles (small arrowheads). Uncorrected tissue type map (lower left) produced by multispectral analysis, showing that mass and normal brain are correctly classified while skin and skull are misclassified as mass; other areas (cyst, ventricles, scattered voxels) are unclassified and shown in black. Tissue type map (lower right) after application of error correction algorithms, showing correct classification of all tissues.

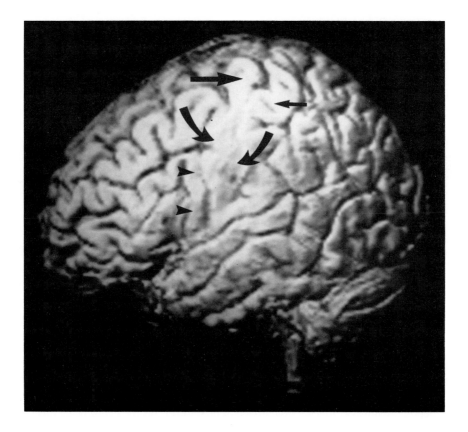

FIGURE 5A. MR-derived model of the brain of a child with intractable seizures due to extensive encephalitis. There is a mild abnormality (curved arrows) of the lower segments of the motor (long arrow) and sensory (short arrow) strips, just posterior to the speech area (small arrowheads). This abnormality, which was confirmed at surgery, was not clearly shown by the original cross-sectional images.

the surface of the brain model, derived from an MR scan of the same patient. The operator was provided with a software "switch" for instantly turning the PET-derived coloring on or off while viewing the rotating 3D model of the brain surface; References 18 to 20 contain colored illustrations of such a 3D display of MR and PET data. Thus, the highly resolved MR images were used to create a 3D atlas of each patient's brain anatomy for the purpose of localizing poorly resolved PET measurements of brain metabolism. Experience with 11 clinical cases suggests that this technique can be used to localize superficial metabolic abnormalities with respect to anatomical landmarks such as the movement, sensory, auditory, and speech areas. We have found that this is particularly useful in planning surgery for patients with intractable epilepsy. The MR scans alone are often unrevealing since these patients usually have normal brain anatomy. Although PET scans frequently demonstrate foci of abnormal metabolism, which are the cause of the seizures, the PET images are too poorly resolved to serve as a basis for surgical planning. The integrated 3D display of MR and PET data demonstrates the anatomical location of the functional abnormality to be resected (Figure 5).

We have also experimented with combined displays of MR and CT data. MR and CT delineate complementary features of anatomy, e.g., MR depicts soft tissues optimally while CT is best for imaging bones and calcifications. We have transferred CT-derived images of the skull and calcifications into a volume of MR data in three clinical cases. After the CT-

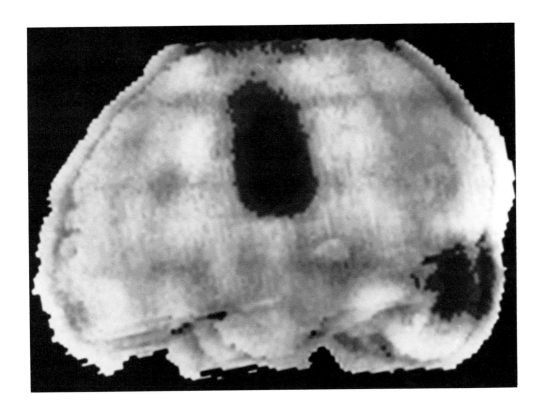

FIGURE 5B. PET-derived 3D model of cortical metabolism of the patient in Figure 5A. A hypermetabolic focus (dark gray) is seen in the left frontal-parietal area. The cerebellar "hot spot" is an artifact of the reconstruction process.

detected features were given distinctive color and opacity attributes, volume rendering was used to create 3D views of the bones and calcifications in the context of an MR-derived rendition of brain soft tissues. The resulting images can give the surgeon a better appreciation of the patient's anatomy in cases which involve complex abnormalities of both soft tissues and bones.

VI. INTERACTIVE DISPLAY OF 3D IMAGES

We have written several interactive programs which enable the Pixar operator to relate 3D views displaying different types of anatomical or functional information. For example, he may instantly toggle between "movies" depicting the rotating opaque surface of the brain and a rotating translucent model of the brain showing internal structures. Another software key can be used to "switch on" (or "switch off") coloring, which encodes functional measurements on the brain surface, derived from PET. It is also possible to peer into the opaque brain through a "window of translucency" which the operator draws with a mouse-controlled cursor.

We have devised a "plane-picker" program[11,12,18–20] for displaying cross-sectional images in the context of 3D views of the brain (Figure 6). As the operator moves a mouse-controlled line-cursor through the 3D model, one or more cross-sectional images of the corresponding slice are updated on the right side of the same screen. The 3D view may be an MR-derived rendition of the patient's brain surface, brain interior, or skin surface; it may also be an integrated 3D display of MR and PET data or MR and CT data. The cross-sectional views may inlcude MR images (T1-weighted and/or T2-weighted), CT scans, and/or PET images. This

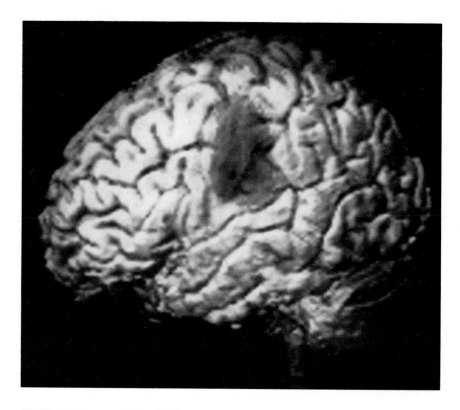

FIGURE 5C. Integrated 3D model of brain anatomy and function, obtained by merging the MRI and PET data in Figures 5A and 5B. It is evident that the anatomical and metabolic abnormalities are located at the same position. Intraoperative electroencephalography showed that this was the location of the most pronounced focus of the patient"s nearly continuous seizure activity.

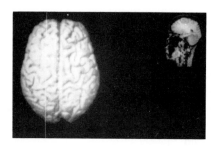

FIGURE 6A. Interactive "plane-picker" display for simultaneous viewing of cross-sectional and 3D views. The sagittal MR image was selected by positioning the line-cursor on the tumbling 3D model of the brain of the patient in Figure 2.

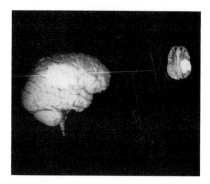

FIGURE 6B. Axial MR image picked by placing the line-cursor on a rotating translucent 3D model of the brain of the patient in Figure 3.

enables the operator to "roam" freely through the brain, examining its internal structure and function on cross-sectional images which intersect its surface at selected positions. Inspection of each type of image (3D, cross-sectional) enhances the interpretation of the other.

Software for staging "electronic surgery rehearsals" has also been written by our group.[11,12,18–20] The program begins with a 3D "movie" of the rotating (or tumbling) brain and skin of the subject, displayed side-by-side on the same screen. The operator stops the movie at a frame which resembles the desired position of the patient at surgery; he uses a mouse-controlled cursor to outline the portion of the brain surface which he wishes to uncover. Then, the machine simulates the necessary craniotomy by removing the corresponding patch of skin and skull on the adjacent view of the subject's external surface and replacing it with a view of the underlying brain surface. This exercise enables the surgeon to predict what exposure will result from each type of incision. Since the "operation" can be "erased" and performed again in a few seconds, the configuration of the incision can be optimized in a few minutes. A film or videotape of the simulated surgery is then sent to the operating room to serve as a guide for placement of the actual incision. We have used this technique to plan neurosurgery on 20 patients with brain lesions (Figure 7).

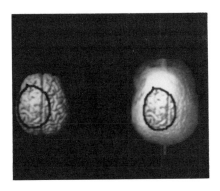

FIGURE 7B. Simulation of surgery performed on the patient in Figures 3 and 6B.

FIGURE 7A. 3D models of the brain and skin of the patient in Figures 2 and 6A, showing the results of a surgery rehearsal.

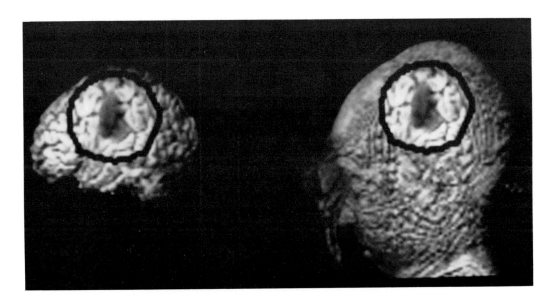

FIGURE 7C. Simulated craniotomy performed on the patient in Figure 5.

In each case intraoperative inspection of the brain surface confirmed the accuracy of the 3D views. Since this technique removes some of the guesswork from surgery, we hope that it will enable surgeons to make smaller incisions. For those cases requiring the greatest precision, stereotaxic surgical methods could be used to transfer the outline of the proposed craniotomy from the 3D image to the patient's head. The surgeon could then duplicate the rehearsed incision by "cutting on the dotted line". The achievement of this type of stereotaxic surgery is now a prime goal of our laboratory.[21]

VII. DISCUSSION

We have demonstrated that it is practical to use MR images to create 3D displays of the surface and internal structure of the brain. These 3D views convey information which cannot be derived by inspection of cross-sectional images; specifically, it is possible to visualize the anatomical relationship between a lesion and important functional centers on the surface of the brain, which control movement, sensation, hearing, and speech. Neurosurgeons can use this information to plan an operative approach which leads to minimal disruption of these critical areas. They have used our surgery simulation software to design incisions which lead to optimal exposure of the underlying brain surface. In addition, we have produced integrated 3D displays of MR, CT, and PET images of the brain. The simultaneous 3D display of MR and PET data enables one to localize PET measurements of brain function with respect to an MR-derived model of brain anatomy. We are now working on techniques for augmenting our 3D display of brain anatomy by adding a rendition of brain vasculature, derived from MR "angiograms".[22]

The software described in this chapter could form the basis of a workstation for 3D brain imaging. Radiologists would be able to use "plane-picker" software on such a device to review cross-sectional images from any modality in the context of a 3D model of brain anatomy or function. A workstation of this kind would be a natural component of picture archiving and communications systems (PACS) which are now being developed by many manufacturers. Neurosurgeons would benefit from the presence of such a workstation in their offices and in the operating room so that surgical procedures could be "practiced" on the screen before they are carried out on the patient. Finally, it is not hard to imagine how radiation therapists could use such a workstation to optimize the configuration of radiation beams.

We are now devising a "point and click" interface for our software so that the entire process of image editing and 3D rendering can be carried out by a technologist without any assistance from programmers, scientists, or physicians. We are also attempting to reduce the overall time for this process so that a complete 3D display can be produced in only 2 or 3 hours instead of 5 or 6 hours. Some time can be saved by software enhancements, e.g., it should be possible to automate most of the image editing process. However, hardware upgrades may also be necessary.

ACKNOWLEDGMENT

This work was supported in part by Siemens Medical Systems, Inc., the Cancer Research Foundation, the Brain Research Foundation, and the Technology Commercialization Center.

REFERENCES

1. Herman, G.T., and H.K. Liu. Three-dimensional display of human organs from computed tomograms, *Proc. Comput. Graph. Imag.* 9:1–21 (1979).
2. Drebin, R.A. Volume-rendering, *Comput. Graph.* 22:65–74 (1988).
3. Harris, L.D., R.A. Robb, T.S. Yuen, and E.L. Ritman. Noninvasive numerical dissection and display of anatomic structure using computerized x-ray tomography, *SPIE Proc.* 152:10–18 (1978).
4. Fishman, E.K., B. Drebin, D. Magid, et al. Volumetric rendering techniques: Applications for three-dimensional imaging of the hip, *Radiology* 163:737–738 (1987).
5. Kennedy, D.N., and A.C. Nelson. Three-dimensional display from cross-sectional tomographic images: An application to magnetic resonance imaging, *IEEE Trans. Med. Imag.* MI-6:134–140 (1987).
6. Cline, H.E., C.L. Dumoulin, H.R. Hart, W.E. Lorensen, and S. Ludke. 3-D reconstruction of the brain from magnetic resonance images using a connectivity algorithm, *Magn. Res. Imag.* 5:345–352 (1987).
7. Bomans, M., K.H. Hohne, U. Tiede, and M. Riemer. 3-D segmentation of MR images of the head for 3-D display, *IEEE Trans. Med. Imag.* 9:177–183 (1990).
8. Koenig, H.A. and G. Laub. Tissue discrimination in magnetic resonance 3-D data sets, in *Medical Imaging II: Image Formation, Detection, Processing, and Interpretation,* v. 914, Schneider, R.H. and S.J. Dwyer (Eds.) (Bellingham, WA: International Society for Optical Engineering, 1989) pp. 669–672.
9. Chuang, K.S., J.K. Udupa, and S.P. Raya. High-Quality Rendition of Discrete Three-Dimensional Surfaces, Technical Report MIPG 130, Medical Image Processing Group, Dept. of Radiology, Univ. of Pennsylvania (July 1988).
10. Levin, D.N., X. Hu, K.K. Tan, and S. Galhotra. Surface of the brain: Three-dimensional MR images created with volume-rendering, *Radiology* 171:277–280 (1989).
11. Levin, D.N., X. Hu, K.K. Tan, S. Galhotra, A. Herrmann, et al. Integrated 3-D display of MR, CT, and PET images of the brain, *Proc. Natl. Comput. Graph. Assoc.* 1:179–186 (1989).
12. Hu, X., K.K. Tan, D.N. Levin, S.G. Galhotra, et al. Three-dimensional magnetic resonance images of the brain: Application to neurosurgical planning, *J. Neurosurg.* 72:433–440 (1990).
13. Vannier, M.W., R.L. Butterfield, D. Jordan, et al. Multispectral analysis of magnetic resonance images, *Radiology* 154:221–224 (1985).
14. Hu, X., D.N. Levin, A. Herrmann, K.K. Tan, and S. Galhotra. Continuous tissue characterization algorithm for 3-D volume-rendering of MR images, *Proc. Soc. Magn. Res. Med.* 1:196 (1988).
15. Pelizzari, C.A., and G.T.Y. Chen. Registration of multiple diagnostic imaging scans using surface fitting, in *The Use of Computers in Radiation Therapy,* Bruinvis, I., P. van der Giessen, H. van Kleffens, and F. Wittkamper (Eds.) (Amsterdam: Elsevier, 1987), pp. 437–440.
16. Pelizzari, C.A., G.T.Y. Chen, D.R. Spelbring, R.R. Weichselbaum, and C.T. Chen. Accurate three-dimensional registration of CT, PET, and MR images of the brain, *J. Comput. Assist. Tomogr.* 13:20–26 (1989).
17. Levin, D.N., C.A. Pelizzari, G.T.Y. Chen, C.-T. Chen, and M.D. Cooper. Retrospective geometric correlation of MR, CT, and PET images, *Radiology* 169:817–823 (1988).
18. Levin, D.N., X. Hu, K.K. Tan, S. Galhotra, et al. Integrated three-dimensional display of MR and PET images of the brain, *Radiology* 172:783–789 (1989).
19. Hu, X., K.K. Tan, D.N. Levin, C.A. Pelizzari, and G.T.Y. Chen. A volume-rendering technique for integrated three-dimensional display of MR and PET data, in *3D Imaging in Medicine,* Hohne, K.H. et al. (Eds.) (Berlin: Springer, 1990), pp. 379–397.
20. Levin, D.N. MR and PET data merge in 3-D images of brain, *Diagn. Imag.* pp. 150–157 (November 1989).
21. Tan, K.K., D.N. Levin, C.A. Pelizzari, and G.T.Y. Chen. Interactive stereotaxic localization of cranial anatomy, *Proc. Soc. Magn. Res. Med.* 1:569 (1990).
22. Hu, X., K.K. Tan, D.N. Levin, S.G. Galhotra, G.T.Y. Chen, C.A. Pelizzari, R.N. Beck, C.T. Chen, and M.D. Cooper, Volumetric rendering of multimodality, multivariable medical imaging data, Proc. of Chapel Hill Workshop on Volume Visualizaton, (Chapel Hill, NC: University of North Carolina), May 1989.

Chapter 11

A HISTORICAL PERSPECTIVE
OF HEART AND LUNG 3D IMAGING

Eric A. Hoffman

TABLE OF CONTENTS

I. VOLUMETRIC CT:
A NEW DIMENSION TO IMAGING OF THE HEART AND LUNGS

The skeleton of the Dynamic Spatial Reconstructor (DSR),[1] boxed and handled as priceless cargo, arrived at the front door of the Medical Sciences Building of the Mayo Clinic in July of 1979. Crowds gathered to watch the riggers inch the 15-ton gantry through a hole in the wall of the building. Over the next year, members of the Biodynamics Research Unit under the leadership of Erik Ritman participated in the assembly, mounting, and alignment of the 14 X-ray guns and juxtaposed television cameras onto the superstructure which, via a high-voltage slip ring assembly, would whirl the imaging chains at 15 rotations per minute allowing for the gathering of X-ray projection data sets which could be reconstructed into volumetric images of the body represented by up to 240 0.9-mm thick computed tomographic (CT) slices. In the limit, each stack of slices from the DSR represent a 1/60-second aperture, stop action volume image of the scanned body region. These temporally sequenced volume images opened the door to the development of computer-based multidimensional biomedical image display and analysis of cardiopulmonary data.

Previous imaging techniques used to study the heart and lungs had included: (1) monoplane and biplane video and cine ventriculography;[2] (2) invasive implantation of ultrasound crystals[3] into the heart walls to measure limited, single, linear dimensions of the cardiac chambers; (3) implantation of metallic markers[4] into the cardiac and lung tissue for subsequent location in three dimensions via biplane fluoroscopy; (4) two-dimensional (2D) ultrasound for the planar assessment of dynamic anatomic detail; and (5) injection of radiopharmaceuticals[5,6] for the assessment of parameters such as regional tissue blood flow, regional pulmonary ventilation, cardiac chamber (limited in practice to the left ventricle) ejection fraction, and, to a limited degree, regional myocardial wall motion. To handle the limitations of 2D image data sets used for the purpose of assessing three- (or greater) dimensional (3D) geometric and functional analysis problems, numerous models,[7,9] such as the assumption that the left ventricle can perhaps be described as an ellipsoid of revolution, have been developed to translate 2D data into an estimate of 3D geometry. Because of the irregularity of the shape of the right ventricle and the right and left atria, in addition to the problem of superposition particularly in the case of the atria, the success of the models was largely limited to the left ventricle where volumes could be estimated to within 15 to 20% of a "gold standard," although a gold standard for "real life" data gathering situations has not been found. Better than 15 to 20% accuracy is achievable when the model is developed on the gold standard, but applicability to subsequent hearts with individual shape variability is questionable. Truly, volumetric CT requires no simplifying assumptions in measuring the geometry of the body's organ systems.

2D techniques have been developed to assess the regional and global functioning of the heart and have been limited in large part to the measurement of regional and global wall thickening and thinning. Inference of 3D parameters from 2D information[10] have included: estimates of the blood volume ejected by the ventricle (cardiac output when integrated over time), fraction of blood volume ejected by the ventricle (relative to the volume in the ventricle at the beginning of the cardiac cycle), calculation of work via the use of pressure-volume loops,[11] calculation of the contractile properties of the muscle via families of pressure-volume loops, etc. Important parameters assessed for regional lung function have included measurements of regional and global parameters related to the relative and absolute distribution of pulmonary ventilation and pulmonary blood flow. Since a primary function of the lungs is to oxygenate the blood and remove carbon dioxide, the appropriate matching of ventilation and perfusion is important. Quantitative assessment of these parameters via imaging techniques has in large part been limited to the arena of nuclear imaging, with projection imaging of radioactive xenon (either inhaled or injected into the bloodstream) being the most common approach. In addition to quantitative analysis of heart and lung function, 2D image data sets have been, and continue to be, used extensively for subjective assessment of the status of these organ systems.

The data format and data flow involved in dynamic, volumetric CT changed the ground rules for image display and analysis dramatically. In planning for the DSR, members of the research team spoke the words related to the image analysis tasks ahead with the megabyte and gigabyte descriptors. However, following the arrival of the scanner, digital 9-track tapes filled with CT sections began to line the laboratory walls, and the true magnitude of the problem was made clear.

A. A NEW TRIP ALONG PREVIOUSLY TRAVELED ROADS: VOLUMETRIC IMAGE DISPLAY

In the course of designing the scanner, early research brought Gabor Herman to Mayo in 1975 on sabbatical with the goal of implementing reconstruction algorithms. It was during this time that Herman and his graduate student Liu, at the traditional Friday noon lab lunch meeting, unveiled the first shaded surface representations derived from CT data. The images showed the ventricular myocardium of an isolated heart.[12] The use of shaded surface displays to depict X-ray-derived, geometric relationships of the heart had their origins in the earlier work at Mayo by James Greenleaf, Craig Coulam, and colleagues[13,14] which estimated the 3D shape of the left ventricular chambers from biplane ventriculograms. This work was preceded by earlier surface displays, by the same group, of isolated lungs, onto which contour maps were generated representing regional distribution patterns of pulmonary blood flow as detected via radioactive microspheres. The earliest version of Liu's surface display algorithm resided on a Control Data Corporation-3500 mainframe computer and had no general access to the "end user." This wasn't much of a problem since in those early days the "end user" him/herself was a rather ill-defined entity.

In 1980 Jay Udupa,[15] having found Gabor Herman's group in Buffalo, traveled to Mayo to implement the first "user friendly" version of the surface display package on the newly acquired ModComp Classic Computer. In the first application at Mayo, applying the data to soft tissue, Hoffman (this author) et al.[16] succeeded in isolating the lung from an intact living dog (see Figure 1A) scanned upright within the DSR predecessor, the Single Source Dynamic Spatial Reconstructor (SSDSR), and the right heart chambers of an anesthetized dog stop action imaged at end-diastole (see Figure 1B). Segmentation was achieved via a simple thresholding scheme and a region growing algorithm.[17] This task, and several subsequent segmentations, including a herculean effort (the end product of which is shown in Figure 2) by Sinak et al.[18] (published years later) to manually segment the heart of a young boy with a common ventricle and a segmentation of the chambers of a young man with idiopathic hypertrophic subaortic stenosis (IHSS)[19] during a 72-hour marathon session, demonstrated the need for a more user friendly approach. Although alluring, the 3D representations of the heart and lungs, in the form of shaded surface displays, have thus far provided little new or important information regarding the normal or pathophysiologic function of these organ systems. What the displays have done is to verify whether or not the structures of interest have been properly detected in the course of a volume determination and, to a limited extent, have provided the ability to interactively separate the chambers one from the other, allowing for separate determination of the atria from the ventricles (see Plate 13*). Surface displays are beginning to show signs of potential clinical utility in aiding the cardiologist to understand the complex malformations associated with various forms of congenital heart disease and in aiding the cardiologist in imparting that understanding to others such as the cardiothoracic surgeon.[18,20,21] As has become the case for craniofacial surgery, it is expected that 3D cardiac imaging will form a fronticepiece to future cardiac surgical planning and postsurgical evaluation.

* Plate 13 appears following page 294.

THREE-DIMENSIONAL SHADED SURFACE DISPLAY OF IN VIVO LUNGS INFLATED TO TLC
(25 cm H₂O TPP-Intact Dog, Head-Up, Morphine-Pentobarbital Anesthesia)

FIGURE 1(A). First X-ray CT-derived shaded surface display of *in vivo* lungs. Dog was held upright and scanned via the SSDSR with lung volume maintained at 24 cm H₂O airway pressure. (From Hoffman, E.A., T. Behrenbeck, P.A. Chevalier, and E.H. Wood. Estimation of regional pleural surface expansile forces in intact dogs, *J. Appl. Physiol.* 55:935–948 (1983). With permission.)

B. BENEATH THE SURFACE

In wrestling, at Mayo, with the questions of volume image analysis, at a weekly Friday noon lunch meeting, the term "pixel" was replaced with "voxel" and new display schemes were developed which did not require prior image segmentation. Such techniques included "projection dissolution"[22] which, in its more complex form, has become known as volume rendering. To avoid the superposition problem (which drove investigators to develop CT in the first place), Lowel Harris et al.[22] proposed a radial projection which in essence unrolled the heart as a map maker unrolls the earth onto a flat surface. An example of this is shown in Figure 3. In his research on the utility of numerical reprojection, Harris worked on the display of these images in stereo pairs and took the concept a step further by working with a vibrating mirror[23] which reflected slices displayed in a fast decay phospor television monitor, thus providing to the viewer the ability to peer into the mirror of a holographic-type representation of the object which could be interacted with via a 3D joy stick.

Numerical biopsy and oblique sectioning[24] (which has been given other names such as volume reformatting) proved to be of particular utility in image quantitation. These functions subserve the requirements of locating within the volume image the specific structure of interest in the orientation of interest so that geometric and density parameters can be appropriately determined. In the case of the heart, examples of measurements of interest are the cross-sectional area of the orifices of the valves separating the atria and ventricles, the cross-sectional area of the coronary arteries (measured from slices perpendicular to the local long axis of the arteries), thickness of the ventricular muscle measured within short axis planes of the heart (planes perpendicular to the axis defined by an imaginary line linking the left ventricular apex with the midpoint of the mitral valve ring), and the brightness changes occurring over time within the region of the myocardium in a temporal scan sequence gathered as an injected bolus of radiopaque contrast agent traverse the muscle of interest, thus indicating regional tissue blood flow. Lawrence (Larry) Sinak, a cardiology fellow who spent several years of his training associated with the DSR project, transported aviation nomenclature to the task of defining commands whereby the end-user could instruct an oblique sectioning program in the search for the appropriate cross-sectional image. "Roll" referred to tilts of the left and right sides of the plane relative to the standard viewing angles of the coronal, sagittal, or transverse imaging planes. As per convention, transverse images are viewed foot to head with the spine down, sagittal images are viewed right to left with the spine to the left, and coronal images are viewed ventral to dorsal with the right side located on the left side of the viewing screen. "Pitch" referred to tilts relative to the top or bottom of the starting plane and "yaw" related to inplane rotations, while elevate related to movement of the plane in parallel steps above or below the starting plane. This nomenclature seemed particularly fitting since the history of the laboratory dated back to World War II when Earl Wood was recruited to Mayo as part of a team investigating pilot blackout during exposure to the high gravitational forces of dive bombing maneuvers.

C. PLATFORMS AND PACKAGES

As lessons were learned through the process of developing the necessary image analysis tools to handle the data emerging from the DSR project, and with the rapidly changing scope

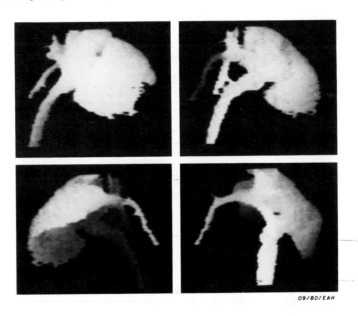

FIGURE 1(B). First X-ray CT-derived shaded surface display of *in vivo* right heart chambers. Dog was scanned supine in the DSR with lungs held at 0 cm H_2O airway pressure, and images represent the end-diastolic volume. Images of heart chambers and lungs were produced in 1980 by Dr. Eric A. Hoffman utilizing a surface display algorithm implemented by Dr. Jay Udupa on a Mod Comp Classic Computer at the Mayo Clinic. (From Sinak, L.J. and E.L. Ritman. Dynamic Spatial Reconstructor, in *Computerized Transmission Tomography of the Heart and Great Vessels — Experimental Evaluation and Clinical Applications,* C.B. Higgins (Ed.) (Mt. Kisco, NY: Futura Publishing, 1983), pp. 61–73. With permission.)

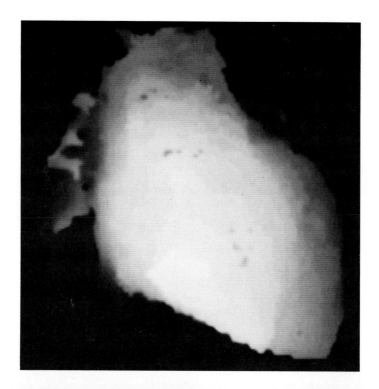

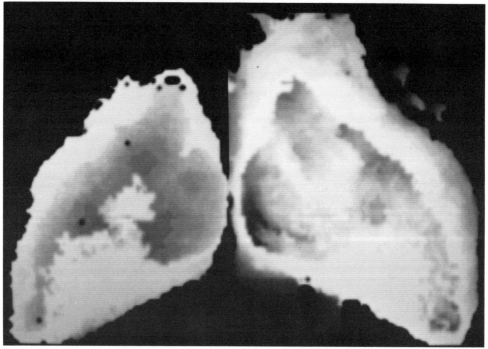

FIGURE 2. 3D shaded surface display of myocardium of a 13-year-old boy diagnosed with a univentricular heart. Upper panel demonstrated an interior view of the heart while in the lower panel the heart is shown "open", demonstrating the incomplete interventricular septum. These images represent one of the earliest displays of X-ray CT-derived images of congenital heart disease and were generated by Dr. Lawrence J. Sinak. (From Sinak, L.J., E.A. Hoffman, P.R. Julsrud, D.D. Mair, J.B. Seward, D.J. Hagler, L.D. Harris, R.A. Robb, and E.L. Ritman. The Dynamic Spatial Reconstructor: Investigating congenital heart disease in four dimensions, *Cardiovasc. Intervent. Radiol.* 7:124–137 (1984). With permission.)

of computing hardware, the analysis platform shifted from a large mainframe approach to the use of individual workstations, the first being the Charles River Universe which ran a UNIX look-a-like operating system called UNOS, and "C" became the programming language of choice. In the background, a development group within the Biodynamics Research Unit, under the direction of Lowell Harris and subsequently Richard Robb, generated the skeleton for an image analysis package, while in the foreground a small core of "end-users" struggled with questions relating to "What are the questions?". The analysis of the unique data sets consisted of 4 or more dimensions (3 spatial, temporal, and functional dimensions such as blood flow, regional wall dynamics, etc.). As the end-users defined important analysis algorithms, programmers incorporated the algorithms within a growing package of analysis tools, and, in turn, an analysis tool would sometimes appear independent of an "end-user" and would be found to be an important missing tool. Patrick Heffernan, in the end, laid the groundwork for the use of the pop-down menu format with a shared memory structure which gives the look, feel, and structure to what has become known as "Analyze".[25] Analyze has now been found to have tremendous utility to not only end-users associated with the DSR project, but also researchers engaged in the evaluation of data sets emanating from the ever-growing list of devices generating multidimensional image data sets.

The Analyze approach is to essentially provide the end-user with a "tool box" of image analysis routines which can be mixed and matched to create individually tailored data manipulation schemes. The package loads a volume image into memory, and that volume can be worked on by the multiple tools within the package. Display approaches include (but are not limited to): (1) the ability to extract the standard orthogonal slices including transverse, sagittal, and coronal orientations; (2) oblique sectioning discussed above; (3) volume rendering which uses ray casting to display the volume in formats including standard summation of the pixels along a ray, brightest voxel projection, and gradient shading.

COMPARISON OF RADIAL AND PARALLEL PATH REPROJECTION IMAGES OF RECONSTRUCTED VOLUME
Coronary Arteriogram of Isolated Canine Heart
(Left Anterior Descending COR ART Ligated)

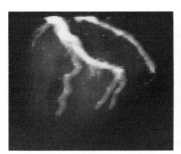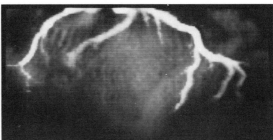

Parallel Reprojection   **Radial Reprojection**   MAYO 1978

FIGURE 3. Demonstration of radial reprojection of volumetric data set of coronary arterial tree. Comparison of parallel (left) and radial (right) path reprojection images of a reconstruction of an isolated canine heart scanned via the SSDSR. Images produced by Dr. Lowell D. Harris. (From Harris, L.D., R.A. Robb, T.S. Yuen, and E.L. Ritman. Display and visualization of three-dimensional reconstructed anatomic morphology: Experience with the thorax, heart, and coronary vasculature of dogs, *J. Comput. Assist. Tomogr.* 3:439–446 (1979). With permission.)

Shaded surface display is also provided, and one can display one or more color-coded object surfaces which have been previously segmented from the volume data set via a process of gray scale range selection, interactive editing of unwanted voxels within the gray scale range, and region growing to select for the voxels connected to a selected seed voxel. Perhaps one of the most powerful tools within Analyze is a module entitled "transform". This module allows the user to write mathematical equations with nested parentheses and to then specify the variables in terms of image slices, volume data sets, or multiple volume data sets such as a temporal sequence. Without the need for any prior knowledge of computer programming, user-definable matrix manipulation algorithms can be tested and set into routine approaches to data analysis. Other quantitative tools provide for the interactive sampling of regions of interest with a reporting of voxel gray scale statistics, measurement of linear dimensions within a slice, calculation of the angles between structures of interest, measurement of regional tissue blood flow based upon brightness changes occurring within a temporal sequence of images gathered as an injected bolus of iodinated, radiopaque contrast agent traverses the imaging field,[26,27] and volume estimation via the use of randomly scattered voxel tags which yield statistical information regarding the likelihood of a structure being a certain volume based upon the number of tags identified to be within the organ structure of interest relative to the total number of tags and the volume into which the tags were scattered.[28]

II. IMAGING: A NEW, INTEGRATIVE LOOK AT CARDIOPULMONARY PHYSIOLOGY

As a first approach to the evaluation of the utility of these image analysis tools to study physiology, we asked the question regarding how well the overall volume of the heart and lungs could be determined. We demonstrated that the lung volumes could be detected to within 3% of excision and water displacement values,[29] cardiac chamber volumes could be detected to within 5% of the water displacement measurements of radiopaque cast scanned *in situ* and subsequently excised,[30] and ventricular muscle mass could be measured to within 5% of postmortem quantitation.[31,32] The visual anatomic fidelity of the surface displays of the heart chambers and lungs correlated well with the structures apparent in the excised chamber casts and excised air-dried lungs.

A. THE HEART

As shown in Figures 4 and 5, via the DSR we have been able to evaluate the volumes of the cardiac chambers and myocardium throughout the cardiac cycle, and the combined chamber volume plus myocardial volume (pericardial sac contents) has been termed "total heart volume". As we began to evaluate how the heart worked under pathologic states, we soon realized that we didn't know how the heart worked in nonpathologic states and sought to answer a rather basic question: "What is the normal, global relationship between the various volumetric components of the heart throughout the cardiac cycle?" We have found that the total volume of the heart remains essentially constant throughout the cardiac cycle. This constant heart volume relationship appears to be attributable in large part to the reciprocal emptying and filling of the atria and ventricles, with a piston-like motion of the valve plane between atria and ventricles and a fixed positioning of the epicardial (outer surface of the myocardium) apex.[33,34] Subsequent scanning of normal human volunteers via magnetic resonance imaging[35] and fast X-ray cine CT[36] has shown that this constant heart volume in dogs also holds in humans. Presumably, this constant heart volume relationship is the most efficient form of functioning whereby little work is wasted in moving extracardiac structures while the majority of the cardiac work load can be expended in moving blood. We have begun to find disease states such as atrial fibrillation where this constant heart volume relation is disrupted.

Simultaneous with the investigation of the constant heart volume relationship, clinically

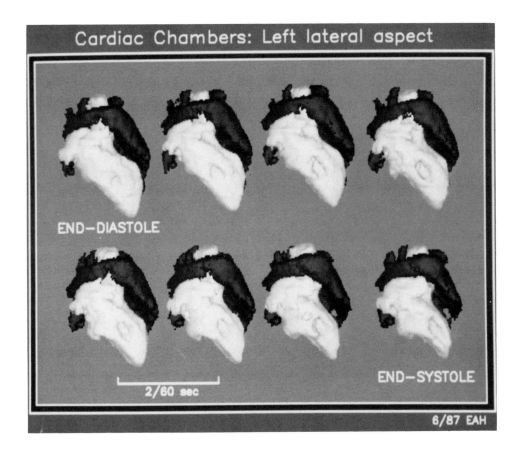

FIGURE 4. Shaded surface displays of cardiac chambers of a dog scanned via the DSR. Images are viewed from the left lateral aspect demonstrating the left heart chambers reconstructed at 2/60 sec scan intervals between end-diastole (upper left) and end-systole (lower right). Note the reciprocal emptying and filling of the atrium and ventricle with a piston-like motion of the mitral valve plane. These sorts of data sets have led to the observation that the total volume of the heart (contents of the periardial sac) remains nearly constant throughout the cardiac cycle. (From Hoffman, E.A. and E.L. Ritman. Invariant total heart volume in the intact thorax, *Am. J. Physiol. Heart Circ. Physiol.* 249:H883–H890 (1985). With permission.)

related protocols involved the use of 3D display of congenital heart disease,[18,37] acquired heart disease,[19] and the evaluation of ventricular ischemia.[27] Erik Ritman[1] has devoted a large amount of his resources to evaluate the validity of volumetric X-ray CT measurements of coronary artery dimensions and regional myocardial blood flow. In addition, efforts have been directed towards the validation of the measurements of vascular dimensions.[37–42]

B. THE RESPIRATORY TRACT

As demonstrated in Figure 6 we have imaged the lungs from a number of species including dogs,[43] sloths,[44] humans,[45,46] and miniature horses.[47] In addition to demonstrating our ability to extract the volume of the lungs accurately, we have shown that, via the DSR, we are able to image the regional lung air content to within 7% of known air content values,[43] and the relative air content values from lung region to lung region appear to be accurate to within 3% in the absence of significant beam hardening artifacts which appear to be consistently located at the dorsal, basal portion of the lung of dogs. In the course of our studies of the distribution pattern of regional ventilation, Patrick Heffernan et al. helped us[48,49] by developing a method of displaying the regional lung air content as a color coding integrated into the shaded surface display. These color-coded data sets are depicted in Figure 7 along with an artists' concept of the compositing of a sampling plane with the surface display whereby statistical analyses of the determinants of regional ventilation can be interrogated.[50] Our studies of regional lung air content distribution have led to the hypothesis that the intrathoracic position of the heart is a major determinant of the regional distribution of ventilation, and the presence of lobar fissures allows the lobes of the lungs to accommodate shape changes of the diaphragm and rib cage without altering the volumes of the alveoli (air sacs comprising the region of gas exchange with the blood). It is interesting to note that horses do not have multiple lobed lungs and the oxygenation of the blood is compromised when recumbent, particularly supine. This poses a particular problem, for example, when the horse breaks a leg or when a newborn foal must be maintained sedated on a ventilator because of complications of premature birth. Contrary to the horse, it is interesting to note that the sloth has lobar fissures at birth, but these fissures fill in with fibrous tissue shortly after birth. Perhaps this is due to the rigid rib cage structure (the sloth has 22 rib pairs) whereby the shape of the chest cavity does not vary as the sloth changes body postures and thus the lung lobes would not be expected to have to slip much relative to each other. Under normal circumstances, the lung blood flow appears to follow the pattern of regional ventilation fairly well. However, recent evidence suggests that blood flow may be, in part, determined by vascular conductance (vascular geometry) and therefore blood flow and ventilation have nonoverlapping parameters determining their relative distribution patterns. Studies are underway to determine what alterations might occur to the determinants

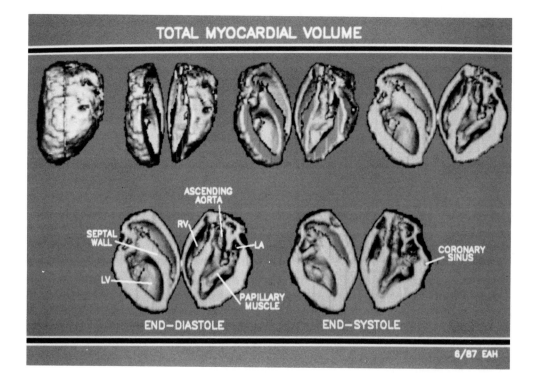

FIGURE 5. Shaded surface display of myocardium of the heart shown in Figure 4. LV = left ventricle; RV = right ventricle; LA = left atrium.

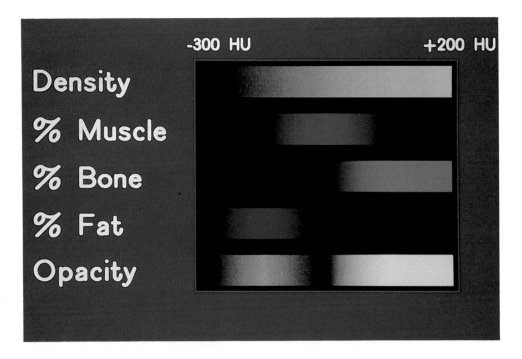

PLATE 8. Volumetric rendering as used on Pixar image computer. Three channels are assigned for muscle, bone, and fat. The opacity or degree of transparency is assigned a fourth channel. A high opacity value will make structures more opaque, while a low opacity value will make the images more transparent. This technique takes advantage of the Pixar's four-channel parallel processors.

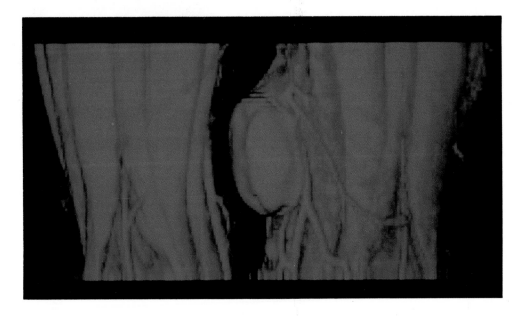

PLATE 9. Melanoma of thigh. Large hypervascular mass in left thigh is malignant melanoma. Note large feeding vessels as well as satellite lesion.

 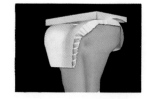 

PLATE 10   PLATE 11   PLATE 12

PLATE 10. View of a model of an intact knee in 60 degrees of flexion. Note the anterior and posterior cruciate ligaments seen in the intercondylar notch. PLATE 11. Surface-shaded rendering of the femoral component sitting atop an originally flat tibial surface which has been dished out. PLATE 12. Wire-frame rendering of the computer-designed femoral and tibial components.

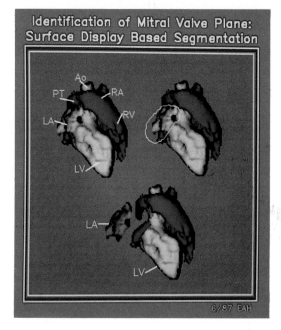

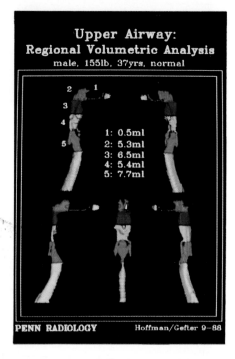

PLATE 13. Shaded surface display of end-diastolic view of the cardiac chambers of a dog scanned via the Dynamic Spatial Reconstructor. An interactive editing program was utilized to identify the mitral valve plane separating the left atrium from the left ventricle. LA = left atrium; LV = left ventricle; PT = pulmonary trunk; AO = aorta; RA = right atrium; RV = right ventricle.

PLATE 14. By segmenting the airway into regional volumes delineated by readily identifiable anatomic landmarks, we are able to evaluate the regional volumes at multiple airway pressures and thus generate regional compliance curves allowing for a characterization of the surrounding soft tissue.

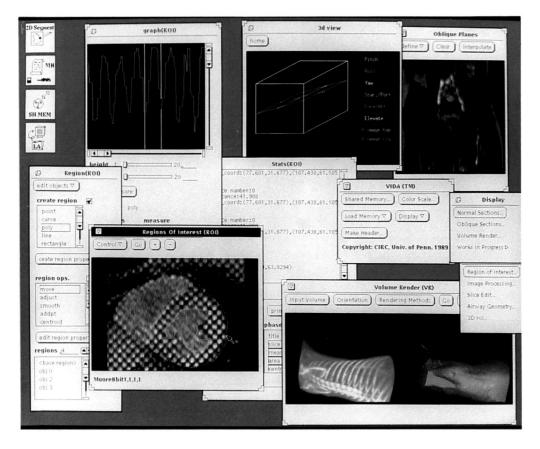

PLATE 15. Photograph of the computer screen showing multiple x-windows of VIDA™. Lower left: region of interest module (ROI) with sub-panels showing statistical analysis of MR tagged heart cross section. Lower right: volume rendering module displaying brightest pixel projection and pixel gradient shaded surface of a sloth scanned via the DSR. In the upper right are sub-panels from a module allowing for selection of oblique sections from volume images in shared memory. Voxels do not have to be cubic for oblique images to be displayed properly. The main VIDA panel is shown in right center with various sub-panels. Iconified modules are shown in the upper left. All volumes depicted reside simultaneously in a unique shared memory structure.

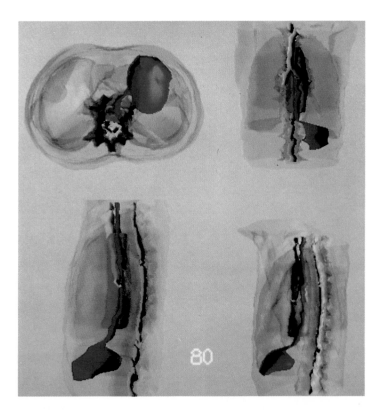

PLATE 16. A 3D dose distribution from a "3-field" esophageal treatment plan displayed using a surface-based display. One of the radiation beams comes in from the front, and two from oblique angles in the back. Shown (from upper left, clockwise) are transverse, anterior, lateral, and posterior oblique views. The 80% radiation isodose surface is in red, the esophagus and stomach in blue, the trachea (seen well in the anterior view) in white, the lungs, vertebral bodies, and external surface in grey scale, and the spinal cord in yellow.

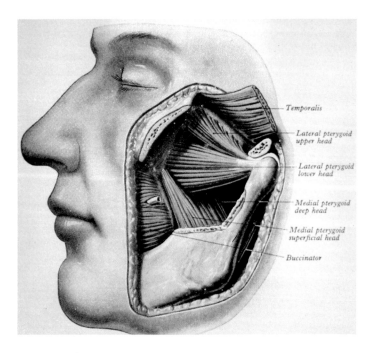

PLATE 17. Illustration of muscles of mastication. (From Warwick, R. and P.L. Williams. *Gray's Anatomy*, 35th British ed. (Edinburgh: Churchill Livingstone, 1973). With permission.)

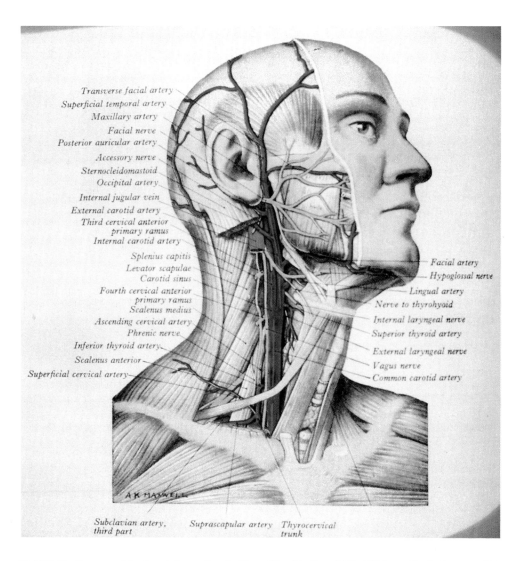

Transverse facial artery
Superficial temporal artery
Maxillary artery
Facial nerve
Posterior auricular artery
Accessory nerve
Sternocleidomastoid
Occipital artery
Internal jugular vein
External carotid artery
Third cervical anterior
primary ramus
Internal carotid artery
Splenius capitis
Levator scapulae
Carotid sinus
Fourth cervical anterior
primary ramus
Scalenus medius
Ascending cervical artery
Phrenic nerve
Inferior thyroid artery
Scalenus anterior
Superficial cervical artery

Facial artery
Hypoglossal nerve
Lingual artery
Nerve to thyrohyoid
Internal laryngeal nerve
Superior thyroid artery
External laryngeal nerve
Vagus nerve
Common carotid artery

A K MAXWELL

Subclavian artery,   Suprascapular artery   Thyrocervical
third part   trunk

PLATE 18. Cutaway of the lateral face and neck. (From Warwick, R. and P.L. Williams. *Gray's Anatomy,* 35th British ed. (Edinburgh: Churchill Livingstone, 1973). With permission.)

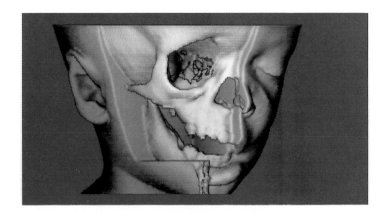

PLATE 19. Cutaway of the facial skeleton. (Provided by ISG Technologies, Inc., Mississauga, Ontario.)

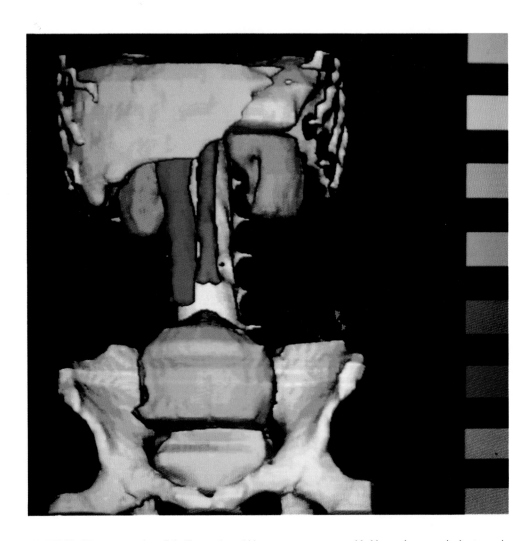

PLATE 20. 3D representation of the liver, spleen, kidneys, aorta, vena cava, bladder, and supravesicular tumor in the abdomen. (Provided by Elscint, Inc., Boston, Massachusetts.)

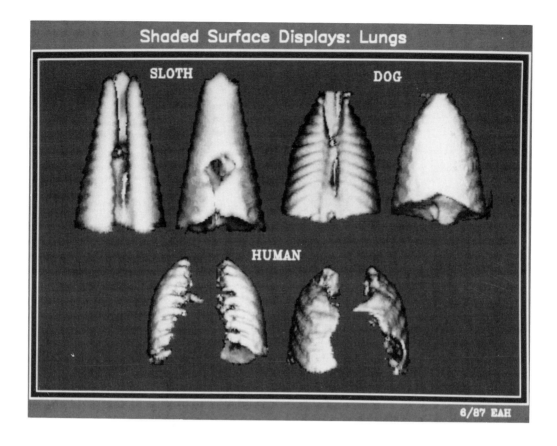

FIGURE 6. Shaded surface displays showing the variations in lung geometry. The left panel of each image pair shows the dorsal lung surface and the right panel of each image pair shows the ventral lung surface of sloth, dog, and human all scanned supine in the DSR. Note the indentations on the lung surface caused by the ribs.

of ventilation or perfusion which may serve to uncouple what is normally a well-matched ventilation/perfusion relationship.

C. THE BLOOD VESSELS

Methods have been developed whereby the course of blood vessels can be followed and oblique sections can be extracted such that the sections are always perpendicular to the local vessel long axis. This allows for the accurate measurement of vessel cross-sectional area and, for instance, rate of tapering of the vessel. As depicted in Figure 8, one approach to the location of the vessels of interest is to display four brightest pixel projection angles and to display the 3D location of a mouse-controlled cursor in all four projections simultaneously. In this way, the investigator can provide the computer with the information regarding which vessel branch is of interest, and the computer can then calculate the appropriate oblique sections of the vessel along the length delineated.

D. THE UPPER AIRWAYS

As a slight diversion from the heart and lungs, we have recently begun to apply our image analysis techniques to another problem in cardiopulmonary medicine known as obstructive sleep apnea. Five to eight percent of the population are afflicted with this syndrome to one degree or another. In this syndrome, as the individual falls asleep, the upper airway collapses, respiration stops, and the individual awakens. Thus, a person with this syndrome gets little rest and is tired all of the time. In the extreme cases, the sleep apneic is seen to fall asleep in the middle of speaking a sentence. There are a number of treatments ranging from diet (many of these individuals are obese), application of continuous positive airway pressure (CPAP) via a face mask worn to bed, tracheostomies, or (in the extreme) surgical removal of tissue from the upper airway region. Of particular interest is to define the pathology more thoroughly to aid in better tailoring the treatment to the individual since the etiology of the disease is most probably multifold and not the same for each person. Causes can range from purely mechani-

V̇/Q̇ EVALUATION

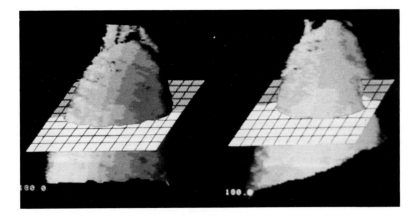

FIGURE 7. Lower panel demonstrates the color coding of the lung surface representing a polynomial fit to regional lung density. The lung on the left represents the supine body posture with a strong ventral-dorsal gradient in regional lung expansion and the lung on the right represents the prone body posture with a more dispersed variation in regional lung density. The highlighted plane demonstrates an artist's concept of an interactive interrogation of the polynomial fit to the data used to generate displays such as demonstrated in the upper panel whereby statistical evaluations can be performed to understand body posture-related variations in regional ventilation and blood flow. (Modified from Hoffman, E.A., R.S. Acharya, and J.A. Wollins. Computer-aided analysis of regional lung air content using three-dimensional computed tomographic images and multinormal models, *Int. J. Math. Model.* 7:1099–1116 (1986). With permission.)

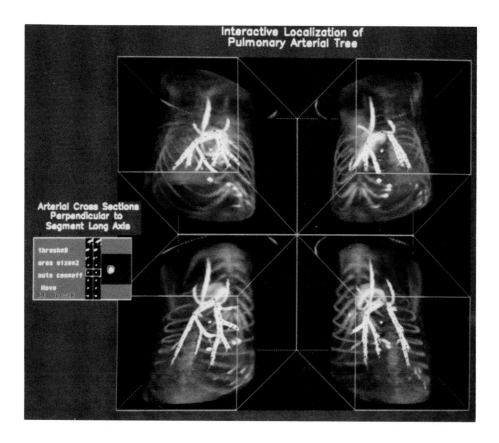

FIGURE 8. Demonstration of a module in Analyze which allows for the interactive tracking of blood vessels (in this case the pulmonary arterial tree of a dog) and the subsequent automated sectioning of the vessels perpendicular to the vessel long axis (panel inserted at left of figure) for cross-sectional area and blood flow measurements.

cal abnormalities to central nervous system abnormalities. As shown in Figure 9 we have devised a 5-minute magnetic resonance scan sequence which allows us to extract a 3D image of the upper airways.[51] Display techniques have been developed to demonstrate the orientation of the upper airway relative to surrounding soft tissue structures. Figure 10 shows a surface display of the airway embedded within a projection image of the soft tissue structures. Through the interactive sectioning (see Plate 14*) of the airway image data set, we are able to locate regions of interest through a series of scans acquired at various levels of positive airway pressure and thus measure the regional compliance of the airway. This approach along with an assessment of the fat and water content of the upper airway soft tissue structures via nuclear magnetic resonance spectroscopy is being applied to a large number of volunteers both from a normal population and from a continuum of non-apneic snorers and sleep apneics.

III. VOLUMETRIC IMAGING BEYOND THE DYNAMIC SPATIAL RECONSTRUCTOR

Throughout this discussion, we have alluded to the fact that volumetric imaging has now gone beyond just the DSR project. A commercially available scanner produced by Imatron

* Plate 14 appears following page 294.

(the Cardiovascular Computed Tomography scanner: known as the CVCT scanner, the C-100, the Ultrafast Cine CT Scanner, or, now that it is marketed by Picker, the "Fastrac")[52] has been placed in a number of clinical sites. Imatron's scanner offers the capability of gathering two 8-mm contiguous slices in 50 msec and 4 contiguous slice pairs (with 4-mm gaps between pairs) in 250 msec or 40 3-mm thick slices in 70 sec with each higher resolution slice gathered in 100 msec. Thus, in the 50-msec scanning mode, with gating techniques, whereby slice pairs at appropriate spatial locations are gathered while maintaining contrast enhancement of the cardiac chamber, it is possible to generate image sets which potentially allow for a 3D analysis of the beating heart. Such an analysis is depicted in Figures 11 and 12 whereby the total volume of the heart of a normal human volunteer was sampled at 16 spatially distributed levels at 10 points through the cardiac cycle (mixing and matching slices sampled over a number of cardiac cycles).[36] Via the use of gray scale selection, region growing, and shape-based interpolation[53] the volume and geometry of the heart was assessed and is displayed in Figure 12. Even with the shape-based interpolation, one can see the 8-mm thick slicing reflected in

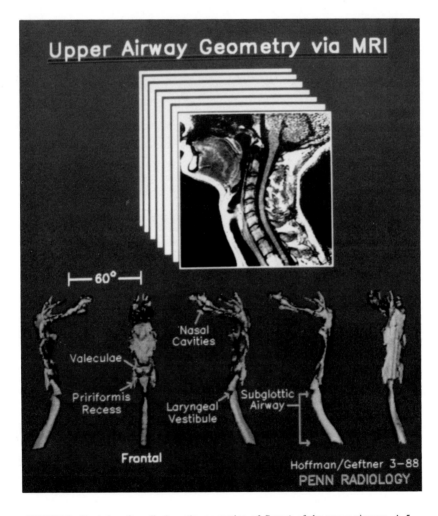

FIGURE 9. Shaded surface displays (lower portion of figure) of the upper airways. A 5-minute protocol has been designed for magnetic resonance imaging whereby 3-mm interleaved sagittal sections are generated. Such images are being used to evaluate the normal airway compared with the airways of patients with obstructive sleep apnea syndrome.

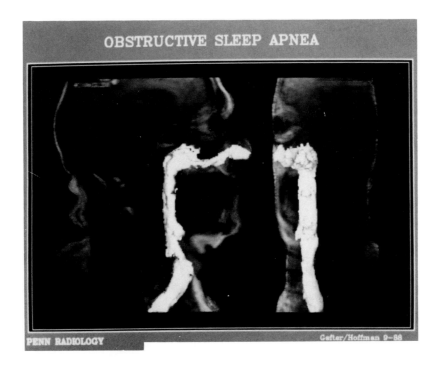

FIGURE 10.   A technique for embedding a shaded surface display of the upper airways into a projection (ray casted) image of the surrounding soft tissue has been developed. These displays allow for a better understanding of the particular soft tissue structures which are associated with geometric defects.

the 3D geometry. Of interest was the finding, from this series of scans, that the center of mass of the heart varies by less than 3 mm in any direction throughout the cardiac cycle,[36] again confirming the notion that the normal functioning of the heart serves to minimize wasted energy expenditure.

Magnetic resonance imaging (MRI) offers the potential of imaging the heart without the need for injection of contrast agents, since the properties of flowing blood within the scanned field can be used to provide for intrinsic contrast between the chambers and the myocardium. Negative aspects of MRI techniques for use with cardiac imaging have related to the long scanning times required to gather a volumetric data set (imaging times can amount to more than 1.5 hours to gather a full cardiac volume spanning the aortic arch to the epicardial apex) and the inconsistent image quality from subject to subject and from time point to time point within the cardiac cycle. Other roadblocks relate to the difficulty of the segmentation task which is complicated by multiple parameters contributing to the final reconstructed voxel value and by image quality inconsistency. Nevertheless, as shown in Figures 13 and 14 (from the work of Weinberg et al.[21]), valuable 3D information is obtainable. Figure 13 demonstrates the segmentation of the 3D information from a scan of a young child in which both the aorta and pulmonary artery arose from the right ventricle and the left ventricle is seen as a thin blind pouch. Figure 14 shows a sequence of shaded surface displays of the heart of a young child scanned after surgical repair whereby a conduit was placed to connect the right ventricle to the pulmonary arteries. The heart muscle is in lighter shades and the blood in the cardiac chambers is depicted in darker shades.

Rezdian and Pykett[54] have recently reported on the success of a technique whereby, with hardware and software modifications to conventional scanners, it is possible to achieve scanning times for a slice which approach the order of 26 msec. Image quality is somewhat degraded. However, if reliability of image quality is related to flow artifacts, gating accuracy, and respiratory motion, one might expect subject to subject reliability to improve. Zerhouni et al.[55] and Axel and Dougherty[56] have demonstrated an MRI technique of labeling the myocardium of the heart through the addition of extra radiofrequency pulses to the imaging sequence. The result is a set of black stripes which can be followed through the cardiac cycle such that stripe motion indicates regional wall function. An example of these stripes is shown in Figure 15. The motion of these stripes has demonstrated that an important contribution to the piston-like motion of the valve plane between the atria and ventricles is the wringing motion of the ventricular myocardium. The apex of the heart appears to move in a counterclockwise direction about the ventricular long axis (viewed apex to base) while the myocardium located more towards the valve plane reverts to a clockwise motion in mid-systole.[36] This differential pattern of motion correlates well with the different pattern of muscle fiber layering and angulation at the apex and base of the heart.[57]

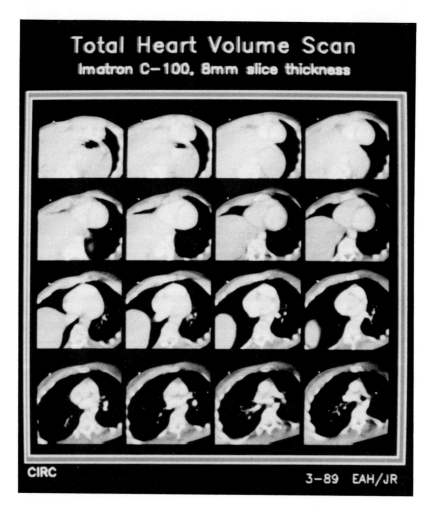

FIGURE 11.   8-mm thick slices of the human heart obtained via Imatron's cine X-ray CT scanner. Sequential slice pairs are contiguous but 4-mm gaps appear between one slice pair and the next. Images represent 50-msec scan apertures and the volumetric data set were obtained via scanning at end diastole over a period of several heart cycles.

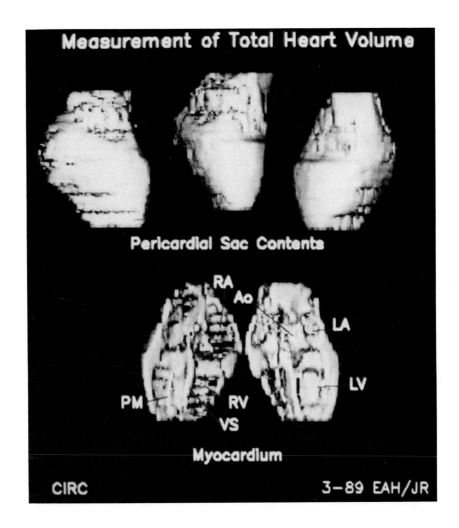

FIGURE 12. Shaded surface display of the heart from Figure 11. A shape-based interpolation algorithm[53] was used to create cubic voxels. PM = papillary muscle; RA = right atrium; AO = ascending aorta; RV = right ventricle; VS = ventricular septrum; LA = left atrium; LV = left ventricle.

Ultrasound has seen the advent of color doppler techniques[58,59] allowing for the assessment of regional flow in the cardiac chambers and directional flow across valves and across septa in the case of defects. Tissue characterization techniques have shown promise for the evaluation of myocardial parameters by use of specular nature of the image in the region of the myocardium,[60] and there has been a recent strong interest in the use of ultrasound to gather volumetric image data sets of the heart,[61] allowing for the extension of the above-mentioned echocardiographic techniques. A sequence of images (derived from the work of McCann et al.[61]) from a normal human volunteer is shown in Figure 16. Noninvasive imaging of metabolic activity of the heart has fallen within the realm of nuclear medicine, and positron emission tomography (PET) has become a prospect for a source of volumetric image data sets.[62,63] Finally, the electrical activity of the heart must also be considered as a form of cardiac imaging with the need for the development of volumetric image display and analysis approaches.

IV. MULTIPARAMETRIC IMAGE DISPLAY
VIA COLOR CODING

Most of the work to date relating to display of cardiopulmonary function and anatomy has concentrated on 3D display techniques (three spatial or two spatial and one temporal) or, to a limited degree, four-dimensional display where three dimensions are spatial and one is temporal. For the most part, displays have been limited to an individual modality. The growing need to link information from multiple imaging modalities and the advancement of a single imaging modality to "sense" multiple parameters has brought about the real need for color display as well as the need for other display modalities.

In general, the biomedical imaging community has viewed the use of color as largely "showmanship". Color mapping of parametric information onto anatomic data sets from the same modality has had some success in specific cases. Utilizing computed ultrasound tomography, Greenleaf et al.[64] mapped green and blue color scales onto cross-sectional images of the breast to represent parameters of attenuation and speed in an attempt to provide added information to aid in the characterization of a detected lesion. This study may have been the first attempt at displaying multiple ultrasound parameters via the use of color. Color flow

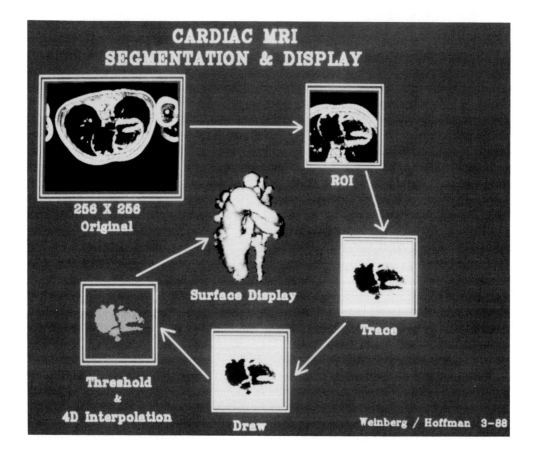

FIGURE 13.   Segmentation sequence to extract the chambers from a multislice spin echo sequence gathered via MRI. Scanning was of a boy with a congenital heart defect consisting of both the aorta and pulmonary artery emanating from the right ventricle. The left ventricle is seen as a blind pouch on the right side of the shaded surface display. (Volumetric data set courtesy of Drs. Paul Weinberg and Alvin Chin.)

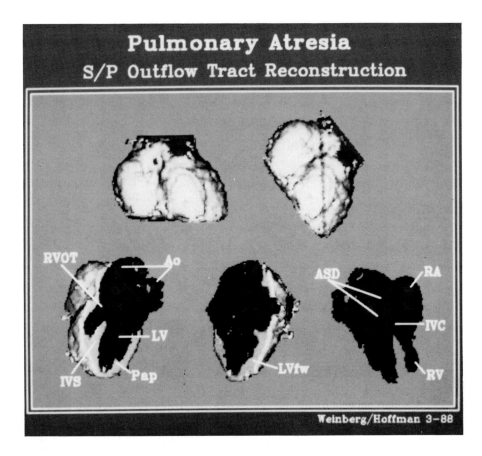

FIGURE 14. Two colors are used to differentiate the myocardium from the blood pool. Images are generated from a magnetic resonance scan of a pediatric patient scanned after a congenital heart repair whereby a conduit (right ventricular outflow track) was placed to connect the right heart chambers to the pulmonary arteries. RVOT = right ventricular outflow tract; IVS = intraventricular septum; PAP = papillary muscle; LV = left ventricle; AO = aorta; LVFW = left ventricular free wall; ASD = atrial septal defect; RA = right atrium; IVC = inferior vena cava; RV = right ventricle. (Scan data courtesy of Drs. Paul Weinberg and Alvin Chin.)

doppler imaging has proven to be a prime example of the utility of color to impart important parametric information.[15,58,59] In this case, it has been possible to combine blood flow and anatomy using a doppler ultrasound dual image and color to create an image of flow. The color assignment depends on the direction of the signal (with respect to the transducer) and its velocity. As discussed by Levkowitz,[65] the generalized lightness hue and saturation (GLHS) based approach is particularly suitable. For instance, echo signals from stationary objects are assigned to gray scale lightnesses relative to their amplitudes, creating an ultrasound image of anatomy. Signals from moving objects can be assigned to colors as follows. The directionality relative to the transducer is assigned to hue. For example, red can be used for movement away from the transducer and cyan for movement towards the transducer. Speed is assigned to saturation. Thus signals from stationary objects (speed 0) will show the anatomy in gray scale (saturation 0), while signals from moving objects will show flow in colors where the hue represents directionality and the saturation represents speed. While direction information in doppler ultrasound is commonly limited to directions to and away from the transducer, to a limited extent, information regarding turbulence can be recovered.

MRI holds strong potential of measuring flow with 360° angle information, and Reichek et al. from our laboratory have demonstrated the ability to extract flow voids within the MR images caused by regurgitant jets across the mitral valve.[66] With such information, the GLHS approach, which Leukowitz has developed, can be used to its full power.

In the field of digital subtraction angiography, numerous investigators[67] have demonstrated the utility of color coding such as to display a temporal progression of the passage of a bolus of contrast agent as it traverses an organ system of interest. In this case, the two parameters being displayed are (1) anatomic information from the projection image as viewed prior to the injection of the contrast bolus and (2) arrival time of the contrast bolus assessed from sequential digitized fluoroscopic projection images.

Recently the use of color has been most prevalent in the field of radioisotope imaging, probably because of the fact that the anatomic detail is so limited, and the boundaries of regions of interest are so fuzzy that color serves as a perceptual aid. For myocardial perfusion imaging using multislice data sets from single-photon emission-computed tomography (SPECT), the bull's-eye plot[68] has been used to represent 3D myocardial perfusion data in a 2D fashion. Here, concentric rings from a short axis view of the heart are displayed with the apical region at the center and the most basal portion of the left ventricle mapped into the outer ring. Although this technique has been shown to be a useful way of depicting perfusion, it provides only limited information about the size and shape of the defects and can, in fact, lead to erroneous perceptions of the dimensions of the defect and its anatomical location. Functional mapping using SPECT has taken the form of phase analysis of gated equilibrium studies. Color-coded data sets attempt to show temporal sequencing of wall motion.[69,70]

A. APPROACHES TO MULTIPARAMETRIC DISPLAY

As early as 1970, Greenleaf and colleagues[13,14] developed methods of estimating the 3D

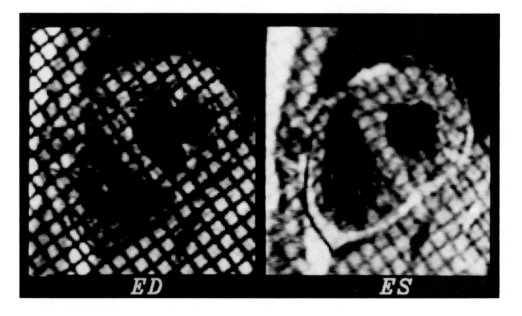

FIGURE 15.   Image data demonstrate a noninvasive, myocardial tagging technique developed by Dr. Leon Axel and Larry Dougherty at the Department of Radiology, University of Pennsylvania. Extra radiofrequency pulses are intermixed into a conventional spin echo, magnetic resonance, cardiac-gated scan sequence. The result is a suppression of signal in selected planes (dark lines). The proton characteristics stay with the myocardium, and thus the line distortion between end-diastole (ED) and end-sytole (ES) reveals regional wall motion. Images are shown in a short axis orientation with the left ventricular chamber at the right of each slice and the right ventricular chamber on the left. This tagging technique has been called spatial modulation of magnetization or SPAMM.

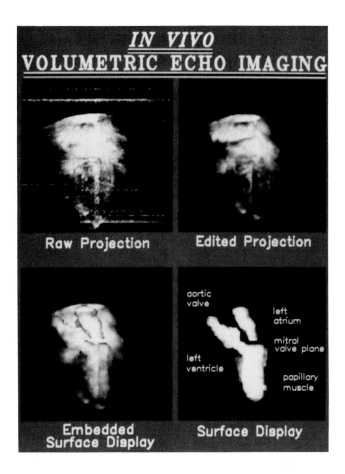

FIGURE 16. Volumetric cardiac data set gathered from a normal human volunteer using ultrasound. The lines in the raw data image (projection upper left) are the result of the writing (patient identification, etc.) on the original 2D sector scan. These are manually erased to produce the edited projection (upper right). The computer-generated surface display can then be combined with this projection (lower right) or displayed alone (lower left). This data represents an example of a multiparametric display which did not require the use of color. (Image modified from McCann, H.A., K. Chandrasekaran, E.A. Hoffman, L.J. Sinak, T.M. Kinter, and J.F. Greenleaf. A method for three-dimensional ultrasonic imaging of the heart in vivo, *Dynamic Cardiovasc. Imag.* 1:97–109 (1987). With permission.)

geometry of an organ of interest (heart or lungs) from biplane projection images and utilized radiopaque markers to align *in vivo* image data with subsequent measurement of regional blood flow data estimated from radioactive microspheres counted after excision of the organ. Regional lung expansion was estimated from the intermarker distances measured via the biplane X-ray system. Change in regional lung expansion estimated in this fashion was also mapped onto the estimated lung surface geometry. Since all of this was accomplished utilizing a monochrome oscillographic display device, only single parameters were displayed at a time. As discussed earlier, subsequent to these early surface displays of Greenleaf, the "art" of surface display was advanced and applied to CT data by Herman and Liu[12] and Udupa.[15] Heffernan and Robb[48] developed and implemented algorithms which allowed for the more rapid display of the shaded surfaces (without the need for special purpose hardware) and then demonstrated how parametric data can be mapped onto the surface using color. However, counting our own work (to be discussed in the next section), little has been done with this type of display technique largely because of the existence of other larger road blocks to data gathering and analysis. Preliminary efforts by others to display parametric data in volumetric formats have appeared sparsely. Perfusion data from SPECT have been demonstrated using 3D shaded surface display,[72] but these are hampered by the noise and limited spatial resolution of SPECT images. Furthermore, this technique relies on functional images to provide anatomic information. Similarly, the color display of single parametric data sets onto shaded surface displays of organ geometry derived from the same volumetric data set has been applied, largely for demonstration purposes, to information from PET.[72,73] It is our feeling that these preliminary attempts at the linking of radionuclide data with anatomic information have met with minimum utility because of the inherent poor anatomic quality of the data sets in the first place and because of the fact that the application has been largely aimed at display rather than linking the display to a method for user interrogation and quantitation of the data set. Preliminary studies have been carried out in the brain by Levine and colleagues[73] to map PET-derived metabolic information onto the anatomic details from MRI studies and the volumetric images have proven to be quite useful.

An area which has not traditionally been considered in the realm of medical imaging, but which is of great interest to us, is the mapping of the electrical activity of the cardiac muscle. In cardiac electrophysiologic studies performed longer than 10 to 20 years ago, mapping was generally carried out using a single electrode which was moved from site to site.[74] Electronic hardware for the simultaneous acquisition of a large number of electrophysiologic channels is now available[75] including commercial systems (Bard Electrophysiology, Billerica, MA). Special socks containing multiple epicardial electrodes[76] and plunge electrodes[77] for recording multiple transmural electrograms have been developed. Despite these advances in the acquisition of electrophysiological data, and though some automation has been incorporated in the display systems, the form of the display, namely idealized, flattened 2D representation of 3D geometry,[71,72,75,78,79] has changed little since studies done early in this century.[80] Ideker and colleagues have described what they believe to be the ideal "simultaneous multichannel cardiac mapping system" and have included within their paper a projection image (2D) derived from epicardial surface data recorded from a dog in normal sinus rhythm. Projections of half of the heart were calculated presumably to eliminate superposition problems. Activation timing was color coded from blue to red with the blue end of the spectrum representing the earlier activation times. An artistic rendering of the position of the coronary arteries was superimposed on the projections. Such a display offers a glimpse at the important information which could be gained by combining the electrode information with cine CT or other methods of acquiring high-resolution, dynamic, 3D anatomic detail allowing for correlation between the anatomy, electrophysiology, and mechanical function of the heart. Information to be gained from such multiparametric displays is expected to be of particular importance to ongoing studies within our group relating excitation/contraction coupling events: events which lead to the normal interchamber interrelationships serving to minimize the wasted mechanical load on the heart in pulling extracardiac structures along with each heart beat. As another example, lessons learned from the normal heart and image display and analysis routines will serve as important tools in efforts within the Department of Surgery to develop an efficient skeletal muscle ventricle.[81]

V. A NEW APPROACH TO MULTIPLE VOLUME DATA SETS AND THE TOOL BOX

To cope with the needs of multiple volume image display and to utilize the powerful new utilities of window-based analysis systems we (The Cardiothoracic Imaging Research Center or CIRC) have begun the development of an end-user-directed software analysis package we call "Volumetric Image Display and Analysis" or VIDA™. This package provides a UNIX/X-windows-based interactive multitasking environment for the display, manipulation, and analysis of multiple multidimensional objects. It is being implemented on a variety of Sun Microsystems workstations. CIRC is a consortium of clinicians and research scientists from numerous disciplines including cardiology, radiology, computer science, bioengineering, and physiology, both from within and outside the University of Pennsylvania. Features of the package which are of particular significance to the display and analysis of multidimensional, multiparametric data sets are the configuration of shared memory to allow for the simultaneous loading of multiple volumetric data sets and parameter passing between data sets, and the organization of data internally to enable specification of volumetric and temporal parameters including volume, slice, time, phase, and imaging source. The window-based design allows for processing of multiple modules and copies of the same module simultaneously. This enables multiple display and manipulation of the same object or of several objects and allows, for example, the selection of an oblique section which can appear within a region of interest module for quantitative evaluation. Window interactions conform to the "open look" graphical user interface specifications.

VIDA™ has been developed as a two-tier process: (1) the core and (2) application-specific modules. The core of VIDA™ is the infrastructure of the package; it provides the foundations that will enable the functional modules to communicate with each other (via a unified shared memory approach) and with the user (via a consistent "look and feel"). This is achieved by providing a complete set of internal functions to access shared memory and file structures and to establish screen layout and interaction. These functions are "packaged" in a small number of libraries that are included with each module that is written. The advantage of the approach is twofold: once the core is developed, implementors within the CIRC do not have to waste time duplicating efforts in order to achieve communication, interaction, and "look and feel" consistency. More importantly, it facilitates the development of additional functional modules by interested parties outside CIRC; anybody interested in developing a module for VIDA™ needs only to concentrate on the details of that module and its interaction with the core of VIDA™. Once this is carried out, the module is ready to be incorporated into the package. Several basic utility modules are also included within the core. These include the ability to read and write data in various formats for various modalities (such as CT, MRI, PET, ultrasound) and various machines within each modality, the display of 2D cross sections in the standard orthogonal sections (transverse, coronal, and sagittal), and a module allowing for the calling of optimized color lookup tables[65] along with windowing.

Functional modules are being developed to carry out various display and analysis tasks, including:

- 3D rendering of objects using shaded surface and volumetric rendering techniques
- Manipulation of data such as filters (2D and 3D), interactive editing, object recognition, and segmentation
- Quantitative analysis such as volume computations, various statistics on regions of interest, assessment of regional image parameters representing tissue metabolism, blood flow, or motion, to name a few; derived numerical quantitative information is stored to a database for subsequent further data reduction and analysis
- Multiparametric representations using color and other coding mechanisms
- Customization and automation of processes according to the needs of particular users

In Plate 15,* a montage of VIDA™ modules is shown intermixed on the workstation

---

* Plate 15 appears following page 294.

screen. Modules shown include volume rendering, oblique sectioning, and region of interest analysis. Data sets of a tagged human heart imaged via MRI, a sloth imaged via the DSR, and a bright blood data set of human pelvis all reside simultaneously with shared memory.

VI. CONCLUSIONS

Noninvasive, multidimensional imaging offers a new frontier to the biologic sciences. While exciting developments in cellular and molecular biology have taken whole departments of physiology into the investigations of cellular and molecular principles, there will remain the need to place these studies into the context of the whole organ and the whole organ system. This capability is available through our ability to image macroscopic anatomy and function (via ultrasound, dynamic X-ray CT, and MRI) along with the 3D mapping of cellular and subcellular activity (via PET, SPECT, magnetic resonance spectroscopy, cardiac electrical mapping, etc.). The key to utilizing the full potential of these imaging modalities lies in the continued development of multidimensional/multiparametric display and analysis techniques linking the unique features of the individual scanning modalities. While much of the commercially based image analysis workstation efforts have been oriented towards dramatic display techniques, this is a short-sited approach. For imaging to serve as a road back from the reductionist approach of the cellular and molecular biologists, efforts will have to focus on development of new innovative approaches to image data quantitation.

REFERENCES

1. Ritman, E.L., R.A. Robb, and L.D. Harris. *Imaging Physiological Functions: Experience with the DSR* (Philadelphia, PA: Praeger, 1985).
2. Ritman, E.L., R.E. Sturn, and E.H. Wood. Biplane roentgen videometric system for dynamic (60/second) studies of the shape and size of circulatory structures, particularly the left ventricle, *Am. J. Cardiol.* 32:180–187 (1973).
3. Rankin, J.S., P.A. McHale, C.E. Arentzen, D.L. Ling, J.C. Greenfield, Jr., and R.W. Anderson. The three-dimensional dynamics geometry of the left ventricle in the conscious dog, *Circ. Res.* 39:304–313 (1976).
4. Carlsson, E., and E.N.C. Milore. Permanent implantation of endocardial tantalum screws: A new technique of functional studies of the heart in the experimental animal, *J. Can. Assoc. Radiol.* 19:304–309 (1967).
5. Loken, M.K., Ed. *Pulm. Nucl. Med.* (Norwalk, CT: Appleton and Lange, 1987).
6. Gerson, M.C., Ed. *Cardiac Nucl. Med.* (New York, NY: McGraw-Hill, 1987).
7. Arvidson, H. Angiographic determination of left ventricular volume, *Acta Radiol. (Stockholm)* 56:321–339 (1961).
8. Davilla, J.C., and M.E. San Marco. An analysis of the fit of mathematical models applicable to the measurement of left ventricular volume, *Am. J. Cardiol.* 18:31–42 (1966).
9. Dodge, H.T., H. Sandler, D.W. Pallew, and J.D. Hard. The use of biplane angiocardiography for the measurement of left ventricular volume in man, *Am. Heart J.* 60:762–776 (1960).
10. Miller, D.D., R.J. Burns, J.B. Gill, and T.D. Ruddy, Ed. *Clin. Cardiac Imag.* (New York, NY: McGraw-Hill, 1988).
11. Sagawa, K., L. Maughan, H. Saga, and K. Sunagawa. *Cardiac Contraction and the Pressure-Volume Relationship* (New York, NY: Oxford University Press, 1988).
12. Herman, G.T., and H.K. Liu. Display of three-dimensional information in computed tomography, *J. Comput. Assist. Tomogr.* 1:155–160 (1977).
13. Greenleaf, J.F., J.S. Tu, and E.H. Wood. Computer generated three-dimensional oscilloscopic images and associated techniques for display and study of the spatial distribution of pulmonary blood flow, *IEEE Trans. Nucl. Sci.* NS-17:353–359 (1970).
14. Greenleaf, J.D., E.L. Ritman, C.M. Coulam, R.E. Sturm, and E.H. Wood. Computer graphic techniques for study of temporal and spatial relationships of multidimensional data derived from biplane roentgen videograms with particular reference to cardioangiography, *Comput. Biomed. Res.* 5:368–387 (1972).
15. Udupa, J.K. Interactive segmentation and boundary surface formation for 3-D digital images, *Comput. Graph. Imag. Proc.* 18:213–235 (1982).

16. Hoffman, E.A., T. Behrenbeck, P.A. Chevalier, and E.H. Wood. Estimation of regional pleural surface expansile forces in intact dogs, *J. Appl. Physiol.* 55:935–948 (1983).

17. Hoffman, E.A., L.J. Sinak, R.A. Robb, and E.L. Ritman. Noninvasive quantitative imaging of shape and volume of lungs, *J. Appl. Physiol.* 54:1414–1421 (1983).

18. Sinak, L.J., E.A. Hoffman, P.R. Julsrud, D.D. Mair, J.B. Seward, D.J. Hagler, L.D. Harris, R.A. Robb, and E.L. Ritman. The Dynamic Spatial Reconstructor: Investigating congenital heart disease in four dimensions, *Cardiovasc. Intervent. Radiol.* 7:124–137 (1984).

19. Sinak, L.J., E.A. Hoffman, R.S. Schwartz, H.D. Smith, D.R. Holmes, A.A. Bove, R.A. Robb, L.D. Harris, and E.L. Ritman. Three-dimensional cardiac anatomy and function in heart disease in adults: Initial results with the Dynamic Spatial Reconstructor, *Mayo Clin. Proc.* 60:383–392 (1985).

20. Laschinger, J.C., M.W. Vannier, S.F. Gutierrez, et al. Preoperative three-dimensional reconstruction of the heart and great vessels in patients with congenital heart disease, *J. Thorac. Cardiovasc. Surg.* 96:464–473 (1988).

21. Weinberg, P.M., E.A. Hoffman, A.J. Chin, L. Axel, and N. Reichek. Three-Dimensional Display of In Vivo Magnetic Resonance Imaging of Congenital Heart Defects, CIRC Technical Report CIRC-100. (Cardiothoracic Imaging Research Center, Dept. of Radiology, Hosp. of the Univ. of Penn., 3400 Spruce St., Philadelphia, PA 19104) (1989).

22. Harris, L.D., R.A. Robb, T.S. Yuen, and E.L. Ritman. Display and visualization of three-dimensional reconstructed anatomic morphology: Experience with the thorax, heart, and coronary vasculature of dogs, *J. Comput. Assist. Tomogr.* 3:439–446 (1979).

23. Harris, L.D., and J.J. Camp. Display and analysis of tomographic volumetric images utilizing a vari-focal mirror, *Proc SPIE* 507:38–45 (1984).

24. Harris, L.D. Identification of the optimal orientation of oblique sections through multiple parallel CT images, *J. Comput. Assist. Tomogr.* 5:881–887 (1981).

25. Robb, R.A., P.B. Heffernan, J.J. Camp, and D.P. Hanson. A workstation for multidimensional display and analysis of biomedical images, *Comput. Methods Programs Biomed.* 25:169–184 (1987).

26. Wu, X., L.A. Latson, T. Wong, D.J. Driscoll, G.J. Ensing, and E.L. Ritman. Regional pulmonary perfusion estimated by high-speed volume scanning CT, *Am. J. Physiol. Imag.* 3:73–80 (1988).

27. Wang, T., X. Wu, N. Chung, and E.L. Ritman. Myocardial blood flow estimated by synchronous, multislice, high-speed tomography, *IEEE Trans. Med. Imag.* 8:70–77 (1989).

28. Bentley, M.D., and R.A. Karwoski. Estimation of tissue volume from serial tomographic sections: A statistical random marking method, *Invest. Radiol.* 23:742–747 (1988).

29. Hoffman, E.A., L.J. Sinak, R.A. Robb, and E.L. Ritman. Noninvasive quantitative imaging of shape and volume of lungs, *J. Appl. Physiol.* 54:1414–1421 (1983).

30. Hoffman, E.A., and E.L. Ritman. Shape and dimension of cardiac chambers: Importance of CT section thickness and orientation, *Radiology* 155:739–744 (1985).

31. Iwasaki, T., L.J. Sinak, E.A. Hoffman, R.A. Robb, L.D. Harris, R.C. Bahn, and E.L. Ritman. Mass of left ventricular myocardium estimated with Dynamic Spatial Reconstructor, *Am. J. Physiol. Heart Circ. Physiol.* 15:H138–H142 (1984).

32. Sinak, L.J., E.A. Hoffman, and E.L. Ritman. Subtraction gated computed tomography with the Dynamic Spatial Reconstructor: Simultaneous evaluation of left and right heart from single right-sided bolus contrast medium injection, *J. Comput. Assist. Tomogr.* 8:1–9 (1984).

33. Hoffman, E.A., and E.L. Ritman. Invariant total heart volume in the intact throrax, *Am. J. Physiol. Heart Circ. Physiol.* 249:H883–H890 (1985).

34. Hoffman, E.A., and E.L. Ritman. Intra-cardiac cycle constancy of total heart volume, *Dynamic Cardiovasc. Imag.* 1:199–205 (1988).

35. Hoffman, E.A., R.L. Ehman, L.J. Sinak, J.P. Felmlee, K. Chandrasekaran, P.R. Julsrud, and E.L. Ritman. Law of constant heart volume in humans: A non-invasive assessment via x-ray CT, MRI and echo, *J. Am. Coll. Cardiol.* 9:38A (1987).

36. Hoffman, E.A., J. Rumberger, L. Dougherty, N. Reichek, and L. Axel. A geometric view of cardiac efficiency, *J. Am. Coll. Cardiol.* 13:86A (1989).

37. Liu, Y.H., E.A. Hoffman, D.J. Hagler, J.B. Seward, P.R. Julsrud, D.D. Mair, and E.L. Ritman. Accuracy of pulmonary vascular dimensions estimated with the dynamic spatial reconstructor, *Am. J. Physiol. Imag.* 1:201–207 (1986).

38. Block, M., Y.H. Liu, L.D. Harris, R.A. Robb, and E.L. Ritman. Quantitative analysis of a vascular tree model with the dynamic spatial reconstructor, *J. Comput. Assist. Tomogr.* 8:390–400 (1984).

39. Block, M., A.A. Bove, and E.L. Ritman. Coronary angiographic examination with the dynamic spatial reconstructor, *Circulation* 70:209–216 (1984).

40. Bove, A.A., M. Block, H.C. Smith, and E.L. Ritman. Evaluation of coronary anatomy using high speed volumetric CT scanning, *Am. J. Cardiol.* 55:582–584 (1985).

41. Liu, Y.H., E.A. Hoffman, D.J. Hagler, J.B. Seward, P.R. Julsrud, D.D. Mair, and E.L. Ritman. Accuracy of pulmonary vascular dimensions estimated with the dynamic spatial reconstructor, *Am. J. Physiol. Imag.* 1:201–207 (1986).

42. Liu, Y.H., E.A. Hoffman, and E.L. Ritman. Measurement of three-dimensional anatomy and function of pulmonary arteries with high-speed x-ray computed tomography, *Invest. Radiol.* 22:28–36 (1987).

43. Hoffman, E.A. Effect of body orientation on regional lung expansion: A computed tomographic approach, *J. Appl. Physiol.* 59:468–480 (1985).

44. Hoffman, E.A., and E.L. Ritman. Effect of body orientation on regional lung expansion in dog and sloth, *J. Appl. Physiol.* 59:481–491 (1985).

45. Krayer, S., K. Rehder, K.C. Beck, P.D. Cameron, E.P. Didier, and E.A. Hoffman. Quantification of thoracic volumes by three-dimensional imaging, *J. Appl. Physiol.* 62:591–598 (1987).

46. Wilson, T.A., K. Rehder, S. Krayer, E.A. Hoffman, C.G. Whitney, and J.R. Rodarte. Geometry and respiratory motion of the ribs, *J. Appl. Physiol.* 62:1872–1877 (1987).

47. Hoffman, E.A., E.L. Ritman, D.E. Mason, and L.E. Olson. Lung/chest wall shapes in miniature horses by x-ray CT, *The Physiologist* 31(4):A222 (1988).

48. Heffernan, P.B., and R.A. Robb. Display and analysis of 4-D medical images, in *Proc. CAR '85* (Berlin: Springer-Verlag, 1985), pp. 583–592.

49. Hoffman, E.A., and P.B. Heffernan. A computer graphics-aided 3-D analysis of heart/lung interaction reconstructed via DSR scanning, *Nat. Comput. Graph. Assoc. Conf. Proc.* III:81–92 (1985).

50. Hoffman, E.A., R.S. Acharya, and J.A. Wollins. Computer-aided analysis of regional lung air content using three-dimensional computed tomographic images and multinomial models, *Int. J. Math. Model.* 7:1099–1116 (1986).

51. Hoffman, E.A. and W.P. Gefter. Multimodality imaging of the upper airway: MRI, MR spectroscopy, and ultrafast X-ray CT, in *Sleep and Respiration,* F.G. Issa, P.M. Suratt, and J.E. Remmers (Eds.) (New York, NY: Wiley-Liss, 1990), pp. 291–301.

52. Boyd, D.P., and M.J. Lipton. Cardiac computed tomography, *Proc. IEEE* 71:298–307 (1983).

53. Raya, S.P., and J.K. Udupa. Shape-Based Interpolation of Multidimensional Objects, *Med. Image Processing Group Tech. Rep.* MIPG129 (Medical Image Processing Group, Department of Radiology, University of Pennsylvania, 1988), pp. 1–32.

54. Rezdian, R.R., and I.L. Pykett. Instant images of the human heart using a new, wholebody MR imaging system, *Am. J. Radiol.* 149:245–250 (1987).

55. Zerhouni, E., D. Parrish, W.J. Rogers, A. Yang, and E.P. Shapiro. Tagging with MR imaging — a method for noninvasive assessment of myocardial motion, *Radiology* 169:59–63 (1988).

56. Axel, L., and L. Dougherty. MR imaging of motion with spatial modulation of magnetization (SPAMM), *Radiology* 171:841–845 (1989).

57. Greenbaum, R.A., S.Y. Ho, D.G. Givson, A.E. Becker, and R.H. Anderson. Left ventricular fiber architecture in man, *B.R. Heart J.* 45:248–263 (1989).

58. Sahn, D. Real-time two-dimensional echocardiographic flow mapping, *Circulation* 71:849–853 (1985).

59. Switzer, D.F., and N.C. Nanda. Doppler color flow mapping, *Ultrasound Med. Biol.* 3:403–416 (1985).

60. Miller, J.G., J.E. Perez, and B.E. Sobel. Ultrasonic characterization of myocardium, *Prog. Cardiovasc. Dis.* 28:85–110 (1985).

61. McCann, H.A., K. Chandrasekaran, E.A. Hoffman, L.J. Sinak, T.M. Kinter, and J.F. Greenleaf. A method for three-dimensional ultrasonic imaging of the heart in vivo, *Dynamic Cardiovasc. Imag.* 1:97–109 (1987).

62. Miller, T.R., J.B. Starren, and R.A. Grothe. Three-dimensional display of positron emission tomography of the heart, *J. Nucl. Med.* 29:530–537 (1988).

63. Gould, K.L., N. Mullani, W.H. Wong, and R.A. Goldstein. Positron emission tomography, in *Cardiac Imaging and Image Processing,* S.M. Collins and D.J. Skorton (Eds.) (New York, NY: McGraw-Hill, 1986), pp. 333–360.

64. Greenleaf, J.F., S.A. Johnson, R.C. Bahn, W.W. Samayra, and C.R. Hansen. Images of acoustic refraction index and of relationship of tissue types with excised female breasts, *Ultrasound Med.* 38:2091–2093 (1977).

65. Levkowitz, H. Color in Computer Graphic Representation of Two-Dimensional Parameter Distributions, Ph.D. thesis, University of Pennsylvania, Philadelphia, (1988).

66. Reichek, N., E.A. Hoffman, D.C. Gnanaprakasam, and L. Axel. Three-dimensional evaluation of aortic valvular jets throughout the cardiac cycle, *Circulation* 78:II-589 (1988).

67. Bursch, J.H., and P.H. Heintzen. Parametric imaging, in *Digital Radiography Radiological Clinics of North America.* G. Hillman and J. Newell (Eds.) (Philadelphia, PA: W.B. Saunders, 1985).

68. Garcia, E.V., K. VanTrain, J. Maddahni, et al. Quantification of rotational thalium-201 myocardial tomography, *J. Nucl. Med.* 26:17 (1985).

69. Goris, M.L. Functional or parametric images, *J. Nucl. Med.* 23:360 (1981).

70. Botvinick, E.H., M.W. Dae, J.W. O'Connell, M.M. Scheiman, R.S. Hattner, and D.B. Faulkner. First harmonic fourier (phase) analysis of blood pool scintigrams for the cardiac contraction and conduction, in *Cardiac Nucl. Med.,* M.D. Gerson (Ed.) (New York, NY: McGraw-Hill, 1987), pp. 109–147.

71. Drebin, R.A., L. Carpenter, and P. Hanrahan. Volume rendering, *Comput. Graph.* 22:65–74 (1988).

72. Gibson, C.J. Interactive display of three-dimensional radionuclide distributions, *Nucl. Med. Commun.* 7:475–487, 1986.

73. Levine, D.N., X. Hu, K.K. Tan, S. Galhotra, A. Herrmann, C.A. Pelizzari, G.T. Chen, R.N. Peck, C.T. Chen, and M.D. Cooper. Integrated 3-D display of MR, CT, and PET images of the brain, in *3D Imaging in Medicine,* J.K. Udupa and G.T. Herman (Eds.) (Boca Raton, FL: CRC Press, 1990), Chap. 10.

74. Josephson, M.E., L.N. Horowitz, A. Farshidi, and J.A. Kastor. Recurrent sustained ventricular tachycardia. 1. Mechanisms, *Circulation* 57:431–439 (1978).

75. Ideker, R.E., W.M. Smith, P. Wolf, N.D. Danieley, and F.R. Bartram. Simultaneous multichannel cardiac mapping systems, *Pace* 10:281–292 (1987).

76. Worley, S.J., R.E. Ideker, J. Mastrototaro, W.M. Smith, H. Vidaillet, Jr., P.S. Chen, and J.E. Lowe. A new sock electrode for recording epicardial activation from the human heart: one size fits all, *Pace* 10:21–31 (1987).

77. Kassel, J., and J.J. Gallagher. Construction of a multipolar needle electrode for activation study of the heart, *Am. J. Physiol. Heart Circ. Physiol.* 233:H312–H317 (1977).

78. Kramer, J.B., J.E. Saffitz, F.X. Witkowski, and P.B. Corr. Intramural reentry as a mechanism of ventricular tachycardia during evolving canine myocardial infraction, *Circ. Res.* 56:736–754 (1985).

79. Ideker, R.E., W.M. Smith, A.G. Wallace, J. Kassel, L.A. Harrison, G.J. Klein, R.E. Kinicki, and J.J. Gallagher. A computerized method for the rapid display of ventricular activation during the interpretative study of arrhythmias, *Circulation* 59:449–458 (1979).

80. Lewis, T., and M.A. Rothschild. The excitatory process in the dog's heart. Part II: The ventricles, *Phil. Trans. Roy. Soc. London* 206:181–223 (1915).

81. Acker, M.A., R. Hammond, J.D. Mannion, S. Salmons, and I.W. Stephenson. Skeletal muscle as the potential power source for a cardiovascular pump: assessment in vivo, *Science* 236:324–327 (1987).

Chapter 12

3D IMAGING IN RADIOTHERAPY TREATMENT PLANNING

Julian Rosenman

TABLE OF CONTENTS

I. INTRODUCTION TO RADIATION THERAPY

Cancer is one of the most dreaded diseases of our times; approximately 1/4 of all Americans will contract this illness sometime in their life. Although great efforts have been made to find more effective cancer therapies, the mainstay of cancer treatment continues to be surgery, radiation therapy, and chemotherapy. Although these treatment modalities are not new (radiation and surgery were used to treat cancer before the turn of the century, and chemotherapy has been in use for 40 years), they have been constantly improved, with radiation therapy, perhaps, undergoing the most dramatic changes. In this chapter we will show how the incorporation of 3D computer graphics into radiation therapy treatment planning may result in another improvement in the effectiveness of this cancer treatment modality.

A. HISTORICAL PERSPECTIVE
1. Early Difficulties

Within weeks of the discovery of the X-ray by Röntgen[1] in 1896, it was noted that some tumors regressed after exposure to high doses of radiation.[2] Some malignancies, such as Hodgkin's disease, were so radiosensitive that their *cure* with X-irradiation was heralded as early as 1903.[3] However, this enthusiasm proved to be premature, and the therapeutic use of radiation became primarily that of palliation of cancer symptoms or, at best, temporary delay of cancer growth.

The problems facing early *radiation oncologists* were formidable. They did not know how much radiation was necessary to destroy a tumor nor how this dose was to be administered; it had already been discovered that the same *physical* radiation dose (measured in *rads*) had different biologic effects depending on the time period over which it was given. Another problem was that of tumor localization. Most tumors could not be directly imaged, so early oncologists had to rely on physical examination to determine the extent and location of the *tumor volume*. To make matters even more difficult, it soon became evident that malignant cells often spread beyond the obvious tumor volume, so the radiation *target volume* had to include all likely local pathways of tumor spread in addition to the palpable tumor volume.

In addition to the biologic difficulties confronting the early radiation oncologists, there were also serious problems generated by the equipment that was available to deliver the radiation. Originally the same (or similar) X-ray machines were used for both diagnosis and therapy. The low energy X-rays (≤ 300 keV) that diagnostic machines produce give remarkable contrast between metal, bone, and soft tissue (necessary for high-quality diagnostic films) as their interaction with matter depends strongly on the atomic number (Z) of the material through which they pass. However, this property of low energy X-rays means that high Z tissues will absorb a disproportionate amount of radiation during therapy. So if a man were being treated for prostate cancer, the pelvic bones would absorb several times the *tumor* dose and possibly be severely damaged before the prostate cancer could be destroyed. In addition to bone damage, low energy X-rays are very damaging to the skin, as their maximum energy is deposited within a few microns of the skin surface. Patients undergoing radiation treatment thus often suffered severe radiation "burns" which necessitated prematurely halting treatment.

A final difficulty with the therapeutic use of low energy X-rays is that they do not penetrate tissues well enough to deliver large radiation doses deep within the body without delivering a large dose to the surface.

To alleviate these problems, radiation oncologists began to treat patients with multiple cross-fired radiation beams so as not to give too high a dose to any particular patch of skin or bone. In treating prostate cancer, for example, one might use right and left lateral portals as well as anterior and posterior portals. This way the skin over the buttocks and anterior pelvis would receive only 1/2 the dose, the other 1/2 being received by the right and left sides of the pelvis.

2. The Need for Radiation Treatment Planning

Determining the radiation dose from a single radiation beam is fairly straightforward, but the use of multiple radiation fields to spare skin (and other normal tissues) required the early radiation oncologists to engage in *radiation treatment planning*. Before actually treating the patient, the physician needed to know how much dose different points of the body would receive. At first, these doses had to be hand calculated, so that the number of alternative treatment plans that could be produced for clinical consideration were extremely limited. As the treatment planning process became computerized (beginning in the 1950s) it became possible to calculate doses not just at a few select points, but throughout an entire plane. Traditionally these points of equal radiation dose are connected together to make *radiation isodose plots*. It is current practice for the radiation dosimetrist to present the clinician with several of these plots so that the most satisfactory one can be chosen.

3. The Need for Higher Energy X-Ray Therapy

While radiation oncologists were struggling with techniques for external beam therapy (teletherapy), a more successful branch of radiation therapy was being developed, that of implanting the radiation sources directly within the patient (brachytherapy). Apparently first suggested by Alexander Graham Bell,[4] by the 1920s patients with cancer of the uterine cervix were being routinely cured by radium insertions. Dosimetry, if done at all, was limited to one or two fixed points, and the complication rate following such treatment was often quite high. Nevertheless, the possibility of curing cancer with some form of radiation became well established, and it became reasonable to assume that, with time, cures of other malignancies would become possible.

By the 1950s it became well recognized that many of the problems of teletherapy could be overcome if the energy of the radiation beam could be raised to the megavoltage range (≥ 1000 keV). At this energy the interaction between X-rays and matter is almost entirely Z independent, and bone and soft tissue would absorb approximately the same dose from a given beam. In addition, the maximum radiation dose would not fall exactly at the skin surface, but rather at a point somewhat under the skin (the exact depth would depend on the beam energy profile). Finally, because of the greater penetrating power of megavoltage X-rays, less dose has to be applied to the surface to produce a given dose at depth. Machines to produce megavoltage X-rays had been built (cyclotrons), but they were extremely clumsy for medical use and prohibitively expensive. However, the development of the 2-MeV Van de Graff generator and the commercial availability of ^{60}Co, a radioisotope that emits γ-rays (by then known to be physically identical to X-rays) of energy approximately 1 MeV, made the use of megavoltage irradiation possible. Megavoltage equipment did not do away with the need for treatment planning, however. On the contrary, the delivery of *tumoricidal* radiation doses, which were very high by traditional standards, required meticulous treatment planning to reduce the risk of severe treatment complications.

B. CURRENT STATUS OF RADIATION THERAPY

1. Biologic and Technical Improvements

In the past 20 years there have been many technical and biologic improvements in radiation treatment planning and delivery. The Cobalt and Van de Graff machines have been replaced with high-energy linear accelerators which can deliver well focused radiation at a high dose rate. Computed tomography (CT) now gives the clinician a high-quality digital image of the patient's tumor and normal anatomy, allowing for a much more precise determination of tumor target volume. In addition, our clinical knowledge base has substantially expanded, so that pathways of tumor spread are much better understood than previously. Because we now have a real chance to cure many cancer patients it is more important than ever to have accurate tumor targeting.

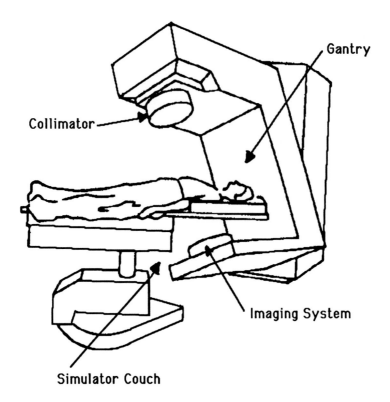

FIGURE 1. A patient being simulated for radiation treatment. The simulator can mimic the geometry of most treatment machines and produces a high-quality radiograph known as a simulator film. A combination of gantry rotation, couch swivel, and collimator angulation suffices to direct the radiation beam from almost any desired angle.

2. Treatment Planning

The treatment planning process in a modern radiation therapy treatment facility typically begins by designing the treatment beams on a special purpose X-ray device known as a radiation treatment *simulator* (see Figure 1). The simulator has the same geometry, machine motions, and controls as the radiation treatment machine, but the high energy source is replaced with a standard X-ray imaging tube so as to create a high-quality image. A combination fluoroscope/film-tray assembly is mounted opposite to the X-ray source. Films are taken in the proposed treatment position with a grid superimposed on them to mark the center and edges of the radiation field (Figure 2). The tumor volume, which usually cannot be visualized on the roentgenograph, must be drawn on by hand. Help locating its position can come from a CT scan and/or physical exam. The tumor *target* volume is then outlined by hand as well. In the case of Figure 2 the target volume includes the prostate and nearby surrounding tissue, such as seminal vesicles and base of the bladder. (The tumor in this patient was judged to be of low malignant potential; had it been more aggressive we might have chosen the entire pelvis as the target volume.) If the clinician judges that the simulation film is not adequate, it may be retaken, with a different field center or size. Once the field is deemed acceptable permanent markings are placed on the patient's skin that will allow the therapy technologists to reproducibly locate the field center and edges.

The radiation dose distribution from the selected beams is then calculated, and an isodose plot in (typically) the center plane is displayed. It is customary for the radiation dosimetrists to prepare several alternative plans with different beam weightings. Figure 3 shows the plan adopted for the patient whose simulation film is shown in Figure 2. Shown on the plan are the femoral heads (proximal end of the thigh bones that articulate with the pelvis) which are known to be particularly radiosensitive. In this plan they will receive about 60% of the total dose which is considered acceptable. The dose across the tumor target volume itself appears to vary by only about 5% (from approximately 93 to 98%), which is also considered acceptable. Notice the beam-modifying "wedges" that have been placed in the lateral radiation beams. Because the wedges are thicker in front than in back, the dose to the anterior part of the patient is reduced. In this patient that was desirable as it nicely compensates for his irregular shape. (If the wedges had been left out, the dosage would have been too "hot" in front and the plan rejected.) It should be emphasized that the entire treatment planning process is done in two, not three, dimensions.

During the treatment course portal or check films will be taken through the treatment machines to be sure the treatment is being adequately reproduced.

3. Limitations of Current Treatment Planning Techniques
Even in the best radiation treatment facilities geometric accuracy (covering the entire tumor

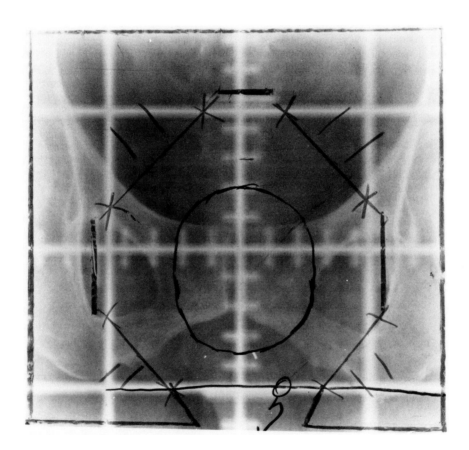

FIGURE 2. A simulation film taken AP (anterior posterior) for treatment of prostate cancer. The prostate, itself invisible on the X-ray, has been drawn in, its location inferred from CT data and from physical examination. The tumor target volume has also been indicated; special shielding blocks will be machined to conform to the drawing and be placed between the X-ray beam and the patient during treatment time.

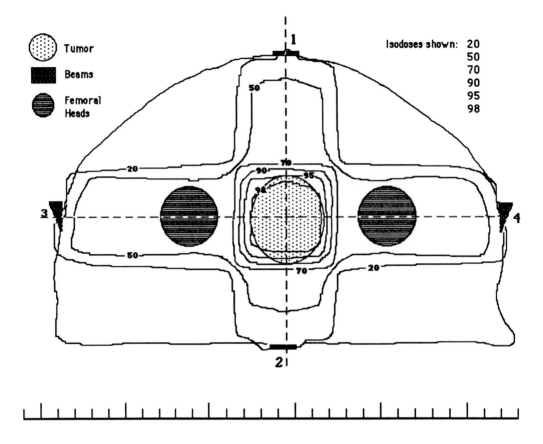

FIGURE 3. An isodose distribution resulting from a "four field" plan for a patient with prostate cancer. Typically the tumor volume and some of the relevant anatomy are drawn on the plan.

volume with all radiation beams) is frequently not achieved. In a review by Tepper and Padikal[5] of six published reports, treatment was found to be altered in 195 of 402 (49%) with the use of CT scans, due to geometric inaccuracies of beam placement during simulation. Suit and Westgate,[6] Coleman,[7] and Suit and Tepper[8] estimate that almost 40,000 cancer patients per year that are currently failing treatment could be cured if local tumor control were always assured (Table 1), a goal that is admittedly unrealistic. However, Goitein has estimated that the elimination of geometric misses alone might increase cure rates by 3 to 4%. For the nearly 300,000 cancer patients treated annually with radiation this would amount to a saving of ~10,000 lives.

It is currently common practice to use CT, magnetic resonance imaging (MRI), and other imaging data to augment the information present on simulation films. The approximate spatial position of tumor volumes, and radiosensitive organs, as inferred from these scans are then routinely drawn on simulator films by the physician. However, accurate registration of the CT data with the simulator film, while allowing for differences in patient position between simulation and CT scanning and for beam divergence, cannot be consistently and accurately achieved by the clinician without the appropriate graphics tools. Errors introduced in the data transfer process are certainly one of the causes of tumor mistargeting.

The probability of eradicating most tumors is directly related to the radiation dose delivered.[10] However, the safety of the treatment depends on how well normal tissue can be excluded from the high-dose region. *Thus, to be safe and effective, the radiation dosage must be carefully matched to the target volume.* This "tightness of fit", or *conformation,* between radiation dose distribution and target volume, then, provides an additional measure of treatment accuracy, for if even a small portion of the tumor receives less than the prescribed dose, the chances of eradicating it are reduced.

Good dose conformation requires the use of multiple, carefully shaped radiation beams whose spatial orientation must be specifically selected for each patient. However, using conventional simulation techniques, it is often difficult to be sure that a given radiation beam is completely covering the tumor target, especially if the orientation of the beam is an unusual one. For that reason, most radiation oncologists employ simplified, standard treatment techniques that are optimized only within narrow limits.

II. 3D RADIATION TREATMENT PLANNING

A. THE NEED FOR 3D GRAPHIC DISPLAYS

CT is now generally considered essential for modern radiation therapy treatment planning because of the wealth of anatomic information provided. Exclusive use of the CT data in the conventional planar slice format, however, would impose important limitations on the treatment planning process. For example, it would not be possible for the treatment planner to specify a radiation beam by drawing its desired projection on each CT slice since parallel CT slices cannot all lie in the plane of a divergent radiation beam. (It is very difficult to predict the projection of a radiation beam onto a CT slice if it is not perpendicular to the z-axis, especially if the beam is irregularly shaped.) Although methods exist to rapidly calculate the appearance of CT slices in arbitrary oblique planes,[12] determination of the incident beam angle (and thus the oblique planes in which to calculate the scans) is, in fact, one of the goals of the treatment planning process. It is likely, therefore, that the treatment planner would consider only plans which used radiation beams that were transverse co-planar if the CT data were available only in the planar format.

Another limitation of the traditional CT format is that by *itself* it cannot adequately convey the spatial distribution of the radiation dose that would result from a given beam configuration. By displaying the radiation isodoses on each CT slice, the merits of multiple competing treatment plans can be compared only in a piecewise fashion. Experience has shown that under these conditions it is not always easy either to recognize the best treatment plan or to suggest useful modifications,[13] as a *global* sense of the dose distribution is lacking.

Finally, for brachytherapy treatment, the conventional CT format may offer ambiguous information as to the location of the implant; it may be impossible to determine whether a

TABLE 1
Estimated Increase in Survival in Local Tumor
Could Always Be Eradicated

Tumor	New cases/year	Local-regional failure rate (%)	Additional survivors if no local failures
Uterine cervix[6]	16,000	23	2,700
Oropharynx[6]	18,700	27	2,000
Colorectum[6]	120,000	30	17,600
Ovary[6]	18,000	40	2,000
Lung (non-oat cell)[7]	102,000	20	7,200
Prostate[7]	75,000	13	3,600
Bladder[7]	38,500	31	3,000
Soft tissue sarcoma[8]	4,250	30	900
Total	392,450	≈24	39,000

radioactive seed seen on one CT slice is the same as that seen on an adjacent slice or whether two seeds are present.

We believe that the problems of tumor targeting, understanding complex radiation dose distributions, and locating the spatial position of radioactive implants can be made substantially easier through the use of high-quality 3D displays. For tumor targeting, suitable displays could effectively portray the 3D spatial relationships between anatomical structures, target(s), and radiation beams, which is needed for proper beam placement. 3D displays may also offer the clinician insight into the global distribution of the calculated radiation dose distribution and greatly facilitate judging the merits of competing treatment plans. Finally, suitable displays might make locating tiny radiation sources a far less laborious process than is currently possible.

B. TREATMENT DESIGN SOFTWARE
1. Integrating Digital Images into the Treatment Planning Process
The concept of tightly integrating CT scans into the treatment planning process is not new;[14] such systems are currently under development at many institutions.[15–22] At the University of North Carolina we have built on the work of these groups, but our techniques are somewhat different in that we have tightly integrated traditional radiation therapy treatment planning and advanced computer technology and display techniques into a coherent system for 3D treatment design and delivery.

Originally the design of radiation portals was done from external landmarks alone (1D planning). In a sense the patient was modeled by his visible and palpable anatomy. With the advent of the radiation simulator it became possible to design radiation portals entirely from plane radiographs (2D planning) which became the new patient model. What we now propose is that treatment planning be done entirely from CT data (3D planning), with plane radiographs (and skin marks) being used mainly as quality control. By adopting this point of view we overcome the difficult issue of how to accurately exchange data from one format (CT scans) to another (plane radiographs) as we can go directly from the CT scans to the patient. Moreover, by adopting this approach, computer-aided design (CAD) of radiation therapy treatment becomes analogous to the technical aspects of industrial CAD of automobiles, airplane wings, or any other complex product. As a result we benefit enormously from the development of ever more powerful, lower cost CAD workstations, produced in response to the needs of corporations with large investments in engineering CAD.

2. Virtual Simulation
"Virtual simulation" is the name given by Sherouse et al.[23–25] to software that fully implements the function of a "physical" radiation simulator. The virtual simulator maintains a relatively high fidelity to the feel of a physical simulator while providing the user with the ability to rapidly explore a wide variety of treatment geometries.[26]

The user of the virtual simulator is provided with a 3D display of the patient model (currently a wireloop model) which consists of the tumor volume and all relevant anatomical structures, as abstracted from the CT data. CT slice "miniatures" (which can be expanded to full resolution) and views of the (physical) simulator gantry and table position are also available (Figure 4). The virtual simulation software runs under the X Window System,[27] which allows these views to be resized and moved anywhere on screen as desired.

Radiation portal design on the virtual simulator usually begins by moving the patient model along any of the three orthogonal axes so as to place the target volume at the treatment machine *isocenter* (intersection of the central ray of the radiation beam with the axis of rotation of the gantry). Beams can then be arranged either by manipulating the patient model *or* the treatment machine icons at arbitrary angles in three dimensions around the target with full knowledge of the intersection of each beam on the defined normal structures. Since all of

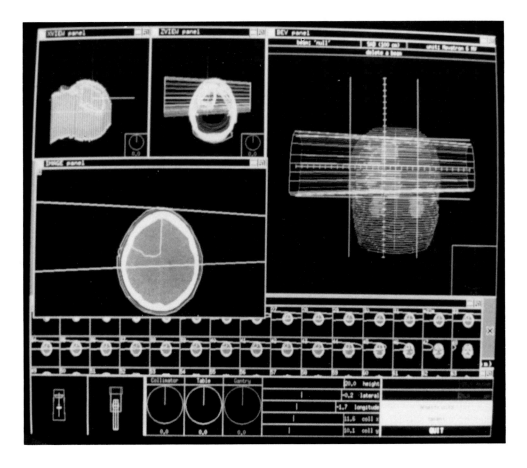

FIGURE 4. A screen from the virtual simulator. The development of a plan for treating a brain tumor is shown. In the upper right a wireloop 3D graphic of the patient's skull is seen with a proposed radiation beam. On the left and below, the physician can see, in 2D, just how the beam intersects with the patient's anatomy and tumor. We have found this combination of 3D *and* 2D display to be necessary for treatment planning.

the explicit concerns of coordinate systems and rotation angles are handled by the software, the user is free to concentrate on the more relevant issues of proper tumor targeting. For example, if the patient model is manipulated, the software automatically updates the gantry and table position. The radiation portal itself is then shaped using a mouse or other pointing device. When the beam outline is completed the user can view the beam overlaid on any or all of the original CT slices.

Once a radiation beam is satisfactorily placed, provisions are available to automatically copy and oppose it or create other beams. When all beams have been designed the virtual simulation is complete and, like its physical counterpart, the virtual simulator can create output which includes digitally reconstructed simulation films with projections of the tumor, target volume, and other structures of interest overlaid. The virtual simulator then produces a list of beam and table parameters (displacements or rotations from the neutral position) needed to set up each specified portal. Other output of the virtual simulator includes construction specifications and templates for beam modifiers such as blocks compensating filters (e.g., wedges).

Following virtual simulation, and the production of the digitally reconstructed radiographs, the patient is fitted into the body cast and placed on the physical simulator. The fields are set up using only the instructions generated during virtual simulation, a process we call "setting up by the numbers". The physical simulation films are then compared to the digitally reconstructed ones as a quality control. Our preliminary experience suggests that we can obtain a consistency between computed and physical radiographs of ≤3 mm for most tumor sites.

It is important to realize that the digitally reconstructed radiographs and physical radiographs are used as a quality control only. The entire treatment planning process takes place from the original CT data.

III. 3D GRAPHIC DISPLAY TECHNIQUES USEFUL FOR RADIATION THERAPY

A. SURFACE-BASED DISPLAYS

Until recently, high-quality display of 3D anatomical structures from CT or MRI data sets has been achieved by surface-based displays employing a *binary* classification as to whether a given pixel or voxel is or is not on a given surface. This surface is then modeled by many small polygons; reflections of ambient light,[28] shadows, textures, and other effects can be added to enhance the 3D effect.[29–32]

Several categories of surface-based displays have been under active investigation. A common approach is to apply binary thresholding directly to the volume data; voxels that are determined to be on the surface are rendered as opaque cubes having six polygonal faces.[29] This approach, known as the *cuberille* model,[30] has been refined to the point that the surface is portrayed with a natural smoothness and texture.[30,31] Another approach to producing a surface-based display relies on edge tracking to yield a set of contours defining features of interest.[32] Then a mesh of polygons can be constructed connecting the contours on adjacent slices using minimal surface area[33] or other techniques. Figure 5 (discussed below) is a 3D rendering using this technique. Variations on these techniques have been developed Lorensen and Cline,[34] who apply surface detectors at each sample location to produce a large collection of voxel-sized polygons which can then be supplemented with analytically defined objects and rendered using conventional algorithms. Further improvements have been developed by Kaufman[35] and others. These display techniques have found some recent success in craniofacial surgery,[36–38] and in detecting and repairing damage to hips[39] and feet,[40] and in studying complex movements such as seen in the wrist.[41]

Yet another approach to displaying surfaces from volume data is the *binary voxel techniques,* in which data samples are mapped directly to image pixels without intermediate geometric constructs.[42,43] Use of these displays in radiation therapy treatment planning, along with special hardware to do interactive display, is under investigation at several institutions.[44–46]

The feasibility of using surface-based rendering for tumor targeting and for display of radiation isodose surfaces has been discussed by Rosenman et al.[47] for external beam and by Chaney et al.[48] for brachytherapy. Figure 5 is an example of a display that might be of use in tumor targeting. The chest of a woman with an obvious cancer in the left upper lung is shown (right side of the picture). Since such displays can now be made fully interactive,[49] one could imagine manipulating this display until, for example, the radiation beam intersected only a minimum amount of the opposite lung.

The ability to display the spatial distribution of radiation dose is an important adjunct to the treatment planning process. In Plate 16* the 80% isodose *surface* is shown that arises when an esophagus tumor is treated with one anterior and two posterior oblique radiation fields. One is able to see at a glance that this surface surrounds the esophagus in a uniform way without

---

* Plate 16 appears following page 294.

much extention into the lung. This simple observation would have required examining 40 to 60 slices if the data had been presented in a format such as shown in Figure 3.

Despite the high quality of presentation, and the evident usefulness provided by surface-based methodologies, they all suffer from a serious weakness; the objects of interest must be defined in a binary fashion *before* rendering can begin. In other words, the objects must be recognized and understood prior to display in 3D form. In addition, at least for radiation therapy treatment planning purposes, these display methods require a large amount of complex preprocessing, including labor-intensive and time-consuming manual contouring.[50]

B. VOLUME RENDERING

Recently a new display approach has been developed, known as *volume rendering* or *volume compositing,* which is done entirely without *binary* decisions.[51-53] In this technique both a grey-scale shade and partial opacity are computed for each voxel in the data set, based on its *surface likelihood.* These values are then blended along the viewing rays to form a translucent gel-like volume image. (Volume rendering, as defined above, should not be confused with earlier *binary decision* 3D displays of the same name.[54])

Such an approach has two theoretical advantages over ones based on binary surface specification. First, minimal predefinition is required; ultimately it may be sufficient for the

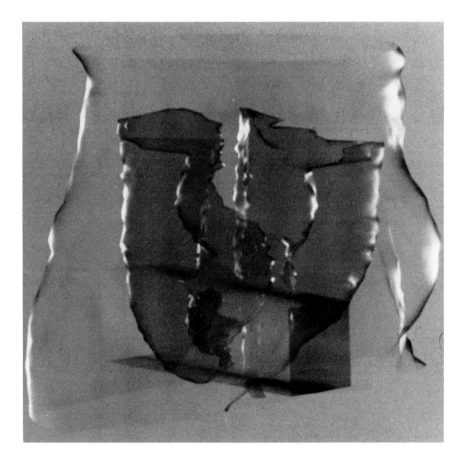

FIGURE 5. A shaded-surface display of a female patient with lung cancer. The external surface is transparent, and the lungs are easily seen. In the left upper lung (right side of the picture) a tumor is seen encompassed by a radiation beam.

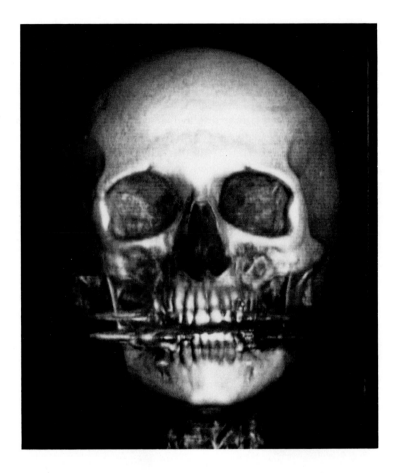

FIGURE 6. A volume rendering of the head of a cadaver. 2-mm slices were taken (107 in all) to produce this extraordinarily detailed image. If this graphic could be made interactive (in near-real time), it would be a wonderful replacement for the wireloop constructs in Figure 4.

user to merely specify *regions of interest* to exclude objects likely to obscure or confuse the portrayal of objects of interest. It is also likely (although not yet convincingly demonstrated) that volume rendering can effectively portray soft tissues in addition to bone. Second, the presentation of the objects is improved, since the method takes into account all the image intensities in a surface region rather than being limited to binarily chosen surface points. As a result, volume rendered images *may* be less prone to false positives (spurious objects) or false negatives (missing objects). Figure 6 shows an example of volume rendering of a cadaver from 107 CT slices, at 2 mm thickness.

1. Use of Volume Rendering in Radiation Treatment Planning

The above attributes of volume rendering make it attractive for radiation therapy treatment planning. Although, to our knowledge, volume rendering has not been used yet for this purpose, it is fun to speculate what the result might be. Figure 7, using the same data set as Figure 6, shows what tumor targeting might be like with volume rendering. We have added a "brain tumor" and "radiation beam" to the data set used in Plate 16. Radiation isodose surfaces could also be incorporated into a volume rendering. In Figure 8 we show a volume rendering of an electron isodensity map taken from an X-ray diffraction study. This same technique could be used to display radiation isodose surfaces. Plate 16 and Figures 6 to 8 were produced and kindly given to us by one of our collaborators, Dr. Marc Levoy.

2. Display of Shaded-Surface Graphics within Volume-Rendered Images

Surface-based rendering which can display user-defined objects has some advantages over volume rendering for radiation treatment planning because some of the objects of greatest importance (for example tumor target volume or radiation beams) are not part of the CT data set. The tumor target volume, especially, is a construct in the mind of the clinician which is formed by integrating the imaging studies with the result of the patient history, physical examination, histologic examination of the tumor, and clinical knowledge about the pathways of tumor spread. Thus it can only be defined by the clinician. At the University of North Carolina we are actively pursuing combining the mixing of geometric and volumetric data and have recently demonstrated the feasibility of this approach.[55] Figure 7 is an early example of this technique applied to radiation therapy.

3. Can Volume Rendering Be Made Interactive in Near-Real Time?

There is no question that volume rendering of CT data provides a better patient model than does the wireloop display seen in Figure 4. However, the wireloop displays are fully interactive in near-real time, and our experience has shown that this is a vital feature for radiation treatment planning. Volume rendering of the quality shown in Figure 6 takes about 5 to 10 minutes to produce on a SUN 4/280™, depending on the number of slices.[56] However, in the next few years graphics engines should become available that can significantly reduce this processing time. Pixel-planes V, a successor to Pixel-planes IV,[49] is being designed and built

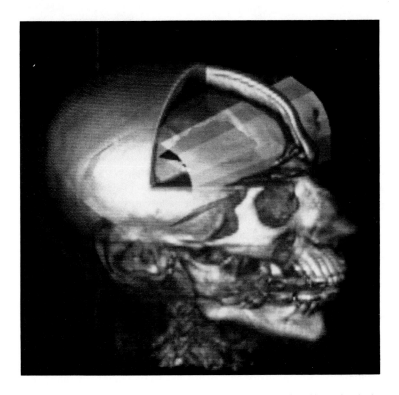

FIGURE 7. How targeting might be done using volume rendering. Shown is a brain tumor and radiation beam.

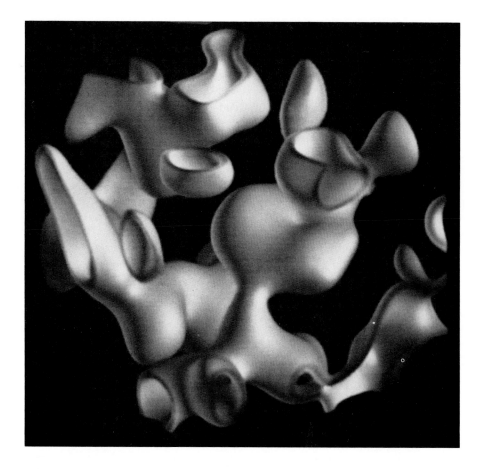

FIGURE 8. An electron isodensity surface shown with volume rendering. The same technique could be used to show a radiation isodose surface.

at our institution.[57] Although not specifically designed for volume rendering, we are estimating that it could render images of the quality of Figure 6 in less than a second.

4. Imbedding CT Data within a Volume-Rendered Image

Sometimes CT data is best displayed in its original 2D format. As previously discussed, it is often difficult for the clinician to register that data with the actual patient. The ability to display the original CT data within the volume-rendered image might then give the clinician both the detail available on the original CT data and the orientation provided by the volume-rendered image. We have been impressed by the displays of McShan and Fraass,[58] using their *axonometric* technique for displaying one or more cross-sectional images in their proper geometric registration. These authors have been investigating the use of these displays for radiation therapy treatment planning for several years. More recently Höhne et al.[59] have demonstrated how MRI data can be imbedded within their version of volume rendering.

5. Motion Correction Algorithms

CT or MR scans of patients take a number of minutes, and some patient motion from slice to slice is inevitable. Our preliminary experience has shown that for volume rendering of the head, severe patient restraint and/or general anesthesia may not be necessary to obtain a satisfactory scan.[60] However, volume rendering of the lungs or chest becomes very degraded because of chest motion. Either a gated CT scan technique or sophisticated motion correction software (or both) will be needed to allow the thorax to be volume rendered in a satisfactory manner. As yet we have little information about the need for motion correction when producing volume renderings of other parts of the body.

6. Future Research in the Use of Volume Rendering for Radiation Therapy Planning
Research at our institution is being directed to developing volume rendering for radiation therapy treatment planning. The following areas are being given special attention:

1. Optimization of the volume-rendering algorithm. By performing a certain amount of precalculation it is possible to increase the speed with which the object can be rotated or moved. In addition, it should be possible to interactively move the light source and highlight and move a 3D region of interest. The use of shadows to provide additional 3D depth cueing is also under investigation.
2. Implementation on Pixel-planes 5. The rendering algorithms will have to be made parallel to take full advantage of multiprocessor architecture.
3. Handling of nonpolygonal geometry. We should be able to incorporate contours and higher order surfaces into the volume rendering.
4. Use of shadows.
5. The use of surfaces having matte or reflective finishes and objects that are partially translucent, not merely partially transparent.

IV. SUMMARY

Radiation therapy remains one of the most important methods of treating cancer that we have. Radiation treatments can be made more effective by methods to display and interactively target tumor in three dimensions. In addition, the enormous amount of data generated when 3D radiation dose calculations are performed can be far better understood when presented in a 3D graphical form.

REFERENCES

1. Röntgen, W.C. On a new kind of ray, *Science* 3:227–231 (1896).
2. Million, R.R., and P.J. McCarty. History of diagnosis and treatment of cancer in the head and neck, in *Management of Head and Neck Cancer*, J.B. Million and N.J. Cassisi (Eds.) (Philadelphia, PA: Lippincott, 1984), pp. 1–18.
3. Senn, N. Therapeutical value of röntgen ray in treatment of pseudoleukemia, *New York Med. J.* 77:665–668 (1903).
4. Sowers, Z.T. The uses of radium, *Am. Med.* 6:261 (1903).
5. Tepper, J.E., and T.N. Padikal. The role of computed tomography in treatment planning, in *Radiation Therapy Planning*, E. Glatstein, J.L. Haybittle, and N.M. Bleehen (Eds.) (New York, NY: Marcel Dekker, 1983), pp. 139–158.
6. Suit, H.D., and S.J. Westgate. Impact of improved local control on survival, *Int. J. Radiat. Oncol. Biol. Phys.* 12:453–458 (1986).
7. Coleman, C.N. Hypoxic radiosensitizers: Expectations and progress in drug development, *Int. J. Radiat. Oncol. Biol. Phys.* 11:323–329 (1985).
8. Suit, H.D., and J.E. Tepper. Impact of improved local control on survival in patients with soft tissue sarcoma, *Int. J. Radiat. Oncol. Biol. Phys.* 12:699–700 (1986).
9. Goitein, M. The utility of computed tomography in radiation therapy: An estimate of outcome, *Int. J. Radiat. Oncol. Biol. Phys.* 5:1799–1807 (1979).

10. Goitein, M., and T.E. Schultheiss. Strategies for treating possible tumor extension: Some theoretical considerations, *Int. J. Radiat. Oncol. Biol. Phys.* 11:1519–1528 (1985).

11. Ling, C.C., C.C. Rogers, and R.J. Morton. *Computed Tomography in Radiation Therapy*, (New York, NY: Raven Press, 1983).

12. Mohan, R., L.J. Brewster, G. Barest, I.Y. Ding, C.S. Chui, B. Shank, and B. Vikram. Arbitrary oblique image sections for 3-D radiation treatment planning, *Int. J. Radiat. Oncol. Biol. Phys.* 13:1247–1254 (1987).

13. Bauer-Kirpes, B., W. Schlegel, R. Boesecke, and W.J. Lorenz. Display of organs and isodoses as shaded 3-D objects for 3-D therapy planning, *Int. J. Radiat. Oncol. Biol. Phys.* 13:135–140 (1987).

14. Reinstein, L.E., D. McShan, B.M. Webber, and A.S. Glicksman. A computer assisted three-dimensional treatment planning system, *Radiology* 127:259–264 (1978).

15. Goitein, M., and M. Abrams. Multi-dimensional treatment planning: I. Delineation of anatomy, *Int. J. Radiat. Oncol. Biol. Phys.* 9:777–787 (1983).

16. Goitein, M., M. Abrams, D. Rowell, H. Pollari, and J. Wiles. Multi-dimensional treatment planning: II. Beam's eye-view, back projection, and projection through CT sections, *Int. J. Radiat. Oncol. Biol. Phys.* 9:789–797 (1983).

17. Dahlin, H., I.L. Lamm, T. Landberg, S. Levernes, and N. Ulso. User requirements on CT-based computed dose planning systems in radiation therapy: Presentation of check lists, *Comp. Programs Biomed.* 16:131–138 (1983).

18. Houlard, J.P., and A. Dutreix. 3D display of radiotherapy treatment plans, in *Proceedings of the 8th International Conference on the Use of Computers in Radiotherapy* (North-Holland: Elsevier Science, 1984), p. 219.

19. McShan, D.L., B.A. Fraass, and A.S. Lichter. Three dimensional portal design and verification, *Med. Phys.* 13:575 (1986).

20. Smith, R.M., J.M. Galvin, and R. Harris. Correlation of CT information with simulation and portal radiographs, *Int. J. Radiat. Oncol. Biol. Phys.* 13 (suppl):176 (1987).

21. Nagata, Y., T. Nishidai, M. Abe, M. Takahashi, Y. Yukawa, H. Nohara, N. Yamaoka, T. Saida, H. Ishihara, Y. Kubo, H. Ohta, and K. Inamura. CT simulator: A new treatment planning system for radiotherapy, *Int. J. Radiat. Oncol. Biol. Phys.* 13 (suppl):176 (1987).

22. Mohan, R., G. Barest, L.J. Brewster, C.S. Chui, G.J. Kutcher, J.S. Laughlin, and Z. Fuks. A comprehensive three-dimensional radiation treatment planning system, *Int. J. Radiat. Oncol. Biol. Phys.* 15:481–495 (1988).

23. Sherouse, G.W., and G. Mosher. User interface issues in radiotherapy CAD software, in *Proceedings of the Ninth International Conference on the Use of Computers in Radiotherapy* (North-Holland: Elsevier Science, 1987), pp. 429–432.

24. Sherouse, G.W., C.E. Mosher, K. Novins, J. Rosenman, and E.L. Chaney. Virtual simulation: Concept and implementation, in *Proceedings of the Ninth International Conference on the Use of Computers in Radiotherapy* (North-Holland: Elsevier Science, 1987), pp. 433–436.

25. Mosher, C.E., G.W. Sherouse, E.L. Chaney, and J.G. Rosenman. 3D displays and user interface design for a radiation therapy treatment planning CAD tool, *Proc. SPIE* 902:64–72 (1988).

26. Rosenman, J., G.W. Sherouse, E.L. Chaney, and J.E. Tepper. Virtual simulation: Initial clinical results, *Int. J. Radiat. Oncol. Biol. Phys.* (submitted).

27. Scheifler, R.W., and J. Gettys. The X window system, *ACM Trans. Graph.* 5:79–109 (1986).

28. Phong, B.T. Illumination for Computer Generated Images, Ph.D. thesis, University of Utah (1973).

29. Herman, G.T., and H.K. Liu. Three-dimensional display of human organs from computer tomograms. *Comput. Graph. Image Proc.* 9:1–21 (1979).

30. Chen, L.S., G.T. Herman, R.A. Reynolds, and J.K. Udupa. Surface shading in the cuberille environment, *IEEE Comput. Graph. Appl.* 5:33–43 (1985).

31. Chaung, K.S., J.K. Udupa, and S.P. Raya. High Quality Rendition of Discrete Three Dimensional Surfaces, Technical Report MIPG 130, Department of Radiology, University of Pennsylvania, Philadelphia (1988).

32. Pizer, S.M., H. Fuchs, C. Mosher, L. Lifshitz, G.D. Abram, S. Ramanathan, B.T. Whitney, J.G. Rosenman, E.V. Staab, E.L. Chaney, and G. Sherouse. 3D shaded graphics in radiotherapy and diagnostic imaging, in *Proc. Computer Graphics '86, National Computer Graphics Association* III:107–113 (1986).

33. Fuchs, H., Z.M. Kedem, and S.P. Uselton. Optimal surface reconstruction from planar contours, *Commun. ACM* 20:693–702 (1977).

34. Lorensen, W., and H. Cline. Marching cubes: A high resolution 3D surface construction algorithm, *Comput. Graph.* 21:163–169 (1987).

35. Kaufman, A. Efficient algorithms for 3D scan-conversion of parametric curves, surfaces, and volumes, *Comput. Graph.* 21:171–179 (1987).

36. Vannier, M.W., J.L. Marsh, and J.O. Warren. Three-dimensional CT reconstruction images for craniofacial surgical planning and evaluation, *Radiology* 150:279–284 (1984).

37. Hemmy, D.C., D.J. David, and G.T. Herman. Three-dimensional reconstruction of craniofacial deformity using computed tomography, *Neurosurgery* 23:534–541 (1983).

38. Zinreich, S.J., D.E. Mattox, D.W. Kennedy, M.E. Johns, J.C. Price, M.J. Holliday, C.B. Quinn, and H.K. Kashima. 3-D CT for cranial facial and laryngeal surgery, *Laryngoscope* 98:1212–1219 (1988).

39. Fishman, E.K., B. Drebin, D. Magid, W.W. Scott, Jr., D.R. Ney, A.F. Brooker, L.H. Riley, Jr., J.A. St. Ville, E.A. Zerhouna, and S.S. Siegelman. Volumetric rendering techniques: Applications for three-dimensional imaging of the hip, *Radiology* 163:737–738 (1987).

40. Adler, S.J., M.W. Vannier, L.A. Gilula, and R.H. Knapp. Three-dimensional computed tomography of the foot: Optimizing the image, *Comput. Med. Imag. Graph.* 12:59–66 (1988).

41. Bresina, S.J., M.W. Vannier, S.E. Logan, and P.M. Weeks. Three-dimensional wrist imaging: Evaluation of functional and pathologic anatomy by computer, *Clin. Plast. Surg.* 13:389–405 (1986).

42. Goldwasser, S.M., R.A. Reynolds, T. Bapty, D. Baraff, J. Summers, D.A. Talton, and E. Walsh. Physicians' workstation with real-time performance, *IEEE Comput. Graph. Appl.* 5:44–57 (1985).

43. Höhne, K.H., M. Bomans, U. Tiede, and M. Riemer. Display of multiple 3D objects using the generalized voxel model, *Medical Imaging II, Proc. SPIE* 914:850–854 (1988).

44. Reynolds, R.A., M.R. Sontag, and L.S. Chen. Algorithm for three-dimensional visualization of radiation therapy beams, *Med. Phys.* 15:24–28 (1988).

45. Goldwasser, S.M. The voxel processor architecture for real-time display and manipulation of 3D medical objects, in *Proc. 6th NCGA Conf.* 3:71–80 (1985).

46. Mott, D.J., C.J. Moore, and J.M. Wilkinson. The creation, manipulation, and display of 3-D images for RTP, in *Proceedings of the Ninth International Conference on the Use of Computers in Radiotherapy* (North-Holland: Elsevier Science, 1987), pp. 33–35.

47. Rosenman, J., G.W. Sherouse, H. Fuchs, S.M. Pizer, A.L. Skinner, C. Mosher, K. Novins, and E.L. Chaney. Three-dimensional display techniques in radiation therapy treatment planning, *Int. J. Radiat. Oncol. Biol. Phys.* 16:263–269 (1989).

48. Chaney, E., J. Rosenman, G. Sherouse, D. Bourland, H. Fuchs, S. Pizer, E. Staab, M. Varia, and S. Mahaley. Three-dimensional displays of brain and prostate implants, *Endo./Hyperthermia Oncol.* 2:93–99 (1986).

49. Fuchs, H., J. Goldfeather, J.P. Hultquist, S. Spach, J.D. Austin, J.G. Eyles, and J. Poulton. Fast spheres, shadows, textures, transparencies, and image enhancements in pixel-planes, *Comput. Graph. Proc. 1985 SIGGRAPH Conference* 19:111–120 (1983).

50. Mills, P.H., H. Fuchs, S.M. Pizer, and J.G. Rosenman. IMEX: A tool for image display and contour management in a windowing environment, *SPIE Med. Imag. '89* (to appear).

51. Drebin, R.A., L. Carpenter, and P. Hanrahan. Volume rendering, *Comput. Graph.* 22:65–74 (1988).

52. Levoy, M. Display of surfaces from volume data, *IEEE Comput. Graph. Appl.* 8:29–37 (1988).

53. Pizer, S.M., H. Fuchs, M. Levoy, J.G. Rosenman, J. Renner, and R. Davis. 3D display with minimal predefinition, in *Computer Aided Radiography (CAR)* (Berlin: Springer-Verlag, 1989, in press).

54. Harris, L.D., R.A. Robb, T.S. Yuen, and E.L. Ritman. Non-invasive numerical dissection and display of anatomic structure using computerized tomography, *SPIE Proc.* 152:10–18 (1978).

55. Levoy, M. Rendering mixtures of geometric and volumetric data, in *Computer Graphics Proc. 1989 SIGGRAPH Conference* (submitted).

56. Levoy, M. Volume Rendering by Adaptive Refinement, Technical Report 88-030, Computer Science Department, University of North Carolina at Chapel Hill (1988) (to appear in *The Visual Computer*).

57. Fuchs, H., J. Poulton, J. Eyles, T. Greer, J. Goldfeather, D. Ellsworth, S. Molnar, G. Turk, B. Tebbs, and L. Israel. A Heterogeneous Multiprocessor Graphics System Using Processor-Enhanced Memories, Technical Report 89-005, Computer Science Department, University of North Carolina at Chapel Hill (1989) (in review, *Comput. Graph.*).

58. McShan, D.L., and B.A. Fraass. 3-D treatment planning: II. Integration of gray scale images and solid surface graphics, in *Proceedings of the Ninth International Conference on the Use of Computers in Radiotherapy* (North-Holland: Elsevier Science, 1987), pp. 41–44.

59. Höhne, K.H., R.L. Delapaz, R. Bernstein, and R.C. Taylor. Combined surface display and reformatting for the three-dimensional analysis of tomographic data, *Invest. Radiol.* 22:658–664 (1987).

60. Davis, R., M. Levoy, J. Rosenman, S.M. Pizer, H.R. Pillsbury, H. Fuchs, J. Renner, and A. Skinner. Three-dimensional high resolution volume rendering (HRVR) from CT and MRI: Present and future applications to otolaryngology-head and neck surgery (in preparation).

Chapter 13

FUTURE DIRECTIONS IN THREE-DIMENSIONAL IMAGING

David C. Hemmy

TABLE OF CONTENTS

I. INTRODUCTION

We have reached a level of clinical acceptance of three-dimensional (3D) imaging: the images provide clinical information in a readily understandable format, they can be provided rapidly, and they are of sufficient quality to be much more than a novel form of viewing. Numerous companies are now marketing software and hardware to provide these images. The product is mainly confined to the provision of osseous images and may permit some viewer interaction with these rendered images. These images have been embraced by those clinical specialties which deal with osseous disorders of varying degrees of complexity requiring multidimensional viewing to permit understanding. Among these specialties are craniofacial surgery, neurological surgery, orthopedics, otolaryngology, and oral surgery. In each of these areas further needs have been expressed based on the good but rudimentary information contained in present-day 3D images. These unmet needs determine many of the future directions investigators and computer scientists must take.

Although other chapters precisely demonstrate what is presently available in 3D imaging, it is necessary to briefly address this present state to show what is lacking and what needs to be accomplished. Presently we have images. These images are of varying quality depending upon scanning factors used in image acquisition, reconstruction algorithm used, and display equipment. These images can be "manipulated". The image manipulation may take the form of varying the viewing or "pose" angle either through the selection of predetermined poses prior to image generation in a "batch" processing program or real or near real-time manipulation of a volumetric rendering. The image can be "dissected" by reducing the portion or portions of the axial slices to be included in the rendered or completed image as well as by removal of voxels from a rendered volume by exclusion of voxels which lie outside a predetermined plane. Furthermore, the character of the image can be altered by the inclusion or exclusion of voxels during image segmentation, the process of determining which voxels are to be included based on their computed tomography (CT) Hounsfield value. Another form of image character change can be made through the use of shading by varying the brightness of voxel faces based on the relative depth of the face or by making the brightness vary in direct proportion to the surface normal described by the voxel face. All these manipulations enhance the image to provide a more pleasing and understandable anatomic representation. They are, however, rudimentary in anatomic complexity because they present only one type of anatomic tissue, bone, with the occasional representation of a second tissue, skin. To be fair, the representation of bone is quite precise, with the major error being found in the exclusion of very thin bone through partial volume averaging of CT attenuation coefficients.[1]

This information, albeit limited, is quite useful to the clinician. In craniofacial surgery it is used to understand complex and rare variations in skeletal anatomy in a fashion which approaches that of a dry skull museum specimen. The object, in this case the skull, can be "handled" as one would handle an unfamiliar object by rotating it through various viewing planes to gather visual information about surface contour, spatial relationships, and hidden surfaces. All this information has been assembled in three dimensions, eliminating the often erroneous mental reconstruction from the two dimensions provided by CT and plain film radiography. The usefulness of these images seems to be acknowledged by younger clinicians, residents, and students and ambiguously embraced and disputed by senior, experienced surgeons.[2] Nonetheless, it remains a tool to permit understanding, facilitating clinical intervention rather than a true aid to planning permitting a prediction of results based on expected surgical manipulation.

In a similar fashion, the anatomic representations provide easily understandable visual information for neurosurgeons and orthopedists wishing to understand the nature of spinal deformities, and for otolaryngologists and plastic surgeons to perceive the extent of involvement of the skull base by a tumor or other destructive process.

In each discipline, there are needs beyond the simple anatomic representation of osseous structures. I propose to discuss these needs based on discussions with surgeons representing these fields, as well as to suggest other areas of potential development.

Inherent in the discussion which follows is the assumption that data are acquired using scanners, both magnetic resonance (MR) and CT, which are currently available on the market with no significant changes to be made in the acquisition hardware. Four general areas will be highlighted: (1) the problems, (2) the image, (3) the workstation, and (4) clinical applications.

II. THE CLINICAL PROBLEMS

As stated earlier, the 3D medical image as we presently know it is a bone or osseous image permitted through the rather sharp demarcation of bone from surrounding tissues because of the rather large difference in attenuation coefficients between bone and the surrounding constituents. This fact made bone the chosen area of interest for early 3D studies.[3] It basically remains the area of interest in most reported clinical studies due to the ease of segmentation of structures as "bone" or "no bone" by including all structures above a selected CT value as bone and eliminating all others (the only "error" being the inclusion of structures such as metals or tissues artificially opacified through the use of contrast agents).[4] The use of contrast agents can occasionally display the relationship of an enhancing tumor to the surrounding bone or the relationship of an anomalous blood vessel to the surrounding skull. The sharp difference between the room air and the skin also provides a segmentation opportunity allowing the production of the *masque mortuaire* which, at present, is of little clinical value.

It is obvious that, for an image to be of clinical use, it must represent something less than all the anatomical parts of that object. CT, for instance, does show most of the anatomical information contained in an object, but this is displayed as a two-dimensional cross-sectional representation rather than as a 3D representation in which only superficial information would be represented at the expense of deep anatomical structures *unless* certain information or structures were selectively removed or given a degree of transparency or translucency.

In craniofacial surgery, it may be desirable to display the entire orbital contents as a single unit without regard for the constituent tissues in contrast to the bony orbit so as to permit volumetric analysis of the soft tissue before and after a surgical procedure. On another occasion, it may be necessary to display the extraocular muscles as they attach to the globe as separate well-defined entities being highlighted from the surrounding bony orbit and the underlying globe without displaying the orbital fat. It may also be necessary to display the globe and attached optic nerve as it courses through the optic foramen to reach the optic chiasm to determine the exact nature of its course and whether or not there is any encroachment upon the nerve or sharp angulation of its course.

Through the use of osteotomies or saw or chisel (osteotome) cuts in bone, large portions of the craniofacial skeleton are made free and movable. These movable segments are positioned to provide a more acceptable facial appearance in individuals with significant deformity. This movement results in a change in the efficiency of muscles, particularly the strong muscles involved in mastication. It is therefore desirable to display these muscles along with the skeletal structures to which they are attached, showing the points of origin and insertion of the muscles.

In neurological surgery there is a need to demonstrate the margins of the brain, the interface between the gray and white matter, the boundaries of the cerebrospinal fluid containing spaces, and the vasculature at an arterial and venous level without the necessity to demonstrate capillaries (very much the same as is displayed by a digital angiogram) and extraneous structures as tumors. There is a similar necessity to that already noted in the craniofacial area to be able to display selected combinations of these structures or perhaps all of these selected structures. For an intracranial, extracerebral structure, such as an aneurysm of the basilar circulation, it necessary to permit a wide range of viewing angles to permit the surgeon to have an understanding of what to him, in his surgical approach, will be hidden features of the aneurysm, but nonetheless important features. Furthermore, he will need to view the aneurysmal field at a viewing angle which will approximate that given him by his limited surgical exposure to permit identification and sparing of blood vessels which maintain normal brain perfusion and have a complex association with the aneurysm. This requires, at the minimum, display of the skull, blood vessels, and the cerebral surface with the ability to vary the skull and brain from opaque to translucent to transparent.

Many tumors of the brain cannot be removed surgically. These tumors are either indiscrete, ramifying through the brain, surrounding vital structures, or they are deeply situated within the brain so that surgical removal would result in disastrous neurologic consequences. Treatment of these tumors is performed by indirect means such as chemotherapy or radiation therapy. In order to treat these tumors, the tissue type of the tumor must be known with certainty as treatment is tailored to the behavior of the tumor. This requires a biopsy for tissue identification. A needle biopsy is the least disruptive form of biopsy. However, this needle must be passed through a hole in the skull (burr hole), with the path of the needle avoiding critical structures and ending in the area to be sampled. The determination of the safety of this pathway requires the display of selected structures for fiducial reference. Although many calculations can be made based on preoperative scan data, verification of the actual needle tract must be made at the time of the surgical procedure and prior to the act of biopsy, which is a somewhat destructive procedure. The verification procedure does require that the imaging information be provided in short period of time, that is minutes rather than hours.

Spinal disorders frequently present problems with abnormal angulation of the spinal column which may result in neurologic deficits, pain, abnormal stature, or interference with respiration through alteration in the architecture of the thoracic cavity. These problems are treated by attempts at restoration of the anatomic configuration to normalcy through manipulation of the vertebral column by surgical osteotomy (cutting) of the vertebrae or through the application of external forces to the vertebral column by means of external orthoses (braces) or internal fixation (rods, loops, and plates). With any mechanical system there are optimum forces and advantages which can be gained by correct siting of the applied system. Inherent is some knowledge of the biologic properties of the materials to which the force is applied as well as an understanding of the pathologic state causing the angulation of the spine. The structural pathology can be understood reasonably well by a display of the vertebrae and the muscles which provide support to the vertebral column. Furthermore, the efficiency of an applied mode of therapy can be determined by serial 3D studies of these structures.

The field of orthodontics presents similar challenges. The problem, simply stated, is abnormal dentofacial structure and the goal is restoration of a normal structure through surgical manipulation and the application of force to the teeth and jaws through external appliances.

The heart has been the subject of many invasive and noninvasive types of studies to determine both the physiology and pathophysiology of an organ which is constantly in motion. With few exceptions, the information gained has heretofore been two dimensional or, at best, biplanar.[5] Furthermore, information obtained from the heart to be useful must contain the dimension of time to delineate the relationships of the cardiac chambers to one another during the entire cardiac cycle.

These problems, then, represent basic clinical problems, many of which can be extrapolated to other areas of clinical endeavor. 3D imaging can be applied to many of these areas to permit some sophisticated clinical solutions.

III. THE IMAGE

The image, to be of greater usefulness than those presently available, must present multiple *selected* tissues. It is only in a cross section or a surface representation that all anatomical parts are displayed. We are all familiar with anatomical texts such as *Gray's Anatomy*.[6] Comprehension of regional anatomy is aided and this anatomy is illustrated by dissection, that is the cutting apart of structures to reveal underlying structures and to permit the demonstration of the relationship of one structure to another. As can be observed in the illustrations taken from *Gray's Anatomy,* some structures have been removed while others are highlighted (Plates 17 and 18*). Although faithful in their representations and representing advanced imaging techniques, electronically "dissected" images presently fall short of gross anatomical displays (Plate 19*).

We must, then, be able to identify discrete tissues such as muscle, blood vessels, and larger organs to permit their faithful display alongside other structures. Examples would be liver and vena cava (Plate 20*); the facial musculature, major craniofacial blood vessels, optic nerves, and the craniofacial skeleton; the heart and coronary arteries; and the chest wall, ribs, lung, major blood vessels, and a lung tumor.

Correct segmentation of tissue other than bone or tissues surrounded by air is difficult but must be improved for the reasons stated previously.

The use of contrast agents permits the thresholds of tissues to be raised so that these tissues can be segmented at the level of bone, providing for the inclusion of such structures as highly vascular tumors or aneurysms (Figures 1 and 2). This can serve an important role in the planning of the surgical approach to such abnormal structures.

MR scanning and CT permit tissues to be distinguished with different degrees of sharpness and contrast. The correct selection of like points in a patient scanned with MR and CT should allow "fitting" and coregistration of data to take advantage of the differences enabling, for instance, the display of the cranium from CT data and the display of the surfaces of the brain from MR data correctly related to one another.

It is possible to identify a discrete tissue such as muscle on a single CT slice. Furthermore it is possible to manually outline this tissue by means of a cursor and trackball. This is a very time-consuming process, many times limited by the eye-hand interaction of the operator and fraught with error at the limits of a tissue or structure, such as the last several slices containing the surface of a spherical structure like the eye, where only one or two pixels represent the structure. Is this not a task which can be assigned to a computer, initially concentrating on tissues with homogeneity but later directing attention to other attributes of a cluster of pixels such as their relative anatomic location and the likelihood that such a cluster is a part of a known anatomical structure?

Images are currently presented as opaque images with hidden surfaces removed or as "transmission images" much like standard plain film X-rays but having the attributes of three dimensionality through shading and depth cueing. There has been some argument as to which type is best. I do not believe there is a "best". There are times when a clinician wishes to encounter an object as it is encountered *in vivo* (the opaque image) and at other times may which to view a structure at depth through a surface structure (the "transmission image"). The planning for a craniofacial reconstruction may require the display of opaque surfaces while the planning for the approach to an intracranial aneurysm of the circle of Willis or a stereotactic biopsy will require a transmission representation.

Considering the accuracy necessary to provide useful images and the need to reliably define tissues or display small structures such as intracranial aneurysms, there can be no compromise

---

* Plates 17 to 20 appear following page 294.

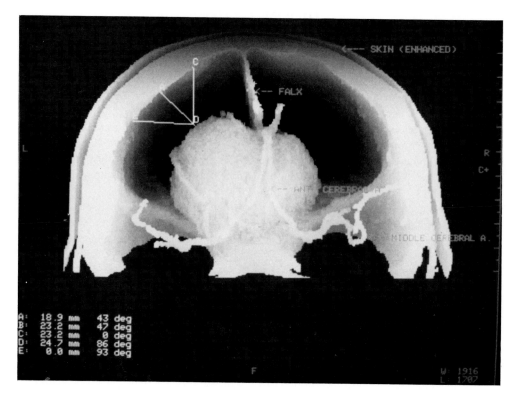

FIGURE 1. Olfactory groove meningioma 3D image obtained by contrast infusion.

on the technique used to acquire the images. The data used must come from the thinnest, abuting slices. Anything else will permit relative distortion of the anatomy. CT scanning is now performed rapidly so that 50 to 100 slices can be obtained without an unnecessary burden on the scanning facility. 3D imaging loses its usefulness with the gross distortion in thick slices.[1]

IV. THE WORKSTATION

The workstation will be a dedicated workstation, dedicated to 3D images and the data contained in an acquired 3D imaging volume. It will permit the display of substrate CT and MR slice data as well as reformatted multiplanar images. 3D images should be presented as previously described to be viewed from any desired pose angle. Magnification and minification of the object to permit close or distant observation of features is desirable. With the inclusion of multiple organ surfaces, a color display becomes mandatory. Color selection may be an option of the operator, may be applied by assigning colors derived from an optimum color scale using the concept of "just noticeable differences"[7] to a range of CT Hounsfield values,[8] or may be assigned based on the tissue type corresponding to colors commonly used in anatomical texts.

With a reduction in hardware costs, multiple screens should be implemented. A monochrome dialogue screen and a high-resolution viewing screen are probably a basic unit. Two high-resolution screens are desirable to contain a "working" image and a series of "work product" images. Failing the presence of two high-resolution screens, the use of windowing to contain the work product is necessary. Further enhancement of the viewed object can be made through the use of animated "movie loops" and stereoscopic viewing.

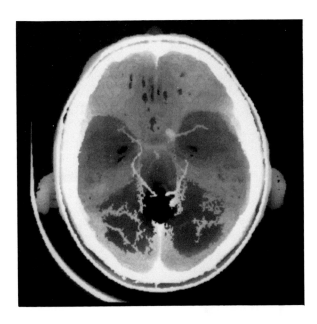

FIGURE 2. Circle of Willis aneurysm obtained with contrast infusion.

Above all, the workstation will be a workstation more than a viewing station. It will permit analysis of the object through the use of measurement and cutting or electronic "dissection". It will allow manipulation by movement of skeletal and soft tissue parts. Lastly, the workstation will facilitate the method of surgical intervention and allow the determination of the results of surgical and radiotherapeutic intervention.

Anatomic points are often difficult to locate accurately by a single inspection of an object. A point is generally identified by noting its correspondence from two or more viewing perspectives. Therefore one must be able to identify a point on a three dimensionally displayed object, fix it, and move the object to determine if it was correctly located. Once points are fixed, linear (point to point through 3D space as well as surface tracking) measurements and volumetric measurements can be made. Volumetric measurements in anatomical structures present a peculiar problem. The volume to be measured is a somewhat poorly defined volume. For instance, the intracranial space is bounded by cranium but also lacks boundaries at points of fenestration such as the optic foramina, foramen lacerum, or foramen magnum. A calculation of volume must define what is to be called the intracranial volume and, if it is to be defined by artificially closing the fenestrations, how and where they are to be closed. There must be agreement by all to permit uniformity in measurement.

Dissection or division of the object into parts may be carried out by using the planar cutting of a volume-rendered object so that voxels "fall off" or are removed from the object, exposing its inner surfaces or by splitting the object along a linear plane so as to open it like a book. However the most realistic method of "cutting the object" is to simulate a saw or knife which penetrates to a selected depth (usually, in the case of bone, completely through the tissue to the inner surface) permitting the description of a free segment.[9]

X, Y, and Z coordinates for deep structures must be readily determined to permit the calculation of the course of passage of a probe or biopsy needle. Furthermore, provision must be made for rapid rendering of a 3D object during a stereotactic procedure to permit corrections in erroneous probe courses.

Not only should the workstation be available in the clinical area but also an interface should be located in the office of the physician. As the physician becomes more facile in the use of computers and powerful programs are made friendlier, there will not only be a need to view images and records as suggested by the purveyors of PACS systems, but also to act on this data by generating sophisticated solutions to clinical problems.

V. CLINICAL APPLICATIONS

Craniofacial surgery, a specialized surgical discipline, carried out in but a relatively few locations world-wide, was one of the first medical disciplines to embrace 3D imaging.[3] 3D imaging was and is particularly useful in this area because the fundamental problem is osseous and the bony structures complex. However, the uses of 3D imaging have barely scratched the surface in this area and many others. At present, we have pretty and useful pictures. What can we do with these pictures in the future? We can, as has been suggested innumerable times, plan surgical procedures. We can, as has also been suggested, design prostheses and implants. We can use it to follow the course of treatment or intervention by comparison of serial studies. In the course of doing these things better, we can design better prostheses and better design surgical procedures through the use of finite element analysis and modeling applied to the 3D substrate. The problem in craniofacial surgery is not simply solved by the movement of bones. The skin and muscle are also shifted. These structures may yield to the change in bone position in different ways and, furthermore, the new positions of points of origin and insertion of the muscles may affect the ultimate stability of the craniofacial skeleton. There may be questions as to whether newly moved skeletal segments should be fixed or allowed to "float" freely, particularly in the infant whose growing brain exerts a significant influence on the calvarium and face. The solutions may lie in the application of finite element techniques to these problems.

Similarly, orthoses, both external (braces) and internal (rods, plates, and springs), are placed to exert forces on the skeleton to reshape or stabilize the face and the spinal column. The location, fixation, and forces applied by these devices are critical. The 3D image can serve as the model to determine the correct application of these devices. We are already using the 3D image of the spine to locate the proper position of fixation screws to make this procedure accurate and semiautomatic. In a similar fashion the image serves to locate the placement of screws for the osteointegrated dental prosthesis.

The efficiency of a fixation device can be determined more effectively with the information provided through reliable imaging coupled with serial studies of the results of fixation devices. I believe that 3D imaging will become routine in orthodontia and scoliosis surgery as the information is present to design more efficient devices.

Many deep-seated tumors are both unresectable and unapproachable by a classical "open surgical biopsy". Rather, directed needle biopsy techniques are used to minimize the damage to uninvolved structures. It is here that 3D images, provided rapidly, can make a significant contribution to the ease and accuracy of passage of a needle. Classic biplanar radiographic data has required the attachment of devices to the patient to convert 2D coordinates to 3D data. It can easily be seen that the location of a "target" within the body and the vector of passage of a needle is readily obtainable from the 3D image provided it is obtained rapidly and that pathologic tissues are distinguishable from their surroundings.

In summary, we must be able to display multiple tissues simultaneously with the same resolution which exists for bone. The future emphasis must lie in the application of this imaging data to the solution of clinical problems rather than solely the demonstration of the nature of these problems.

REFERENCES

1. Hemmy, D.C., and P.L. Tessier. CT of dry skulls with craniofacial deformities: Accuracy of three dimensional reconstruction, *Radiology* 157:113–116 (1985).
2. Tessier, P.L., and D.C. Hemmy. 3-D imaging in medicine…a critique by surgeons, *Scand. J. Plast. Reconstr. Surg.* 20:3–11 (1986).
3. Hemmy, D.C., G.T. Herman, E.A. Millar, and V.M. Haughton. Three-dimensional reconstruction of the spine and calvarium in children utilizing computed tomography, in *Proceedings of the Second International Child Neurology Congress,* Sydney, Australia (1979), p. 81.
4. Hemmy, D.C. Three dimensional imaging, the first decade, *Proc. IEEE Special Symp.* pp. 61–64 (1988).
5. Laschinger, J., M.W. Vannier, R.H. Knapp, and S. Gronemeyer. Three dimensional reconstruction of the heart using EKG-gated magnetic resonance images, *Proc. Natl. Comput. Graph. Assoc.* III:86–91 (1987).
6. Warwick, R., and P.L. Williams. *Gray's Anatomy,* 35th British ed. (Philadelphia, PA: W.B. Saunders, 1973).
7. Lefkowitz, H., and G.T. Herman. Towards an optimal color scale, *Proc. Natl. Comput. Graph. Assoc.* III:92–98 (1987).
8. Farrell, E.J., and R. Zappulla. Imaging tools for interpreting two and three dimensional medical data, *Proc. Natl. Comput. Graph. Assoc.* III:60–68 (1987).
9. Yokoi, S., and T. Yasuda. A craniofacial surgery planning system, *Proc. Natl. Comput. Graph. Assoc.* III:152–161 (1987).

INDEX

S

Sacroiliac, 135—140
Sagittal split osteotomy, 175
Scaling, 78—79
Scatter plot, 155
Scene, 5, 146
Scene filtering, see Filtering
Scene interpolation, see Interpolation
Scene masking, see Masking
Scene processing, 7—11
Scene region, 5—6, 9
Scene sampling, 11
Scene segmentation, see Segmentation
Scene space, 3, 35
Scene space coordinates, 7
Scene-space shading, 40—42
Scene transformations, see Transformations
Schizophrenia, 154
Segmentation, 14—18, 61, 73—75, 273—274,
     299, 302
  boundary-based, 14—16
  feature plot method, 17
  region-based, 14, 16—18
  volume rendering, 45—47
Segment repositioning, 55
Selected tissues, 335
Semilobar holoprosencephaly, 107
Shaded surface displays, 287—296, 298—299,
     301, 305—306, 323
Shading, 35, 38—43
Shadowing effect, 34
Shape-based interpolation, 13—14, 147, 155, 298
  masking and, 19
Shape descriptors, 183
Shoulder and shoulder joint, 242—246
Simulator, 316, 320
Simultaneous multichannel cardiac mapping system,
     306
Single-photon emission-computed tomography
     (SPECT), 304
6d digitizer, 185
Skull images, 155, 248—251
Slice display, see Display
Slice editing, 272
Slice imaging, 3—4, 21—26
Slice location, 7
Slice thickness, 7, 194—196
Sloths, 293, 308
Software, 59—60, 69, 81, 224
Solids models, 81
Spatial enumeration, 78
Spatial modulation of magnetization (SPAMM),
     304, 306
Specular reflection, 39
Spine, 246—248, 334
Stereo, 39, 42, 150—151
Stereo pairs, 149
Stereophotogrammetry, 176—177, 183
Stereoscopic viewing, 336

Stereotaxic surgical methods, 282
Sternoclavicular injuries, 244
Subregioning, 146, 155
Sun Microsystems, 307
Surface-based cephalometrics and statistics, 181—
     183
Surface-based displays, radiation therapy, 322—323
Surface detection, 33, 149—150
Surface displays, see Shaded surface displays
Surface formation, 25, 27—33
  contouring, 27—28
  patching, 27—30
  quantitation, 147
  surface tracking, 27, 30—33
Surface imaging, 77—78, 80, 85, 146—148, 159
Surface likelihood, 323
Surface reconstruction, 77
Surface reconstruction images, 75, 79
Surface rendering, 3, 25, 30, 33—43, 61—62
  brain, 272
  hidden part removal, 35—38
  shading, 35, 38—43
Surfaces, 30—32, 37—38, 147—148, 155
  measurement on, 58
Surface shading, 11, 14, 194, 201—203
Surface tracking, 27, 30—33
Surface volume rendering, 58
Surgery rehearsals, 280
Surgical planning, 55—57, 171, 176, 186, 273,
     278
  craniofacial, 76
Surgical robotics, 183—187
Surgical simulation, 282
  hip stems, 259
Syndromology, 171, 183

T

T2-weighted, 155
Talocrural joint, 240—242
Target volume, 314, 316, 319
Temporal lobes, 154, 157
Terminology, 5—20, see also specific terms
3D CT, 81, 103—144, 151
  congenital craniofacial deformities, 109—112
  craniofacial anomalies, 164—168
  craniofacial trauma, 112—124
  dynamic ''low-dose,'' 104—105, 131, 136
  examination technique and principles, 225
  extracranial neoplasms, 125, 130—133
  image assessment, 106—109
  intracranial tumors, 119, 122—129
  pelvic fractures, 131, 134—143
  preoperative assessment, 112, 118, 134
  radiation dose, 108
  radiation factors, 106
  technique, 104—105
3D-CT reconstruction, 76
3D83, 149—150, 159, 193
3D98, 148—149, 192
3D reformation, 192—200

Thresholding, 17, 73, 226
Tiling, 29—30, 180
Tilt, 148
Tissue characterization
  brain, 272, 276
  cardiopulmonary imaging, 301
  continuous, 272—273
Tissue mixture, 45
Tissue opacity and color, 45
Topologically connected surfaces, 180
Topologically continuous surface model, 181
Torsion, 181
Total heart volume, 292
Total knee replacement (TKR) designs, 263
Transaxial CT scanning, 241—242, 248
Transformations, 3—4, 7, 19
Transmission image, 335
Transparency, 39, 43, 81
Triangulation techniques, 58
Trigeminal neurofibroma, 125—126
Trilinear interpolation, 12—14
Tumors, 338
  brain, 274—275
  extent, 122
  survival rate, 319
  target volume, 314, 316, 319

U

Ultrafast Cine CT Scanner, 298
Ultrasonic scanners, 177
Ultrasounds (US), 72, 301—302
Ultrasound scans, 179
Unilateral interfacetal dislocation, 206—207
Upper airways, 295—299

V

Validation, 63
Validation studies, 160
Variability, 155—156, 183
Vascular dimensions, 293
Vascular geometry, 294
Ventricular ischemia, 293
Vibrating mirror, 3, 288
Viewing conditions, 33—34
Viewing geometry, 34
View space, 3
View-space shading, 40—44
Virtual simulation, 320—322
Visualization issues, 61—63
Visualization techniques, 80
VOI, see Volume of interest
VOI operation, 7, 9
Volume, 155, 165
Volume calculation, 146, 149—150
Volume compositing, 323, see also Volume rendering

Volume imaging, 3—4
Volume of interest (VOI), 7—9
Volume reformatting, 288
Volume rendering, 3, 30, 43—54, 61—62, 226, 279
  binary scenes, 49—51
  brain, 272—273
  cardiopulmonary imaging, 288, 291
  classification, 45
  future research in use of, 327
  grey scenes, 51—54
  masking, 47
  octrees, use of, 50—51
  opacity functions, 46—47
  preprocessing, 43—47
  radiotherapy, 323—327
  rendering, 43—54
  segmentation, 45—47
  tissue mixture, 45
  tissue opacity and color, 45
Volumetric analysis of brain, 146, 150, 154—160
Volumetric CT, cardiopulmonary imaging, 286—292
Volumetric image display, cardiopulmonary imaging, 287—289
Volumetric Image Display and Analysis (VIDA), 307
Volumetric imaging, 80
  beyond DSR, 297—301
Volumetric measurements, 337
Volumetric visualization, 81, 85
Voxel, 9, 14, 81, 150, 155, 288
  brightest projection, 291
  color, 274, 276, 279
  connectivity, 32
  neighborhood for, 10
  opacity, 274, 276, 279
Voxel-based surface, 30
Voxel-based surface rendering, 61—62
Voxel density, 5, 7, 10, 21
Voxel faces, 37, 147
Voxel projection, 48—49

W

Window-based design, 307
Windowing, 23, 25, 307
Window of translucency, 279
Winged edge library, 180
Wire-frame representations, 81
Working image, 336
Work product images, 336
Workstation, 282, 307, 336—338

Z

z-buffer technique, 37—38